HD
9502
U52
D8
1975

Duchesneau, Thomas D
    Competition in the U.S. energy industry . a report to the
Energy Policy Project of the Ford Foundation / Thomas D.
Duchesneau. — Cambridge, Mass. : Ballinger Pub. Co., 1975.
    xiii, 401 p. : ill. ; 24 cm.
    ISBN 0-88410-337-4. ISBN 0-88410-338-2 pbk.

    l Energy policy—United States. 2 Power resources—United States. 3.
Competition. l Ford Foundation. Energy Policy Project. II. Title.

HD9502.U52D8 1975   338.4   74-22179
                                             MARC

Competition in the U.S. Energy Industry

A Report to the Energy Policy Project of the Ford Foundation

Competition in the U.S. Energy Industry

Thomas D. Duchesneau
University of Maine at Orono
Orono, Maine

Ballinger Publishing Company • Cambridge, Mass.
A Subsidiary of J.B. Lippincott Company

Published in the United States of America by Ballinger Publishing Company, Cambridge, Mass.

First Printing, 1975

Library of Congress Catalog Card Number: 74-22179

International Standard Book Number: 0-88410-337-4
0-88410-338-2 Pbk.

Printed in the United States of America

Library of Congress Cataloging in Publication Data

Duchesneau, Thomas D
    Competition in the U.S. energy industry.

    Sponsored by the Energy Policy Project of the Food Foundation.
    Includes bibliographical references.
    1. Energy policy—United States. 2. Power resources—United States.
3. Competition. I. Ford Foundation. Energy Policy Project. II. Title.
HD9502.U52D8   338.4   74-22179
ISBN 0-88410-337-4
ISBN 0-88410-338-2 pbk.

Contents

INTRO

List of Tables

Preface

The Energy Policy Project was initiated by the Ford Foundation in 1971 to explore alternative national energy policies. This book, *Competition in The U.S. Energy Industry*, is one of the series of studies commissioned by the Project. It is presented here as a carefully prepared contribution by the author to today's public discussion about energy companies. It is our hope that each of these special reports will stimulate further thinking and questioning in the specific areas it addresses. At the very most, however, each special report deals with only a part of the energy puzzle; the Energy Policy Project's final report, *A Time to Choose*, which was published in October 1974, attempts to integrate these parts into a comprehensible whole, setting forth the energy policy options available to the nation as we see them.

This book, like the others in the series, has been reviewed by scholars and experts in the field not otherwise associated with the Project in order to be sure that differing points of view were considered. With each book in the series, we offer reviewers the opportunity of having their comments published in an appendix to the volume. (See page 395.)

Competition in The U.S. Energy Industry is the author's report to the Ford Foundation's Energy Policy Project and neither the Foundation, its Energy Policy or the Project's Advisory Board have assumed the role of passing judgment on its contents or conclusions. We have expressed our views in *A Time To Choose*.

S. David Freeman
Director
Energy Policy Project

Author's Note

This book is one of several background studies financed by the Energy Policy Project of the Ford Foundation. It is concerned with the extent of competition in the production of primary energy in the United States. The study was conducted during a time of unprecedented change in both domestic and world oil markets; the effects of such change can only be partially understood at this time.

The Energy Policy Project Staff provided valuable guidance during the course of the study. The study benefited immensely from the contribution of Walter Mead. Special studies were prepared by Thomas Hogarty, John Lichtblau, Edwin Mansfield, Thomas G. Moore, Reed Moyer, Lester Salamon, and John Sigfried. The authors of the special studies provided valuable insights into specific topics and were instrumental in the completion of this report. The special studies comprise the appendixes to this book.

The manuscript was substantially improved with the advice of a review committee and of a symposium held in June 1973. Many helpful comments were received. I especially want to acknowledge the comments provided by John Blair, Thomas Hogarty, Paul MacAvoy, Reed Moyer, John Siegfried, Milt Russell, and John W. Wilson. The contributions of John M. Ryan of Exxon Corporation and John Lichtblau of the Petroleum Industry Research Foundation provided important insights to many aspects of the industry.

A special debt of gratitude is owed to Sylvia K. Coupe for typing and proofreading of the manuscript.

While many contributed to this study, the author accepts full responsibility for the interpretations and conclusions contained in the book.

Thomas D. Duchesneau
University of Maine at Orono

Chapter One

Introduction

SOURCES OF ENERGY

The extensive use of energy is a fundamental characteristic of the U.S. economy. The United States accounts for slightly less than 6 percent of the world's population but consumes approximately 33 percent of its energy.[1] On a per capita basis, the U.S. is by a wide margin the world's largest consumer of energy. While these figures indicate the importance of energy, the complete role of energy in the U.S. economy is not adequately represented by quantitative measures, because energy use also contains an important qualitative dimension.

The immense energy requirement of the U.S. is met mainly by fuels: crude oil, natural gas, coal, and uranium. Some 95 percent of total energy supply is accounted for by coal, oil, and natural gas. Nuclear energy, while holding promise for the future, presently supplies a very small percentage of the total. It is traditional to distinguish among energy sources by referring to primary and secondary sources. Primary sources include coal, oil, natural gas, and uranium; electricity, the production of which requires the conversion of primary sources, is considered to be a secondary source of energy.[a]

The technological ability to convert primary energy sources into electricity creates a unique interrelationship between electricity and primary sources of energy in end uses where electricity is consumed as an indirect source of energy. This type of relationship led the *Energy Study Group* to state:

> To the extent that electricity can be substituted for other forms of energy in any use, substitution among all primary fuels is at once possible in that use. . . . Even when direct use of some fuels would

---

[a]Hydropower is also a primary source of energy. Presently, electricity is the most important secondary source but the development of a commercial process for the liquefaction and gasification of coal would add an additional secondary source.

not meet special requirements, all the primary fuel sources, through conversion to electricity, are alternative substitutes.[2]

The subject of this paper is the state of competition in the production of primary energy in the U.S. and the policies necessary to promote and preserve effective competition. Emphasis is on determining the structure of energy production and the competitive implications of its institutional features. Conduct and performance are also analyzed. The book is designed to provide a framework for evaluating various policy options. Hopefully, it provides a basis for a rational long-run policy towards competition in the energy industry.

ROLE OF COMPETITION IN A MARKET ECONOMY

The role of competition in a market economy based on private property rights has been widely discussed since the time of Adam Smith. The meaning of competition in economic theory has not always been clear. One modern author states: "There is probably no concept in all of economics that is at once more fundamental and pervasive, yet less satisfactorily developed, than the concept of competition."[3]

A complete discussion of the concept of competition is not appropriate here, but it is informative to note that the concept has evolved over time.[4] In modern economic analysis, competition describes a situation where sellers and buyers view price as a parameter determined by market forces and recognize that the actions of any single seller or buyer have no perceptible impact on market price.[b]

Viewing competition as a process in which price is thought of as a parameter determined by the individual seller has important implications for public policy. Businessmen commonly describe competition as the process by which they engage in a struggle with other firms for patronage. The attempt to gain patronage may be on a price or a nonprice basis. There is no assurance that the latter type of rivalry is synonymous with competition as the term is used in economic theory, i.e. where price is viewed by the seller as a parameter. For instance, it could be argued that the U.S. automobile industry is characterized by a substantial amount of rivalry but relatively little competition. The focus of public policy is on competition rather than on rivalry.

The Case for Competition

Confusion as to the meaning of competition arises largely because the case for competition rests on both political and economic grounds. It may very well be that, in the U.S., political rather than economic considerations have made competitive markets a social goal. Such a position has been taken by Prof.

---

[b]Effective seller competition requires the existence of a sufficient number of independent sellers to prevent excessive seller influence over price, quantity, and product policies.

Kaysen. He states: "If the regime of competition and the arguments of laissez-faire ever commended themselves widely, it has been primarily on political rather than economic grounds."[5]

Political Factors. Two important attributes of a competitive market system argue politically in favor of competition and, as a corollary, against monopoly or the existence of market power. One is competitive decision-making, which tends to decentralize power. Competition severely limits the discretionary authority of any individual seller or buyer over allocation decisions; the individual seller is compelled to act in accord with the dictates of the market. Secondly, under competition, resources are allocated by impersonal forces. Market forces direct resources toward consumer priorities without bringing personal elements into the decision-making process.

To summarize, the political arguments favoring competitive markets rest upon the existence of decentralized and impersonal decision-making under competitive conditions.

Economic Factors. Economy theory provides some powerful arguments in support of competitive markets. It demonstrates that competitive markets have greater allocative efficiency under definable if rather restrictive conditions.[6] A second economic argument in favor of competitive markets is that they ensure technical efficiency in the production of goods and services. A firm is technologically efficient when it produces a given output with the least input of resources or, in other words, when costs are minimized. Cost minimization is a condition of profit maximization, and competition puts maximum pressures on firms to be technically efficient. The technically inefficient firm simply cannot survive in a competitive market.

It can be argued that a lack of competition in intermediate goods markets can reduce living standards faster than a lack of competition in consumer goods markets. Primary energy sources can, depending upon their use, be either consumer or intermediate goods. Market power or the lack of competition in a consumer good market causes price to exceed the opportunity costs of production. This, in turn, causes consumers to alter their consumption pattern so that fewer units of the good are purchased, which leads to the lower living standards associated with monopoly power. In the case of a producer good, an uncompetitive market not only distorts a product's price but, because the product is an input to a production process, also leads to changes in the production function of the buying industries and a further loss of efficiency. Thus, the effects of market power in an intermediate goods market are transmitted via the market system to all users of the input and may cause a greater loss of allocative efficiency than would result in a consumer good market.[7] This emphasizes the importance of maintaining effective competition in intermediate goods markets.

STRUCTURE, CONDUCT, AND PERFORMANCE

The study of competition in an industry usually rests upon an analysis of market structure, conduct, and performance. Structure refers to the external environment within which the firm's decisions are made. How a firm's policies, especially price policies, are determined is the measure of market conduct, and market performance describes the end results of market processes. Performance involves an assessment of the extent to which the economic results of an industry's market behavior deviate from the optimum contribution it could make to achieving the accepted goals of society. Knowledge of the impact of market structure and conduct on market performance provides a basis for evaluating public policy designed to promote competition. Antitrust laws, regulatory commissions, and legislation affecting competition indirectly influence market performance by changing either the structure of markets or the conduct of sellers in those markets.

Public Policy and Industrial Structure

Many modern econometric studies of industrial organization have been concerned with the various relationships between market structure and market performance. This emphasis reflects the importance of public policy questions in industrial organization. Government policies seldom focus directly on performance, but rather are applied to structural or behavioral features of markets. As a result, the structure-performance relationship spurs considerable research that attempts to serve the needs of policy makers.

The structure-performance relationship remains obscure without a generally accepted theory of oligopoly.[c] Various models of oligopolistic behavior exist, but they fall short of explaining actual behavior. Researchers are forced to take a more inductive approach, attempting to link certain structural variables with aspects of market performance.

The greatest amounts of research have been devoted to establishing the relationship between levels of seller concentration in an industry and allocative efficiency measured by long-run profitability. Lesser amounts of research have been devoted to the relationship between entry conditions and efficiency. It is possible to generate oligopoly theories that lead to almost any prediction, but most researchers have hypothesized that the probability of successful collusion among sellers increases as the level of seller concentration rises in an industry. As a result, high concentration is predicted to be associated with high, long-run profit rates.[d] In an opposite manner, low concentration

---

[c]Strictly speaking the relationship flows from structure to conduct to performance. Given the difficulty of measuring and interpreting conduct, studies of competition generally state their hypotheses in terms of structure-performance relationships.

[d]Assuming that entry barriers exist.

levels, by making successful collusion less likely, should lead to competitive rates of return.

The hypothesized relationship between concentration and profit rates has been heavily researched. Professor Scherer, after a critical review of various empirical tests of this relationship, notes the general consistency of results:

> Although doubts remain and more research with better data is needed, the bulk of the evidence supports the hypothesis that profit increases with the degree to which market power is concentrated in the hands of a few sellers.[8]

Seller concentration is not sufficient by itself, however, to lead to a misallocation of resources. In a competitive market with easy entry conditions, new firms can enter the industry regardless of successful collusion and high profit rates, causing an increase in industry output and the eventual restoration of competition. The situation changes where entry barriers exist. Entry barriers tend to keep new firms out of an industry with excess profitability. In order for high concentration to cause a misallocation, entry barriers must exist. There is evidence to indicate that entry conditions, at least in the case where barriers are very high, are associated with allocative inefficiency.[9] The theoretical reasoning and empirical research linking concentration and entry conditions to allocative inefficiency have important implications for public policy that intends to preserve and promote competition.

Policy formulation does not flow in a simple manner from the existing body of knowledge. One, policy makers are confronted with the reality that the theoretical linkage between aspects of structure and allocative efficiency is not complete nor is it deterministic. Scherer recognizes the dilemma for public policy makers by stating, "It is in part because of this complexity and randomness that oligopoly poses such difficult problems for the economic analyst."[10]

Two, a given level of seller concentration can be consistent with significant variations in the degree of competition in different industries. Economists are not interested in concentration levels *per se*. Concentration measures are indirect measures of what cannot be directly observed; namely, the degree of interdependency among sellers. The degree of interdependency is a determinant of performance. In a given industry, institutional features and arrangements such as joint ventures among firms can create considerable interdependency among sellers. The impossibility of fully determining the degree of interdependency among sellers on the basis of concentration levels is particularly obvious in the case of crude oil production. Various institutional features such as demand prorationing and joint bidding and production arrangements all combine to raise interdependency far above what is indicated by concentration levels.

Thus, interpreting concentration ratios in different industries is not a simple task. The same figure, say 70 percent for the top eight firms, may have very different meanings in different industries. In the energy industry, competition may be determined more by institutions than by concentration levels. An interesting question is raised by the discussion above concerning concentration and certain institutions. It is possible that an industry's market structure is determined to an important degree by certain institutional features. If so, policy makers, when considering institutional changes for whatever reasons, must be aware that such changes affect market structure and the degree of competition. The impact on structure could either help or hinder competition, but it should be kept in mind that economic theory is relatively clear on one point: if an industry's structure is noncompetitive, in the long run performance will suffer.

Three, the relationship between market structure and conduct is not well established in economics. Relatively little research has been conducted on the relationship of a firm's behavior and internal organization to market structure. Certainly, additional knowledge of this relationship is essential to a more complete understanding of market processes.

Despite this shortcoming, an analysis of market structure remains a necessary step in evaluating competition in an industry. Such an analysis, in combination with an awareness of how institutional features can affect conduct and performance, provides a sound basis for a rational policy of competition in the energy industry.

Past studies have generally analyzed structure from the perspective of a single fuel. This emphasis is not incorrect, but it tends to obscure the fact that competition in energy production takes two forms: (1) interfuel and (2) intrafuel. Complete analysis requires that both forms be included.

Interfuel Competition

Interfuel competition occurs in those energy end uses where coal, oil, natural gas, and uranium are interchangeable enough to support the conclusion that they trade in a single economic market. The greatest degree of interchange exists in the electric utility sector of the economy, where the structure of a combined fuels market, an energy market, is a proper concern.

Intrafuel Competition

In those end uses of energy where a single source dominates and no potential substitutes exist, intrafuel competition is the only source of competition. Perhaps the best illustration of this is the transportation sector, where the disappearance of the coal-burning locomotive made oil dominant in total energy consumption.

Policy Implications

Policy must recognize the difference between the two types of competition discussed above. The two examples used, the electric utility and transportation sectors, are in a real sense polar cases. In most end uses there is at least some degree of interchangeability among the various products. In these cases, the distinction between the two sources of competition is not as easily made. This is even more so when one considers that, while coal or uranium may not be directly substitutable for fuels currently being used, electricity is at least a potential substitute; and since fossil fuels and uranium can be converted into electricity, all of the fuels are substitutable in an indirect sense.

In any case, the distinction is important because of the policy implications. Actions such as mergers may have different impacts on competition depending upon the type of competition involved. Some activities may be anticompetitive in terms of intrafuel competition but at the same time have little impact on interfuel competition. At this stage, it is necessary to point out that an analysis of market structure must recognize that two types of competition exist, and that policy should be designed to foster both types of competition.

THE ENERGY COMPANY

The energy company concept is a relatively recent development in the primary fuels sector of the economy. An energy company can be defined as a firm with substantial reserve and/or production positions in the various primary *fuels areas.* Generally, energy companies have resulted from the expansion of oil firms into other fuels. Entry has been accomplished by a variety of ways, including: (1) acquiring fuel reserves; (2) acquiring production facilities; (3) *de novo* entry into production and/or reserve ownership; and (4) the use of joint venture arrangements.

As a result of this type of expansion, most of the large oil companies in the United States have been transformed into energy companies. The extent of this development is indicated by the evidence presented below in Table 1-1. The evidence indicates the fuels in which each of the 25 largest oil firms have established a position. In addition to coal and uranium, Table 1-1 also indicates the companies with interests in oil shale and tar sands. With current crude oil prices in the $9-10 per barrel range, shale and tar sand production appears to be profitable. The Colony Development group was ready to begin production from shale when mid-continent oil was priced at $3.50-4.00 per barrel. In addition, during January, February, and March 1974, three different groups bid very large sums of money to begin oil production from shale. Oil company activity in other fuels is indicated in Table 1-1. Among the 25 largest oil companies, 18 are involved in oil shale, 11 in coal, 18 in uranium, and 7 in tar sands. These findings, based on a search of trade journals, annual reports, and Moody's Industrials, are subject to some error. However, they are substantiated by the results of a survey conducted by Continental Oil. The results are summarized in Table 1-2.

Table 1-1.   Diversification in the Energy Industries by the Twenty-Five Largest Petroleum Companies, Ranked by Assets, as of Early 1970

Petroleum Company	1969 Assets ($ thousand)	Rank in Assets	Energy Industry				
			Gas	Oil Shale	Coal	Uranium	Tar Sands
(1)	(2)	(3)	(4)	(5)	(6)	(7)	(8)
Standard Oil (N.J.)	17,537,951	1	X	X	X	X	X
Texaco	9,281,573	2	X	X	X	X	
Gulf	8,104,824	3	X	X	X	X	X
Mobil	7,162,994	4	X	X		X	
Standard Oil (Calif.)	6,145,875	5	X	X			
Standard Oil (Ind.)	5,150,677	6	X	X		X	X
Shell	4,356,222	7	X	X	X	X	X
Atlantic Richfield	4,235,425	8	X	X	X	X	X
Phillips Petroleum	3,102,280	9	X	X		X	
Continental Oil	2,896,616	10	X	X	X	X	
Sun Oil	2,528,211	11	X	X	X	X	X
Union Oil of California	2,476,414	12	X	X		X	
Occidental[a]	2,213,506	13	X		X		
Cities Service	2,065,600	14	X	X		X	X
Getty[b]	1,859,024	15	X	X		X	
Standard Oil (Ohio)[c]	1,553,591	16	X	X	X	X	
Pennzoil United, Inc.	1,356,532	17	X			X	
Signal	1,258,611[d]	18	X				
Marathon	1,221,288[e]	19	X	X			
Amerada-Hess	982,157	20	X			X	

Ashland	846,412	21	X	X		X	X
Kerr-McGee	667,940	22	X		X	X	X
Superior Oil	494,025	23	X	X			
Coastal States Gas Producing	490,190	24	X				
Murphy Oil	343,914	25	X				

[a]Includes Hooker Chemical Company.

[b]Includes Skelly and Tidewater.

[c]Includes British Petroleum Assets.

[d]As of June 30, 1969.

[e]As of September 30, 1969.

Source: Bruce C. Netschert, Abraham Gerber, and Irwin M. Stelzer, *Competition in the Energy Markets: An Economic Analysis* (Washington: National Economic Research Associates, May 1970).

Table 1-2.   Oil Company Involvement in Other Energy Sources:
Response to Continental Oil Co. Survey

Energy Source	Number of Oil Companies		
	Active or Planned Production	Positions	Total
Coal	7	9	16
Uranium	6	18	24
Oil shale	3	14	17
Tar sands	3	13	16

Source: L.C. Rogers, "Oil Finding Talent Pours into Broad Minerals Drive," *The Oil and Gas Journal*, February 24, 1969, p. 37.

Characterization of the Energy
Company Development

The transformation of oil companies into energy companies is a major development in the primary fuels sector of the U.S. economy. The energy company development has been characterized in various ways. Critics of the industry see it as a drive by oil companies to monopolize energy production in the U.S. Entry, by itself, certainly does not warrant this charge. In defending themselves, oil firms depict such entry as the result of the decision-making process that allocates investment in response to after-tax profit differentials. The question remains of why, starting in 1965, oil companies launched an expansion into coal and uranium. An even more fundamental question concerns the impact of the expansion on interfuel competition. This can, in part, be determined by an analysis of the effect of these mergers upon market structure. In particular, it is necessary to determine the extent to which oil companies have become dominant suppliers of coal and uranium, and to see how this has affected levels of seller concentration and entry conditions.

SUMMARY

The U.S. relies heavily upon a system of competitive markets to allocate resources among alternative uses. Competition in energy production includes both interfuel and intrafuel competition. Concern for the continued existence and health of competition in the production of primary energy has recently increased as a consequence of energy company development. Public policy designed to preserve effective competition must be concerned with both aspects.

An analysis of competition focuses heavily upon the structure of the particular market being studied. This book is concerned with the structural setting of coal, oil, natural gas, and uranium production in the U.S.

NOTES

1. U.S. Department of the Interior, *Report to the Secretary of the Interior by the Advisory Committee on Energy*, Washington, D.C., June 30, 1971, p. 1.
2. Report prepared for the Interdepartmental Energy study by the Energy Study Group under the direction of Ali Bulent Cambel, *Energy R&D and National Progress* (Washington: Government Printing Office, 1964), p. xxv.
3. Paul J. McNulty, "Economic Theory and the Meaning of Competition," *The Quarterly Journal of Economics*, November 1968, p. 639.
4. The evolution of thought in the nature of competition can be found in: P.W.S. Andrews, *On Competition in Economic Theory* (London: MacMillan & Co., 1964), George J. Stigler, "Perfect Competition, Historically Contemplated," *Journal of Political Economy*, February 1957, and P.J. McNulty, "A Note on the History of Perfect Competition," *Journal of Political Economy*, August 1967.
5. Carl Kaysen, "The Corporation: How Much Power? What Scope?" or Edward S. Mason, *The Corporation in Modern Society* (Cambridge: Harvard University Press, 1960), p. 98.
6. In technical terms, competition tends to create a Pareto optimum allocation of resources: A discussion of the conditions necessary for Pareto optimality and the ability of competitive markets to meet these conditions can be found in: T. Scitovsky, *Welfare and Competition* (Homewood: Richard D. Irwin, 1971) and J. deV. Graff, *Theoretical Welfare Economics* (Cambridge: University Press, 1967).

    While perfect competition creates an efficient allocation of resources, it does not follow that the allocation is also equitable.
7. Kaysen and Turner, *op. cit.*, pp. 32-34 and I.M.D. Little, *The Price of Fuel* (Oxford 1953), pp. x-xiv; Prof. Scherer notes that, once vertical price distortions are included estimates welfare loss due to monopoly power increases by about 40 percent. See: F.M. Scherer, *Industrial Market Structure and Economic Performance* (Chicago: Rand McNally and Co., 1970), p. 404.

    It is not necessarily true that monopoly price distortions are always undesirable. Consider the case where negative externalities and monopoly power exist in combination. The market power would tend to offset the impact of the externalities on resource allocation. The net effect may be an improvement in allocative efficiency.
8. Scherer, p. 185. For a comprehensive review of the current state of

empirical research see: Leonard Weiss, "Quantitative Studies of Industrial Organization," in M.D. Intriligator, ed., *Frontiers of Quantitative Economics* (Amsterdam: North Holland Publishing Co., 1973).

9. *Ibid.*
10. Scherer, p. 212.

Chapter Two

Market Structure

SIGNIFICANCE OF MARKET STRUCTURE

Economic theory reasons that market structure, defined as the organizational characteristics of a market, exerts a strategic influence on pricing and on the nature of competition within the market. Among the structural dimensions of a market area: (1) degree of interdependency among sellers and buyers; (2) condition of entry; (3) degree of product differentiation; and (4) demand elasticity. All of these factors, according to economic theory, have an impact on competition in a market.

Seller Interdependency

The extent of competition in an industry depends to an important degree upon the structural features of the market. The closer the structure is to the monopoly-like structure, the greater the likelihood that the industry will be characterized by monopoly-like performance. In an opposite manner, as structure approximates the conditions of the competitive model, the oligopoly market is more likely to display competitive-like performance.

In analyzing the state of competition in a market, economists are interested in observing the degree of interdependency among the sellers.[1] Interdependency is, however, a qualitative factor and cannot be directly measured. As a result, economists have employed measures of seller concentration as indirect measures of the degree of seller interdependency in a market. Seller concentration is defined as the proportion of an industry's economic activity accounted for by the N largest sellers in the industry. Economic activity is often represented by value of shipments, assets, or employment.[2] High levels of concentration, the upper limit being a monopoly where a single firm controls the entire output, represent high degrees of interdependency, while low concentration levels would indicate less interdependency. It should be emphasized that concentration measures are a proxy variable for the degree of seller interdependency, and that a measure of concentration in a given market may not fully indicate the degree of interdependency among sellers. Other factors, such as joint venture arrangements, create interdependency but are not reflected in measured concentration levels.

High levels of seller concentration increase the probabilities that sellers will be able to reach and maintain agreements on a collective course of action designed to achieve either joint profit maximization or qualified joint profit maximization.[a] To attain either of these goals, sellers must be able to coordinate their intrinsically rival actions. High concentration, an indication of a high degree of interdependency, increases their ability to act collectively. In an opposite manner, low levels of concentration make it difficult for sellers to reach agreement on a collective course of action because, as the number of participants to any agreement designed to coordinate activities for group interest increases, the ability to reach and maintain the necessary agreement is seriously weakened. Consequently, low levels of seller concentration promote independent profit maximization and competitive performance in terms of resource allocation.

This reasoning leads to the hypothesis that, in the long run, the tendency is for high concentration levels in an industry to be associated with allocative inefficiency evidenced by relatively high profitability. An opposite tendency exists in industries with low concentration. Low concentration levels should be associated with normal long-run rates of return, an indication of allocative efficiency.[3]

Condition of Entry

High concentration, by itself, is not sufficient to lead to allocative inefficiency or excess profitability. In order for inefficiency to result, there must be some restriction on the ability of new firms to enter the industry. If entry conditions were easy, a joint profit maximization strategy leading to excess profitability would attract new firms to the industry and the eventual erosion of any market power. Thus, entry barriers must be present in order for high concentration to result in allocative inefficiency. In this way, entry barriers reduce the influence of potential competition on established sellers and thus weaken competition in an industry.

The Evidence

The hypothesized relationship between levels of seller concentration and allocative efficiency, described briefly above, has been subjected to substantial amounts of empirical research designed to test its validity. Without it, there could be no basis for policy makers to focus upon market structure as a means to promoting competitive performance.

---

[a]The incentive for firms to coordinate their action in this manner is the monopoly profit resulting from successful coordination.

The results tend to support the theory, which reasons that the degree of seller interdependency, as measured by concentration levels, is inversely related to the strength of competition in a market. While not universal, the results are impressive in light of the variety of time periods, profit and concentration measures, and industry samples used in the different research projects.[4]

This book is designed to address policy questions concerning the preservation and promotion of competition in the domestic production of primary energy. While economic theory is essential to formulating public policy, Prof. Baumol has noted some inherent limitations of theory in this role. Theory, according to Baumol, is better suited to the task of recommending against, rather than supporting, a particular policy action. This results from the necessity to construct models which, by their nature, oversimplify reality. Baumol notes that

> . . . since we have good reasons to suspect the representativeness of the model, we are not likely to have much confidence in policy proposals for which it serves as primary justification.[5]

Thus, it is the nature of model building that weakens the ability of theory to take a positive position in formulating policy.

In spite of the weakness of theory, the available evidence linking market structure to performance has been sufficient to cause U.S. antitrust policy to take a clear turn toward a structural approach and away from a conduct approach. The net effect of this change in relative emphasis is almost certain to produce a much more balanced antitrust policy.

INDUSTRY DEFINITION

Concentration measures are only meaningful if constructed on the basis of properly defined economic markets. This involves defining product market boundaries in a manner that yields a close approximation of an economic market. In addition, the proper geographic dimensions of the market must be determined. Unless this is accomplished, the resulting measures of concentration are of little, if any, value in deducing the strength of competition in a market.

Economic theory provides some rather general guidelines for establishing market boundaries. Markets are defined in terms of the degree of substitutability among various products. An economic market, properly defined, includes all products considered to be good substitutes and excludes those products considered to be poor substitutes for the included products. In more specific terms, two products belong in the same economic market if a small change in price (product) causes a significant diversion in a relatively short time of the buyers' purchases or the sellers' production from one product to another. Beyond this point, economic theory is unable to provide additional guidance in defining markets. For instance, theory can not indicate how substitutable products have to be in order to be considered as trading in the same market. The principle upon which market boundaries are determined is clear, but the application of the principle to a real situation is often a difficult task.

The concern of this book is with the primary sources of energy. Consequently, the immediate task is to determine the market boundaries between coal, natural gas, crude oil, and uranium. Do these products trade in the same economic market or in separate and distinct markets? Such a determination is complicated because (1) there are several distinct end uses of energy with differing degrees of interfuel substitutability, and (2) in certain end uses electricity, produced by converting primary fuels into electricity, competes with the primary fuels.

Interfuel Substitutability—Demand

In assessing the degree of interfuel substitutability on the demand side of the market, energy use can be divided into two general categories: (1) nonutility and (2) utility uses. In 1970, utility consumption of primary fuels accounted for some 24 percent of total gross energy inputs in the U.S. economy.

Nonutility Uses. Among nonutility uses of energy, the major purposes of energy consumption include heating, lighting, and power. Measurements of substitutability are difficult because there is little readily available data to indicate consumption and relative fuel prices in specific end uses. Available consumption data is usually presented in terms of broadly defined economic sectors such as household and commercial, and within each sector there are usually several different end uses of energy. Some indication of the extent of substitutability can be obtained by an examination of historical consumption patterns within these sectors.

Household and Commercial. Fuel use patterns during the 1947-1969 period in the household and commercial sectors are indicated by the data presented in Table 2-1. In 1969, fuel consumption, heavily influenced by home heating uses of fuel, is dominated by natural gas and petroleum. There have been dramatic shifts in relative importance among the various fuels. For example, in 1947, coal accounted for over 47 percent of total BTU consumption. Since that time, there has been a virtual elimination in the use of coal due to increased use of natural gas and electricity.

Industrial Sector. Fuel use patterns in the industrial sector are presented in Table 2-2. Again, substantial amounts of interfuel substitution has occurred. Coal's share has declined, and the shares of natural gas and electricity have increased.

Transportation. Fuel use patterns in the transportation sector, described in Table 2-3, have evolved to the point where petroleum accounts for

Table 2-1. Distribution of Total Consumption of Energy Resources, by Major Sources and Consuming Sectors: Household and Commercial Sector

(Percent of Total)

	Coal[a]	Natural Gas Dry[b]	Petroleum[c]	Hydro[d]	Nuclear[d]	Electricity Purchased[e]	Total	Total Sector Energy Input[f] (trillion BTU)
1947	47.4	15.7	31.4	—	—	5.5	100.0	7,165
1949	41.0	18.8	33.5	—	—	6.6	100.0	7,373
1951	31.3	23.7	37.8	—	—	7.3	100.0	8,471
1953	24.4	27.0	39.9	—	—	8.6	100.0	8,490
1955	18.7	30.1	42.2	—	—	9.0	100.0	9,479
1957	12.9	34.8	41.8	—	—	10.5	100.0	9,730
1959	9.2	36.7	43.1	—	—	11.0	100.0	10,952
1961	7.7	37.9	42.6	—	—	11.7	100.0	11,802
1963	6.1	39.6	41.4	—	—	12.9	100.0	12,704
1965	5.2	39.9	40.8	—	—	14.1	100.0	13,815
1967	4.1	40.6	40.5	—	—	14.7	100.0	15,311
1968	3.6	41.3	39.3	—	—	15.8	100.0	15,615
1969	2.7	42.1	38.6	—	—	16.8	100.0	16,357
1970	2.5	41.8	38.0	—	—	17.7	100.0	16,988
1971[P]	2.3	42.2	37.4	—	—	18.0	100.0	17,467

[P]Preliminary.

[a]Includes anthracite and bituminous coal and lignite.

[b]Excludes natural gas liquids.

[c]Includes still gas, LRG, and NGL.

[d]Represents outputs of hydropower (adjusted for net imports and net exports) and nuclear power converted to theoretical energy inputs calculated from prevailing average heat rates at central electric status. Specific values can be found in *Mineral Yearbook*.

[e]Conversion of electricity to energy equivalent by sectors was made at the value of contained energy corresponding to 100 percent efficiency using a theoretical rate of 3,412 BTU per kilowatt-hour.

[f]Direct fuels plus electricity.

Source: U.S. Department of the Interior, Bureau of Mines, *Minerals Yearbook*, Various Years.

Table 2-2.   Distribution of Total Consumption of Energy Resources by Major Sources and Consuming Sectors: Industrial Sector

	Coal[a]	Natural Gas Dry[b]	Petroleum[c]	Hydro[d]	Nuclear[d]	Electricity Purchased[e]	Total	Total Sector Energy Input[f] (trillion BTU)
			(Percent of Total)					
1947	55.1	22.7	18.8	—	—	3.5	100.0	13,254
1949	46.4	28.1	20.8	—	—	4.1	100.0	11,854
1951	44.1	29.3	21.0	—	—	4.5	100.0	14,354
1953	42.1	31.4	21.3	—	—	5.3	100.0	14,515
1955	38.7	32.6	22.0	—	—	6.7	100.0	15,121
1957	37.1	33.7	22.0	—	—	7.2	100.0	15,800
1959	30.9	38.6	22.5	—	—	7.9	100.0	15,340
1961	29.3	39.9	22.7	—	—	8.1	100.0	16,198
1963	28.7	40.5	22.6	—	—	8.3	100.0	17,689
1965	29.9	40.0	21.6	—	—	8.5	100.0	19,184
1967	27.7	42.1	21.1	—	—	9.2	100.0	20,408
1968	25.8	42.6	22.2	—	—	9.4	100.0	21,756
1969	24.5	43.4	22.4	—	—	9.7	100.0	22,172
1970	22.2	45.1	22.8	—	—	9.8	100.0	22,546
1971[P]	19.0	47.4	23.0	—	—	10.4	100.0	22,316

PPreliminary.

aIncludes anthracite and bituminous coal and lignite.

bExcludes natural gas liquids.

cIncludes still gas, LRG, and NGL.

dRepresents outputs of hydropower (adjusted for net imports and net exports) and nuclear power converted to theoretical energy inputs calculated from prevailing average heat rates at central electric status. Specific values can be found in *Minerals Yearbook*.

eConversion of electricity to energy equivalent by sectors was made at the value of contained energy corresponding to 100 percent efficiency using a theoretical rate of 3,412 BTU per kilowatt-hour.

fDirect fuels plus electricity.

Source: U.S. Department of the Interior, Bureau of Mines, *Minerals Yearbook*, Various Years.

Table 2-3. Distribution of Total Consumption of Energy Resources, by Major Sources and Consuming Sectors: Transportation

				(Percent of Total)				
	Coal[a]	*Natural Gas Dry*[b]	*Petroleum*[c]	*Hydro*[d]	*Nuclear*[d]	*Electricity Purchased*[e]	*Total*	*Total Sector Energy Input*[f] *(trillion BTU)*
1947	34.4	Neg.	65.3	—	—	0.3	100.0	8,819
1949	23.3	N.A.	76.0	—	—	0.3	100.0	8,099
1951	16.5	2.2	81.1	—	—	0.3	100.0	9,229
1953	8.8	2.6	88.4	—	—	0.2	100.0	9,224
1955	4.8	2.6	92.4	—	—	0.2	100.0	9,855
1957	2.7	3.0	94.1	—	—	0.1	100.0	10,251
1959	1.0	3.5	95.3	—	—	0.2	100.0	10,408
1961	0.2	3.6	96.1	—	—	0.2	100.0	11,007
1963	0.2	3.7	96.0	—	—	0.2	100.0	11,983
1965	0.1	4.1	95.6	—	—	0.1	100.0	12,733
1967	0.1	4.2	95.6	—	—	0.1	100.0	14,167
1968	0.1	4.0	95.8	—	—	0.1	100.0	15,320
1969	0.1	4.1	95.7	—	—	0.1	100.0	15,970
1970	—	4.5	95.3	—	—	0.1	100.0	16,489
1971[P]	—	4.4	95.4	—	—	0.1	100.0	17,119

[P]Preliminary.

[a]Includes anthracite and bituminous coal and lignite.

[b]Excludes natural gas liquids.

[c]Includes still gas, LRG, and NGL.

[d]Represents outputs of hydropower (adjusted for net imports and net exports) and nuclear power converted to theoretical energy inputs calculated from prevailing average heat rates at central electric status. Specific values can be found in *Minerals Yearbook*.

[e]Conversion of electricity to energy equivalent by sectors was made at the value of contained energy corresponding to 100 percent efficiency using a theoretical rate of 3,412 BTU per kilowatt-hour.

[f]Direct fuels plus electricity.

Source: U.S. Department of the Interior, Bureau of Mines, *Minerals Yearbook*, Various Years.

over 95 percent of total sector inputs in 1970. Coal's relative importance has declined to insignificance due to the substitution of the diesel for the coal locomotive. The main use of gasoline is for the automobile. Technological alternatives to the gasoline engine exist in the form of natural gas and electrically-powered autos. Whether these become economic depends in part on the movement of relative prices. It is certainly true that higher relative prices of gasoline raise the possibility of those substitutes being actually used. Other factors, such as requirements for clean air, also favor these alternatives and can alter the patterns of interfuel competition in the transportation sector.

Interchangeability among the various energy sources, both primary and secondary, in the various nonutility uses is affected by several factors. One, coal is at a basic disadvantage because of air pollution restrictions. Two, consumption decisions in many nonutility uses are generally not responsive to movements in relative prices except at the initial consumption point because of the large capital costs associated with any change in fuel use. For instance, the type of heating system to be installed in a new home is somewhat sensitive to relative fuel prices but, once installed, this sensitivity is reduced substantially. In light of the costs of converting a heating system to a different fuel, very large movements in relative fuel prices are necessary to create an economic incentive for consumers to switch fuels. Balestra, in his study of the demand for natural gas in the U.S. states:

> ... given the particular characteristics of the gas market, it is unrealistic to assume that the consumer's choice is formed according to the principles of the traditional demand theory. In the gas market, for instance, a change in relative prices may not induce the consumer to revise his choice, because of the high transfer costs involved in the shift to a different type of fuel.[6]

As a result, interfuel competition is most properly evaluated in terms of the new or incremental demand for the various fuels.

Three, the statistical technique employed to measure substitutability among products, the estimation of cross elasticity coefficients, is of limited value in defining market boundaries. This procedure yields a number that indicates the responsiveness of consumers' purchases of a product to changes in the prices of other products. The major problem in this approach is with interpreting the results.[7] A priori, there is no way to identify the critical level of the cross elasticity coefficient, which if exceeded warrants the conclusion that the products trade in the same economic market. As a result, there is a strong subjective element in any interpretation of cross elasticity coefficients. Another weakness is that consumption patterns may change for reasons other than movements in relative prices. Tastes are a highly volatile element and have certainly affected consumption levels of coal. Institutional factors, such as air pollution standards, have a significant impact on fuel use patterns. In spite of these difficulties, numerous research results are available that indicate the magnitude of cross elasticity coefficients. But no one has directly addressed himself to the question of whether cross elasticity coefficients can be used to establish market boundaries and, more specifically, to the question of whether the fuels trade in a single market or a different market.

Four, the competitive position of natural gas relative to other fuels has been substantially enhanced because the price of interstate gas has been held below a market clearing level. As expected, this has led to a widespread substitution of gas for other fuels. If the extent of FPC regulation were reduced and gas prices were to rise, changes in relative prices could have an impact on fuel use patterns.

In spite of the impreciseness of the techniques employed to measure product substitutability, there is little doubt that consumers, in attempting to satisfy their demand for fuel in nonutility uses, have several choices in most end uses of energy. While coal is not considered to be a good substitute and uranium is not directly consumed in nonutility uses, gas and electricity are competitive in many end uses. In addition, in markets such as home heating, oil is also competitive in certain geographic areas. The fact that all of the primary fuels can be used to generate electricity in nonutility markets means that, in an indirect sense, all of the primary fuels are substitutable. In a 1963 study of Energy R&D, the Energy Study Group recognized the increasingly important role of electricity as a factor that tends to increase the extent of interfuel competition. They stated:

> The extent to which energy is consumed as electricity is an important determinant of interfuel substitution. While there are some markets for which only one energy form is now economical, as much as 95 percent of total U.S. energy is consumed for purposes in which several or all of the primary sources are potential substitutes (directly or through conversion).[8]

Electric Utility Sector. The electric utility sector of the economy provides the best opportunity to evaluate the economic relationship among the primary fuels. Utilities use large volumes of fuels and, most importantly, the fuels are used for a single purpose, i.e., to produce the heat used to create steam. From the economist's view, one of the benefits of assessing interfuel competition within the electric utility sector is the availability of relatively reliable consumption and price data. The combination of these factors suggests that substitution patterns in the utility sector should provide a very clear indication of the economic relationship among the various fuels.

FTC Study. The FTC's Bureau of Economics has undertaken a series of economic studies concerned with competition in the production of primary fuels in the U.S. The initial report in this series has been released.[9] The purpose of the report is " . . . to analyze the extent of interfuel substitution in order to establish, as precisely as possible, the product boundaries among the various energy sources."[10]

Fuel consumption and cost data for all electric utilities, as reflected in the FTC study, are updated and presented in Tables 2-4 and 2-5. As can be seen, in 1972 coal accounts for some 54 percent of total BTUs consumed by utilities, followed by natural gas with a 27 percent share, and oil with 19 percent of the total. It is important to note that coal's share has declined some 19 percent, while total BTU consumption by electric utilities increased by 273 percent between 1952 and 1972.

Fossil fuel prices, reported to the FPC on an "as burned basis," are presented in Table 2-5. In 1972, the average price of natural gas was 30.3¢ per million BTU. In light of its relative price advantage, mainly due to the ability of utilities to obtain gas on an interruptible contract basis, it is not surprising that the relative importance of gas as a boiler fuel increased over the period. The reason the increase was not greater lies in the simple inability of utilities to obtain larger quantities of gas.

The price of oil to utilities jumped substantially during 1971 and 1972, but in spite of the increase oil's relative importance increased significantly. Coal prices also increased by a large amount.

The extent of potential competition among the fossil fuels is measured in part by the fuel capabilities of installed generating capacity. In 1973, according to the data presented in Table 2-6, slightly more than 50 percent of installed capacity was limited, in a technological sense, to using a single fuel, usually coal. This does not rule out the possibility that this capacity could be converted to multifuel capability in the future. The results indicate that in 1973 nearly 45 percent of installed capacity had some degree of multifuel capability. This allows the utility to take advantage, with a minimum of delay and additional costs, of any movement in relative costs of the fuels. It should be pointed out that the data in the FTC report may contain a bias in the estimation of multifuel capabilities. The bias results from the procedures followed by the National Coal Association in classifying the original data. If a boiler is capable at whatever level of efficiency of burning an alternative fuel, the boiler is listed in the multifuel category. There is no way of determining whether more multifuel capacity exists in the smaller, older, and less efficient portion of the total generating plant. If this were the case, the multifuel capacity, as reported by the NCA, would be biased upward. Another influencing factor is that it is much easier to convert a boiler from coal to oil or gas than to convert from oil or gas to coal.

The final indicator of potential competition is the extent to which utilities view nuclear fuel as a substitute for fossil fuels. An analysis of the type of fuel to be used in the generating capacity either under construction or being

Table 2-4. Distribution of Fuel Use: Conventional Fuel
Plants—U.S.[a]

	Percent accounted for by:			*Total BTU*
Year	*Coal*	*Oil*	*Gas*	*(billion)*
1952	67%	10%	23%	3,793,787
1953	66	11	23	4,170,507
1954	65	9	26	4,260,880
1955	68	9	23	4,831,273
1956	70	8	22	5,280,485
1957	69	8	23	5,566,308
1958	68	8	24	5,422,751
1959	66	8	26	6,030,213
1960	66	8	26	6,419,110
1961	65	8	27	6,742,796
1962	65	7	28	7,107,487
1963	65	7	28	7,719,048
1964	65	7	28	8,339,775
1965	66	8	26	8,762,623
1966	65	8	27	9,716,440
1967	64	9	27	10,144,066
1968	62	10	28	11,114,942
1969	59	12	29	12,068,281
1970	56	14	30	12,959,071
1971	54	16	30	13,424,600
1972	54	19	27	14,179,455
Percent increase 1952-72	−19.4%	90.0%	17.4%	273%
Percent increase 3 year average[b]	−17.2%	63.0%	20.8%	232%

aContiguous U.S., except Idaho.

b1970-72 and 1952-54.

Source: National Coal Association, *Steam-Electric Plant Factors, 1952-73*, Washington, D.C.

planned should provide some insight into this relationship. As can be seen from Table 2-7, nuclear power is emerging as a real substitute for fossil fuels. Nuclear accounted for 27.6 percent and fossil fuel for over 51 percent of the capacity schedule for completion in 1973, but by 1979 the situation will change to the point where nuclear fuel is to be used in 59.2 percent of new capacity and fossil

Table 2-5. Utility Fuel Costs: As Burned Basis—U.S.

CENTER OVER COLUMN:		(Cents per Million BTU)	
Year	Coal	Oil	Gas
1952	27.3	33.1	14.5
1953	27.3	32.3	16.7
1954	26.1	32.8	17.3
1955	25.2	33.2	18.0
1956	28.1	37.9	18.5
1957	27.5	44.4	19.5
1958	27.4	39.6	20.7
1959	26.5	35.2	22.3
1960	26.0	34.5	23.8
1961	25.8	35.5	25.1
1962	25.6	34.5	26.4
1963	25.0	33.5	25.9
1964	24.6	32.6	25.3
1965	24.4	33.1	25.0
1966	24.7	32.4	25.0
1967	25.2	32.2	24.7
1968	25.5	32.8	25.1
1969	26.6	31.9	25.4
1970	31.1	36.6	27.0
1971	36.0	51.5	28.3
1972	38.2	58.8	30.3
Percent increase 1952-1972	17.0%	77.6%	108.9%
Percent increase 3-year average[a]	30.5%	49.8%	75.9%

[a]1970-72 and 1952-54.

Source: National Coal Association, *Steam-Electric Plant Factors, 1952-71*, Washington, D.C.

fuels in 21.7 percent. This clearly indicates that these two types of fuels are competitive.

The conclusion of the FTC Report as to the product boundaries is summarized below in Table 2-8. The report states:

> For the entire U.S., the evidence indicates that coal, oil, natural gas, and uranium are sufficiently substitutable in their use by electric utilities to support the conclusion that they trade in the same economic market.[11]

Table 2-6. Installed Generating Capacity: Distributed by Fuel Capability—U.S.

Fuel Design	1960	1963	1966	1969	1973
Single fuel:					
Coal	48.0%	46.9%	41.1%	40.8%	39.3%
Oil	2.0	2.0	2.4	2.3	3.7
Gas	6.7	7.2	7.0	7.2	7.4
Subtotal	56.7	56.1	50.5	50.3	50.4
Multiple fuel:					
Coal, oil	7.7	4.1	9.3	9.9	8.6
Coal, gas	11.9	12.7	11.3	10.4	9.5
Oil, gas	16.0	21.0	19.3	19.2	20.2
Coal, oil, gas	7.4	5.7	9.2	8.3	6.3
Subtotal	43.0	43.6	49.0	47.9	44.6
Nuclear:	0.2	0.3	0.5	1.8	4.7
Total[a]	100.0	100.0	100.0	100.0	100.0
Total capacity[b] (thous. kws.)	127,355.1	146,889.6	194,653.9	247,611.1	311,891.7

[a]May not add to 100 due to rounding.
[b]Maximum generator nameplate rating.
Source: National Coal Association, *Steam-Electric Plant Factors*, Washington, D.C.

**Table 2-7.   Fuel Capability of New Generating Capacity:
Planned or Under Construction—U.S.**

		Scheduled Year of Completion		
Fuel Capability[a]	*1973*	*1975*	*1977*	*1979*[b]
		(Percent)		
Conventional Fuels plus geothermal	51.7	61.9	58.9	21.7
Nuclear	27.6	24.6	29.9	59.2
Hydro	9.9	9.7	9.3	17.8
Other[c]	10.8	3.8	1.9	1.3
Total	100.0	100.0	100.0	100.0
Total Capacity[d] (thous. kw.)	50,734	44,121	39,504	38,643

[a]Fuel capability, as reported in *Steam-Electric Plant Factors*, is subject to change for any given plant.

[b]Because of differences in construction lags, a direct comparison of conventional vs. nuclear capacity in 1977 and 1979 can not be made.

[c]Includes gas turbines, internal combustion, and capacity for which fuel capability is currently not known.

[d]Name plate rating in case of nuclear, capacity is as reported by *Electrical World*, June 1, 1973.

Source: National Coal Association, *Steam-Electric Plant Factors, 1973*, Washington, D.C.

Given the unevenness with which fuel reserves are distributed throughout the U.S. and the fact that transportation costs for certain of the fuels are high, market boundaries may differ at the regional level. There are several differences at this level, with the major difference occurring in the West South Central Area, where gas captures the entire market.[b]

Interfuel Substitutability: Supply Side—Long Run

In defining market boundaries, the degree of product substitutability is traditionally assessed from the demand side of the marketplace. It is however equally important to analyze the substitution patterns on the supply side of the market. In terms of supply, two products belong in the same economic market if a small change in price (product) causes producers to change their output mix from one product to the other in a relatively short time.

The importance of supply substitutability for interfuel competition lies mainly in the future, when most likely commercial processes for the liquefaction and gasification of coal will be available. The significance of this is that coal, via conversion to oil or gas, becomes an indirect substitute for natural gas and oil in those end uses of energy where, in its natural form, coal is not presently interchangeable with oil and gas. Coal producers will be able to decide,

---

[b]Natural gas has such an economic advantage that other fuels are not even considered to be potential substitutes.

Table 2-8. Interfuel Competition and Market Boundaries: Electric Utility Sector

Region	Type of Interfuel Competition — Actual	Type of Interfuel Competition — Potential	Market Definition
1) Total United States	Coal Oil Gas	Uranium	Coal Oil Gas Uranium
2) New England	Coal Oil	Uranium	Coal Oil Uranium
3) Middle Atlantic	Coal Oil Gas[a]	Uranium	Coal Oil Gas[a] Uranium
4) East North Central	Coal	Gas Uranium	Coal Gas Uranium
5) West North Central	Coal Gas	Uranium	Coal Gas Uranium
6) South Atlantic	Coal Oil Gas	Uranium	Coal Oil Gas Uranium
7) East South Central	Coal	Gas Uranium	Coal Gas Uranium
8) West South Central	Gas	–	Gas
9) Mountain	Coal Gas Oil[b]	–	Coal Gas Oil[b]
10) Pacific	Oil Gas	Uranium	Oil Gas Uranium

[a]Gas is competitive only in certain metropolitan areas.
[b]Oil is competitive only in certain areas of the State of Utah.
Source: Thomas D. Duchesneau, *Interfuel Substitutability in the Electric Utility Sector of the U.S. Economy*, Economic Report to FTC, Feb. 1972, p. 81.

in response to relative price movements, whether to produce coal or synthetic oil and gas. At present, coal has little consumer appeal other than to electric utilities and even in this case, coal's future is questionable given the present inability to control sulfur emissions. Consequently, the ability to produce synthetic fuels from coal should be a strong procompetitive development in the energy market.

This development, in terms of its impact on interfuel substitution, is similar to the electricity example discussed above.

GEOGRAPHIC BOUNDARIES OF THE MARKET

The proper geographic boundaries, as well as product boundaries, of a market must be established before the construction of measures of market structure. In the case of primary energy sources, the unevenness of the distribution of energy reserves and the importance of transportation costs raise the real possibility that competition may occur within but not between regions. Professor Thomas Hogarty has analyzed the geographic trading patterns of primary energy sources. His report, presented in Appendix A, serves as the basis for the following discussion.

There is substantial theoretical literature on the question of geographic markets but relatively little research has been conducted at the empirical level. This may, in part, be due to the relatively high degree of uncertainty in interpreting empirical results in this area. Past attempts to delineate the geographic dimensions of a market have often erred because of incorrect procedures. The proper method is one that incorporates both supply and demand elements in the analysis of trade patterns. The failure to include both elements can lead to serious error.

Professors Elzinga and Hogarty have developed a technique to define geographic markets. Their procedure examines trade flows into and out of a possible market area. To be a distinct geographic market, two criteria must be satisfied simultaneously: (1) relatively small amounts of the product flow into the region from outside; in Hogarty's terminology, the little in from outside standard (LIFO),[c] and (2) relatively small amounts of the product flow out of the region, the little out from inside standard (LOFI)[d]. It should be emphasized that only by satisfying both criteria simultaneously can a region be considered to be a distinct geographic market. In interpreting the results of the analysis there is an element of subjective judgement. This mainly arises because economic theory, while able to indicate the principles by which geographic boundaries should be established, cannot a priori indicate the exact criterion that, if exceeded, warrants a conclusion that the market is geographic rather than national in scope. In the study of energy markets, 75 percent and 90 percent were adopted as weak and strong standards respectively. In addition, an analysis of trading flows at a given point in time is sensitive to the existing relative prices. Changes in relative prices can lead to different geographic boundaries.

The steps of the procedure are as follows:[12]

---

[c]Little In From Outside, signifying a market area relatively free of "imports."

[d]Little Out From Inside, signifying that market area was not merely an "export" base.

1. Identify the product's major producing centers (areas). If only one major producing center exists and the product is sold nationally, then the market is at least national in scope.
2. Next, organize the product's shipments data in terms of both production origin and consumption destination. Typically, the latter will be available in political units such as states, and for most purposes this is adequate.
3. Taking each producing area one at a time, calculate the minimum area (e.g., minimum number of states) required to account for at least 75 (90) percent of shipments from the producing area. This will satisfy the weak (strong) form of the LOFI criterion. Designate this area as a hypothetical market area (HMA).
4. Of total shipments to destinations within the HMA, do 75 (90) percent or more of the shipments originate from the designated producing area? If so, the weak (strong) form of the LIFO test is met. So long as only one producing center exists within the HMA, the LOFI test is met through step 3. Given more than one producing center, however, the LOFI test is met through step 3. Given more than one producing center, however, the LOFI test must be repeated. If the LIFO test is not met, then redraw the HMA, determining the minimum area necessary to absorb 75 (90) percent of the shipments from producing centers located within the new HMA. If there is no subnational area satisfying this criterion, the market is (at least) national in scope.
5. If both the LOFI (step 3) and LIFO (step 4) criteria are satisfied, the HMA comprises a distinct geographic market area. Assuming this area is less than national in size, the final step consists in calculating total consumption within this area to get market size in terms of volume.

The Findings

Natural Gas. A market in which all producers are located at a single geographic point and with customers nationwide can be considered to be at least national in scope. Production of natural gas is clearly centralized in the U.S. Most of the states have some marketed production of natural gas but, as indicated in Table 2-9, some 90 percent of marketed production originated in only 5 states. Texas and Louisiana, in combination, account for more than 70 percent of total production. Hence, the five-state area composed of Texas, Louisiana, Oklahoma, New Mexico, and Kansas represents the producing area and in combination with nationwide sales of gas suggests that the natural gas market is national in scope.

Presently there are some imports of liquified natural gas (LNG), but at a cost far above domestic gas. In spite of a price differential, imports occur because of extremely limited supplies of new domestic gas. If gas prices are

Table 2-9. Marketed Production of Natural Gas, Five Largest Producing States, 1967-1971 (million cubic feet)

Year	Texas	Louisiana	Oklahoma	New Mexico	Kansas	Total (5 states)	Total (all states)
1967	7,188,900 (39.56)	5,716,857 (31.46)	1,412,952 (7.77)	1,067,510 (5.87)	871,971 (4.79)	16,258,190 (89.47)	18,171,325
1968	7,495,414 (38.79)	6,416,015 (33.20)	1,390,884 (7.19)	1,164,182 (6.02)	835,555 (4.32)	17,302,050 (89.54)	19,322,400
1969	7,853,199 (37.94)	7,227,826 (34.92)	1,523,715 (7.36)	1,138,133 (5.49)	883,156 (4.26)	18,626,029 (89.98)	20,698,240
1970	8,357,716 (38.12)	7,788,276 (35.52)	1,594,943 (7.27)	1,138,980 (5.19)	899,955 (4.11)	19,779,870 (90.23)	21,920,642
1971	8,550,705 (38.01)	8,081,907 (35.93)	1,684,260 (7.48)	1,167,577 (5.19)	885,144 (3.93)	20,369,593 (90.55)	22,493,012

Numbers in parentheses are percentage figures. Detail may not add to total due to rounding.

Source: U.S. Department of the Interior, Bureau of Mines, *Mineral Industry Surveys: Natural Gas Production and Consumption* (various years).

deregulated and increase substantially, geographic market boundaries may change. When the price of domestic natural gas reaches a level equal to the price of imported LNG, there will probably be a rapid development of an international market for gas.

Crude Oil. Refinery receipts of crude oil, classified by state of origin and destination, serve as the basis for analyzing trade flows of crude oil. Initial attention focused upon Texas and Louisiana, where in 1971 76 percent and 84 percent, respectively, of refinery receipts of crude oil were intrastate, thus satisfying the weak form of the LIFO test. But because only 64 percent of Texas crude shipments and 38 percent of Louisiana shipments were made intrastate, the weak form of the LOFI standard was not met.[e] Thus, neither Texas nor Louisiana represents distinct geographic markets.

Further tests based on varying configurations of states were conducted. Table 2-10 contains the results of our geographic market area tests. Examining the 1971 data, we find that: (1) shipments from Area I to five states[f] absorbed 76.22 percent of shipments from Area I (LOFI criterion); and (2) of total shipments to these five states, 92.60 percent originated in Area I (LIFO criterion). On the other hand, while nine states[g] absorbed 92.40 percent of Area I shipments, only 84.73 percent of total shipments to these nine states originated in Area I. Thus, Area I is a distinct market area only on the weak form of our test. The second area failed to pass even the weak form. Since previous tests demonstrated that no subarea could be used to delineate a distinct market area, the market for crude oil is at least national in scope.

Of total domestic refinery receipts of crude oil, only about 64 percent originates in Area I. Thus, crude oil from other sources (both domestic and foreign) constitutes a substantial share of refinery receipts and, unlike the situation in natural gas, the five states composing Producing Area I do not (even approximately) supply the nation's demands. In addition, the numerous other sources of crude oil cannot be used as the basis for a second geographic area. Finally, the market based on Area I shipments is not distinct under the strong form of our test. These results suggest that the market for crude oil is at least national and probably international in scope.

The results reported by Prof. Hogarty are based on a methodology that matches crude oil supply areas with refinery demand areas. By comparing crude oil production areas with consumption areas, the analysis neglects intermediate markets such as pipelines. Hogarty's finding of a national market

---

[e]For example, if we combine Texas and Louisiana, then we find that only 63 percent of total shipments of crude oil from these states is destined for refineries within these states.

[f]Texas, Louisiana, Illinois, Oklahoma, and Indiana.

[g]In addition to the five cited in the previous footnote, we had Ohio, Pennsylvania, Kansas, and New Jersey.

Table 2-10. Geographic Scope of Market Areas for Crude Oil, 1967 and 1971

	1971	*1967*
Minimum number of states[a] required to absorb 75 (90) percent of shipments from Area I[b] (LOFI)	5 (9)	5 (9)
Percent of shipments to these states originating in Area I (LIFO)[c]	92.60 (84.73)	96.09 (84.55)
Minimum number of states[d] required to absorb 75 (90) percent of shipments from Area II[e] (LOFI)	13	9
Percent of shipments to these states originating in Area II (LIFO)[f]	62.21	59.39

[a]These states (in order of importance, 1971) were Texas, Louisiana, Illinois, Oklahoma, Indiana, Ohio, Pennsylvania, Kansas, and New Jersey.

[b]Producing Area I comprised the states Texas, Louisiana, Oklahoma, New Mexico, and Kansas.

[c]Denominator includes imports.

[d]These states (in order of importance, 1971) were California, Pennsylvania, New Jersey, Indiana, Ohio, Illinois, Wyoming, Utah, Montana, Michigan, New York, Minnesota, and Wisconsin.

[e]Area II comprised shipments of crude oil from foreign sources plus all producing states except the 5 in Area I.

[f]Producing Area II failed to meet even the weak form of the test, hence a test involving the strong criteria was not conducted.

Source: U.S. Department of the Interior, Bureau of Mines, *Mineral Industry Surveys: Crude Petroleum, Petroleum Products, and Natural Gas Liquids* (1967, 1971).

between crude producers and refinery user is not necessarily relevant in the case of a pipeline buyer of crude oil. A crude producer may be in a situation where there is only a single buyer in the form of a pipeline operator. In this case, the geographic dimensions of the market are quite restricted.

In the case of a specific refinery, options in buying crude oil may be limited by its location and the characteristics of the pipeline delivery system. The lack of a delivery system from certain crude oil fields to the refinery restricts the refinery's choice of suppliers and results in a less than national market. The Hogarty study, by working with crude oil flows at the state level, does not address this type of situation. But even though a refinery may appear to face a regional crude oil market, its options are often much wider. A given refinery can often buy crude in any region and proceed to use it as trading stock with other firms who have crude available in the area of the refinery. In this way, the relevant trading area becomes much wider.

Coal. An analysis of geographic trading patterns of coal contains some ambiguity. Several factors account for this, including: (1) the unevenness of the distribution of the nation's coal resources; (2) the variations in the quality of coal, especially in terms of sulfur content; (3) controls on sulfur emissions; and (4) innovations in coal transportation.

The analysis of geographic trading patterns of coal is based upon the aggregate producing areas reported in Table 2-11. The results of the tests are presented in Table 2-12. The Eastern area, composed of nine states, represents a distinct trading area on the basis of the weak form of the test, but not on the strong form. The remaining areas, the Central and Western region, are not separate markets in 1971 but, when combined, they represent a distinct market on grounds of the weak standard. Thus, the results for 1971 indicate the existence of two market areas, an Eastern and combined Central and Western market.

The 1971 results conflict with the findings of earlier years. In the years 1965, 1969, and 1970, there are distinct market areas; and in 1961, there were two distinct market areas plus a "no-man's land." Hogarty's analysis reports that the 1961 results may reflect the impact of the severe recession in 1961, during which Eastern coal producers cut prices sufficiently so that Eastern coal was competitive with Central coal. The 1971 results appear to be heavily influenced by air pollution controls. Enforcement of emission controls has broadened the market areas to the point where the Western and Central areas represent a single trading area.

The interpretation of the results is somewhat ambiguous. However, the important point is that the results for 1971 are more consistent with the presumption of a nationwide market for coal than those for earlier years. In

Table 2-11. Coal Producing Areas and Corresponding Bureau of Mines Producing Districts

Coal Producing Area	Bureau of Mines Districts[a]
Eastern[b]	Districts 1-8, 13
Central[c]	Districts 9-12, 14-15
Western[d]	Districts 16-23

[a]Complete descriptions of these districts may be found in the source cited below. In some cases the districts comprise entire states (e.g., District 5 is Michigan); in other instances only part of a state is included (e.g., District 2 is Western Pennsylvania).

[b]Comprises Pennsylvania, Ohio, West Virginia, Michigan, Eastern Kentucky, Virginia, Tennessee, Alabama, and Georgia.

[c]Comprises Western Kentucky, Illinois, Indiana, Iowa, Arkansas, and Oklahoma.

[d]Comprises Colorado, New Mexico, Wyoming, Utah, North and South Dakota, Montana, Washington, Oregon, and Alaska.

Source: U.S. Department of the Interior, Bureau of Mines, *Mineral Industry Surveys: Bituminous Coal and Lignite Distribution* (various years).

Table 2-12.   Geographic Scope of Market Areas for Bituminous Coal and Lignite, Various Years

	1971	1970	1969	1965	1961
Minimum number of states required to absorb 75 (90) percent of shipments from eastern area (LOFI)	9[a] (15)	8[a] (14)	9[a] (15)	9[a] (16)	10[a] (15)
Percent of shipments to these states originating in eastern area (LIFO)	92.90 (81.55)	87.29 (85.31)	86.21 (86.48)	89.82 (86.01)	92.81 (85.17)
Minimum number of states required to absorb 75 (90) percent of shipments from central area (LOFI)	7	6[a] (10)	5[a] (10)	5[a] (10)	5 (9)
Percent of shipments to these states originating in central area (LIFO)	69.17	77.25 (69.19)	80.08 (72.89)	76.70 (66.68)	71.51 (58.49)
Minimum number of states required to absorb 75 (90) percent of shipments from western area (LOFI)	9	7[a] (13)	7[a] (12)	7[a] (10)	6[a] (11)
Percent of shipments to these states originating in western area (LIFO)	42.55	81.17 (42.36)	99.46 (77.79)	98.97 (73.09)	96.83 (71.78)
Minimum number of states required to absorb 75 (90) percent of shipments from central and western area combined (LOFI)	10[a] (19)				
Percent of shipments to these states originating in central and western areas combined (LIFO)	77.75 (53.38)				

[a]Denotes distinct geographic market area.

Source: See source, Table 13, p. 48.

addition to reflecting the impact of air pollution controls, this trend may reflect the delayed impact of transport innovations such as the unit train, large hopper cars, and improved barge transport of coal. A second additional factor—at least for the future—would be technological advance in producing synthetic gas from coal. Finally, a sign of expansion in market scope is the increase in exports of coal, especially to the steel industries of Europe and Japan.

Uranium. Enriched uranium is a highly valued product, and transportation costs represent a minor part of its costs. Consequently, the trading area for uranium fuels, by nearly anyone's analysis, would be at least national in scope and, no formal application of the Hogarty methodology was conducted.

SELLER CONCENTRATION: LEVELS AND TRENDS

Measures of seller concentration in the production of crude oil, natural gas, coal, and uranium are presented below. It should be emphasized that concentration levels mean little by themselves, but are of interest because they allow an inference about the degree of interdependency among sellers in a market. In addition, concentration measures do not reflect seller interdependency due to factors other than the existence of only a few sellers, i.e., institutional factors.

Crude Oil

The petroleum industry in the U.S. encompasses: (1) crude oil production, (2) transportation by pipeline or tanker, (3) refining, and (4) marketing of refined products. The industry comprises a large number of firms, but a relatively small number of very large, fully integrated companies, often referred to as majors, dominate economic activity at each production stage. In general, independents are small firms and are not fully integrated across the different functions. In 1967, crude oil production accounted for some 63 percent of profits earned by integrated companies; in addition 6 percent came from transportation, 21 percent from refining and marketing, and 10 percent came from petrochemicals.[13] This book is primarily concerned with crude production because events at the crude level are a major determinant of market conditions in downstream markets.

Concentration measures for domestic crude oil production are reported below. Obtaining a consistent set of data has been complicated by substantial variation in the definition of crude production as reported by individual firms. Some firms report net crude production defined as total crude production less royalty oil production. Royalty oil is oil produced and usually controlled by the company but in a legal sense owned by the landowner. In other cases, companies report gross production defined as net production plus royalty production. In terms of determining the extent of dominant firm control over production, gross production is obviously a more relevant statistic. Where net production is used, the resulting figure is biased downward, the general size of the bias being indicated by Cookenboo's estimate that royalty oil was equivalent to about 18 percent of total production and de Chazeau and Kahn's estimate of 12.5 percent.[14] Other sources of inconsistency in the data are the manner in which natural gas liquids are treated and, in limited cases, the inability to separate U.S. from Canadian production. The production figures reported in this paper were obtained from company annual reports, *Moody's Industrial Manual,* and *Standard and Poors Industry Surveys* and, more often than not, indicate a company's net crude oil production.

The top domestic producers of crude oil for 1955 and 1970 are listed in Tables 2-13 and 2-14 respectively. The concentration measures are summarized in Table 2-15. In 1955, the four largest firms accounted for some 18.8 percent of total domestic crude oil production; by 1970 the figure for the four firms was 32.5 percent, an increase of approximately 73 percent. At the

Table 2-13.   The Twenty Largest Producers of Crude Oil In
The U.S.: 1955

	Net Domestic Crude Production 1955[1] (000 Barrels)	Percentage of Total	Cumulative Percentage
Standard Oil of N.J.	143,175	6.1	6.1
Texas Co.	115,920	4.9	11.0
Shell Oil Co.	95,220	4.1	15.1
Standard Oil of Calif.	86,595	3.7	18.8
Standard Oil of Ind.	83,145	3.5	22.3
Gulf Oil Corp.	79,695	3.4	25.7
Socony Mobil	79,350	3.4	29.1
Continental Oil Co.	46,230	2.0	31.1
Phillips Petroleum	43,125	1.8	32.9
Sinclair	41,055	1.7	34.6
Sun Oil	35,190	1.5	36.1
Union Oil of Calif.	35,190	1.5	37.6
Ohio Oil Co.	33,465	1.4	39.0
Cities Service	32,775	1.4	40.4
Tide Water Assoc. Oil Co.	30,705	1.3	41.7
Atlantic Refining	29,670	1.3	43.0
Sunray Mid Continent	27,600	1.2	44.2
Skelly Oil Co.	23,460	1.0	45.2
Pure Oil	23,115	1.0	46.2
Total: Top 80	1,084,680		
Total: U.S.	2,348,415		

[1] Thousands of barrels per day multiplied by 345. Barrels per day from: Melvin de Chazeau and Alfred E. Kahn, *Integration and Competition in the Petroleum Industry*, (New Haven: Yale University Press, 1959), pp. 30-31.

Source: Melvin de Chazeau and Alfred E. Kahn, *Integration and Competition in the Petroleum Industry* (New Haven: Yale University Press, 1959).

Table 2-14.   The Twenty Largest Crude Oil Producers U.S.:
1970

	Crude Oil Production[1] (000 Barrels)	Percent of Total	Cumulative Percentage
Standard Oil of N.J.	376,614	10.7	10.7
Texaco	319,676	9.1	19.8
Gulf Oil Corp.	214,718	6.9	26.7
Shell Oil Co.	204,085	5.8	30.5
Standard Oil of Calif.	177,331	5.0	37.5
Standard Oil of Ind.	159,838	4.5	42.0
Atlantic Richfield	151,503	4.3	46.3
Getty	134,456	3.8	50.1
Mobil	132,055	3.8	53.9
Union Oil of Calif.	95,902	2.8	56.7
Sun Oil Co.	78,632	2.2	58.9
Marathon	63,820	1.8	60.7
Continental Oil Co.	60,368	1.7	62.4
Phillips	47,677	1.4	63.8
Cities Service	45,001	1.3	65.1
Amerada Hess	30,879	0.9	66.0
Tenneco	29,576	0.8	66.8
Louisiana Land & Exploration	22,617	0.6	67.4
Superior Oil	18,607	0.5	67.9
Standard Oil of Ohio	10,497	0.3	68.2
Total: Top 20	2,398,900		
Total: U.S.[2]	3,517,450		

[1] Based on average daily production of crude oil as reported in annual reports and Moody's. In all cases, it was not possible to separate gross and net production.

[2] Bureau of the Mines, *Minerals Yearbook* (Washington, D.C.: Government Printing Office, 1972), p. 817.

Source: Annual Reports and Moody's *Industrial Manual.*

eight firm level, concentration in 1955 and 1970 was 31.1 percent and 50.1 percent respectively. Concentration in domestic crude oil production is not excessive relative to many manufacturing industries, but the presence of an upward trend in the concentration levels is significant. To some extent, the upward movement in concentration reflects the easing of market demand prorationing.

Concentration ratios based on production data are an indication of past events in the industry and may not be an adequate reflection of the future. A more appropriate statistic would be concentration figures based upon the extent of dominant firm control of domestic crude oil reserves. For obvious

Table 2-15.   Seller Concentration: U.S. Crude Oil
Production, 1955-1970

| | Concentration Level | | | Increase |
	1955	*1960*[b]	*1970*	*1970-55 %*
Top four	18.8%	26.5%	30.5%	62.2%
Top eight	31.1	43.8	50.1	61.1
Top twenty	46.2[a]	63.0	68.2	48.0

[a]Represents share of top nineteen firms.

[b]Obtained from: Permanent Subcommittee at Investigations of the Committee on Government Operations, "Preliminary Federal Trade Commission Staff Report on Its Investigation of the Petroleum Industry," *Investigation of the Petroleum Industry* (Washington: Government Printing Office, July 12, 1973), p. 7.

reasons, firms do not make this information public and, as a result, reliable information is seldom available. Estimates, based on public data sources are presented in Table 2-16.

Independent Crude Producers.   Some indication of the relationship between the majors and independent crude producers can be obtained from data collected by an FTC survey of independent crude oil producers during the 1967-1971 period.[15]   Data was obtained from some 50 firms and, on the average, the responding firms were larger than the other independent firms.

The FTC survey results indicate a close relationship between independent crude oil producers and major petroleum firms. Of total crude oil sales by independents in 1971, 44.4 percent were to the eight largest majors and 25.8 percent to other majors;[h] in other words, 70.2 percent of independent crude oil production was purchased crude from independents; the crude oil gathering lines were owned by the majors. In 1971, the eight largest majors owned 36 percent of the gathering lines in the crude oil fields operated by independents. When combined with the 24.8 percent of gathering lines owned by other majors, the results indicate that majors control 60.8 percent of the gathering lines used to collect crude production by independents. In addition, the FTC survey reports that once gathering lines are in place, the number of buyers of crude from that field are substantially reduced. Thus, the eight largest majors, by purchasing some 44 percent of independently produced crude and owning 36 percent of the gathering lines, are closely related to independents.

The World Oil Market.   There is every indication that crude oil imports are going to be an increasingly important source of energy in the U.S. Domestic crude oil capacity reached a peak in 1970 and will probably continue to decline until Alaskan production begins to flow. In addition, the substitution of a modest fee system for the import quota program has resulted in a substantial reduction in the restrictions on importing oil.

---

[h]The FTC Report fails to identify the companies in this category.

Table 2-16. Company Shares of Proved Domestic Crude Reserves, 1970

Rank	Company	Share of Domestic Proved Reserves (Percent), 1970
1	Standard (N.J.)[1,2]	9.92
2	Texaco[1,3]	9.31
3	Gulf	8.97
4	Standard (Calif.)[1,3]	8.97
5	Standard (Ind.)[4,3]	8.46
6	ARCO[3,2]	7.48
7	Shell[1]	5.90
8	Mobil[1,3]	4.87
9	Getty[5]	3.85
10	Phillips[1,3,6]	3.55
11	Signal	3.28
12	Union[3]	3.18
13	Continental[3,6]	2.77
14	Sun[1,3]	2.67
15	Amerada Hess[3,7]	2.49
16	Cities Service[1,3]	2.49
17	Marathon[3,5]	2.37
18	Skelly	1.09
19	Superior[1,8]	1.03
20	Tenneco	.90
	Top 4	37.17
	Top 8	63.88
	Top 20	93.55

[1] Includes natural gas liquids.

[2] Includes Alaska.

[3] Includes Canada.

[4] Official, including natural gas liquids.

[5] Official.

[6] Excludes controlled reserves.

[7] Includes equity in Canadian affiliate.

[8] Excludes "probably additional" reserve.

These explanatory notes accompanied the reserve estimates.

Note: Individual company reserves obtained from estimates of Rice, Kerr & Co., Engineers. Universe data obtained from "Report on Crude Oil and Gasoline Price Increases," cited supra. This figure was taken as approximately 39,000,000,000 barrels, including Alaska. Concentration data would be lower the more this figure understates actual proven reserves. Concentration data should be taken to be, at best, only rough estimates. It is not known to what extent both the universe estimate and the individual company reserve data include crude production from secondary recovery techniques. The omission of Standard Oil of Ohio from the table suggests that Alaskan North Slope reserves are not included in the data.

Source: Permanent Subcommittee on Investigations of the Committee on Government Operations, "Preliminary Federal Trade Commission Staff Report on Its Investigation of the Petroleum Industry," *Investigation of the Petroleum Industry* (Washington: Government Printing Office, July 12, 1973), p. 7.

Professor Adelman, in an analysis of the world petroleum market,[16] has constructed measures of seller concentration for the world market. His figures represent the extent of dominant firm control of production shares in countries other than Communist countries and North America. As such, the figures do not represent actual production levels, but the amount firms are entitled to produce out of a field under the provisions of various joint ventures. To the extent that actual crude liftings by firm deviate from the firm's proportionate share, a bias is introduced.

The results, based on 1950, 1957, 1966 and the first half of 1969, are presented in Table 2-17. The evidence indicates a continual decline in concentration levels since 1950. The decline during the 1966-1969 period is somewhat less than the earlier period in part because of the June 1967 war. The decline in concentration has been dramatic but still remains significantly higher than in the U.S. Adelman argues that the major reason for the dramatic decline

Table 2-17. Concentration Ratios and Number of Firms-equivalent in World Oil Market, Selected Years

		1950	*1957*	*1966*	*First Half 1969*
	A. Production Shares				
Largest four:	Esso	30.4	22.8	18.0	16.6
	BP	26.3	14.4	17.0	16.1
	Shell	13.8	17.5	12.9	13.3
	Gulf	12.1	14.8	10.8	9.8
	Subtotal	82.6	69.5	58.7	55.8
Lesser four:	Texaco, SoCal, Mobil, CFP	17.4	22.2	24.7	25.0
All other:		0.0	8.3	16.5	19.2
	Total	100.0	100.0	100.0	100.0
	B.				
Herfindahl measure of concentration		0.2039	0.1319	0.1116	0.1046
Number of firms-equivalent		4.9	7.6	9.0	9.6

The Herfindahl measure is the sum of the squared market percentages of the firms in the market. "Firms-equivalent" is the reciprocal of H-measure; see M.A. Adelman, "On the H-measure as a Firm-Equivalent," *Review of Economics and Statistics*, Feb. 1969.

Note: Table excludes data on Communist countries and North America.

Source: M.A. Adelman, *The World Oil Market* (Baltimore: The Johns Hopkins Press, 1972), p. 81.

has been the high growth rate for petroleum products in the world. Small markets prior to World War II effectively raised entry barriers and resulted in a small number of sellers at each stage.[17]

Crude oil production and, most importantly, the geographic distribution of production have changed dramatically. Several factors account for this, including (a) expansion of capacity by dominant firms, (b) entry of new oil-producing countries, and (c) the entry of new companies, both foreign national companies and smaller U.S. companies, such as Phillips Petroleum, Occidental Petroleum, and Standard Oil of Indiana.[18] The entry of host governments, as sellers of participation oil or nationalized production, has materially affected historical concentration measures.

These factors have combined to bring about fundamental change in the structure of the international oil market. Because of import quotas their impact on U.S. markets was reduced, but with the lifting of restrictions on imports to the U.S., these developments, along with the formation of OPEC, are major influences on the U.S. market.

Effective seller competition requires independence among sellers. The absence of independence, or the existence of interdependency, creates a setting in which sellers recognize that self-interest supports the adoption of collusive price and product policies. Concentration measures are an indirect measure of the degree of interdependency in a market, thus allowing an inference about the strength of competition in the market. The concentration measures for domestic crude oil production presented here are only moderate in comparison with many manufacturing industries, but any conclusion about the effectiveness of competition would be premature because numerous joint ventures among large oil companies raise substantial doubts about their competitive independence. Firms that are parties in significant joint ventures may not be truly independent competitors.

The structure of domestic crude oil production is intimately related to the world market and to the extent to which domestic producers are protected from import competition. The leading producers in the world market are also leading domestic producers. Concentration in the world market has historically been significantly above domestic levels, but there has been a dramatic decline in world levels where concentration at the four-firm level has fallen from 82.6 percent in 1950 to 55.8 percent in early 1969. Whether this trend will continue in light of OPEC's strength remains to be seen. In any case, the 1969 levels still exceed domestic levels. Consequently, as the relative importance of imported crude oil increases, concentration of crude oil sales in the U.S. will increase.

Another factor influencing crude oil concentration will be the entry of oil from the North Slope. The impact will depend upon the composition of the leading firms in Alaskan production in relation to the dominant producers in the lower 48 states. Data from a 1969 lease sale indicates that lease ownership is heavily concentrated among the leading domestic crude producers. A review of the sale results indicate that major oil companies successfully bid, usually via joint bidding among themselves, for over 75 percent of the acreage offered for lease.[19] It should be pointed out that many of the leases may contribute very little actual production, and concentration in terms of Alaskan production will most likely differ from the concentration based on lease sales. There is at least a possibility that Alaskan production will reduce the upward concentration trend in domestic production, but the dominance of lease sales by leading domestic producers suggests otherwise.

Petroleum Refining

Petroleum refining involves the production of various petroleum products from crude oil inputs. The major products of domestic refineries are gasoline and distillable fuel oil and account for about 45 and 22 percent, respectively, of total refinery output. On January 1, 1971, there were some 253 operating domestic refineries representing 129 refining companies. During the 1966-1972 period, domestic refining capacity increased by some 27 percent, with a substantial proportion of the increase occurring during the first four years of the period.

The top refining companies for 1955 and 1972 are listed in Tables 2-18 and 2-19. Concentration ratios are summarized in Table 2-20. In 1972, the top four firms accounted for 33.1 percent of domestic capacity and the top eight firms controlled 59 percent of total capacity. The 1970 concentration levels represent only slight increases over the 1955 levels. Concentration in petroleum refining in 1970, while slightly higher than in crude oil production, can only be considered moderate.

Entry barriers do not appear to be prohibitively high. Refinery construction requires large amounts of capital and a guaranteed access to crude oil supplies. The former, while large in a total sense, does not represent a serious barrier to entry. In Bain's study of entry conditions, a minimum efficient size refinery contained less than 2 percent of the industry output. Important obstacles to new domestic refinery construction in recent years have been due to an inability to obtain low cost, foreign crude supplies and siting limitations due to environmental concerns. The former was, to an important degree, due to the oil import quota program. With the recent movement from a quota to a fee system and special exemptions for new refineries, barriers to new refinery construction appear to have been reduced substantially.

Under the new regulations, new refining capacity is allowed to import 75 percent of its crude oil needs free of the 21¢ per bbl license fee for five years. It was recently reported that several companies are planning new expansions in the Northeast and Midwest areas.[20] Among the companies are Mobil, Standard Oil of California, and Ashland. Reversing the usual pattern, Pittston Co., a large coal producer, has announced plans to construct a 250,000 bbl per day refinery in Maine.

Table 2-18. Top Domestic Refining Companies: 1955

	Domestic Refinery Runs (thousands of barrels per day)	Share of Total	Cumulative Percentage
Standard Oil of N.J.	784	10.5%	10.5
Standard Oil of Indiana	576	7.7	18.2
Texas Co.	562	7.5	25.7
Socony Mobil	541	7.2	32.9
Gulf Oil Corp.	506	6.8	39.7
Shell Oil Co.	465	6.2	45.9
Standard Oil Co. of Calif.	453	6.1	52.0
Sinclair Oil Corp.	408	5.5	57.5
Cities Service	284	3.8	61.3
Phillips Petroleum	229	3.1	64.4
Sun Oil	227	3.0	67.4
Atlantic Refining	196	2.6	70.0
Tidewater Associated	167	2.2	72.2
Union Oil of California	157	2.1	74.3
Pure Oil Co.	155	2.1	76.4
Continental Oil Co.	146	2.0	78.4
Standard Oil of Ohio	129	1.7	80.1
Richfield Oil	122	1.6	81.7
Ashland Oil & Refining	110	1.5	82.2
Total: Top twenty	6217		
Total: U.S.	7480		

Source: de Chazeau and Kahn, *Integration and Competition in the Petroleum Industry* (New Haven: Yale University Press, 1959), pp. 30-31.

Gasoline Marketing

Leading gasoline marketers in terms of domestic sales for 1972 and 1973 are listed in Table 2-21. In 1973, Texaco, with about 8 percent of total gasoline sales, is the largest seller on a national basis. Following Texaco in relative importance are Exxon and Shell with market shares of 7.6 percent and 7.5 percent respectively.

Concentration figures based on national sales of gasoline are summarized and presented in Table 2-22. Concentration levels are only moderate and have generally been stable throughout the 1954-1973 period.

There are significant variations however when concentration is viewed at a regional level. The market share of the top marketers in various regions are indicated for 1973 in Table 2-23. The regions used in this table should be viewed as only approximations of actual geographic market boundaries. The regional concentration ratios are summarized in Table 2-24. At the

Table 2-19.   Top Domestic Refining Companies: 1972

	Crude Capacity Barrels per Day	Percent of Total	Cumulative Percentage
Standard Oil of N.J.	1,237,000	9.1	9.1
Texaco	1,150,000	8.4	17.5
Shell	1,101,500	8.1	25.6
Standard Oil of Calif.	1,030,378	7.5	33.1
Standard Oil of Indiana	1,022,500	7.5	40.6
Gulf	848,500	6.2	46.8
Mobil	836,800	6.1	52.9
Atlantic Richfield	831,600	6.1	59.0
Sun Oil	482,000	3.5	62.5
Standard Oil of Ohio & BP	467,000	3.4	65.9
Union Oil	463,880	3.4	69.3
Phillips	419,458	3.1	72.4
Continental	337,700	2.5	74.9
Ashland Oil	331,130	2.4	77.3
Cities Service	288,000	2.1	79.4
Total: Top 15	10,847,546		
Total: U.S.	13,656,000[a]		

[a]Estimate.

Source: Standard and Poors Industry Surveys, *Oil*, p. 62.

Table 2-20.   Concentration Levels in Domestic Refining

	1955	1972	Increase 1955-1972
Top four	32.9%	33.1%	0.6%
Top eight	57.5	59.0	2.6
Top fifteen	76.4	79.4	3.9

national level, the top four marketers account for 30 percent of gasoline sales in 1973; at the regional levels, the share of the top four ranged from a low of 25.2 percent in the mountain region to a high of 51.5 percent in the Pacific region. In terms of the eight-firm concentration, the lowest value, 43.3 percent, occurred in the Mountain area and the highest, 78.5 percent, is reported for the Pacific region.

An indication of the extent to which the top domestic producers of crude oil also dominate refining and gasoline marketing stages can be seen from

Table 2-21.   Leading Gasoline Marketers: 1973 and 1972

	Market Share of Gasoline Sales		*Change in Percentage Points*	*Cumulative 1973*	*Cumulative 1972*
	1973	*1972*			
Texaco	7.97	8.15	− 0.18	7.97	8.15
Exxon	7.63	6.86	0.77	15.60	15.01
Shell	7.48	7.20	0.28	23.08	22.21
Amoco	6.88	6.96	− 0.08	29.96	29.17
Gulf	6.75	6.50	0.25	36.71	35.67
Mobil	6.48	6.41	0.07	43.19	42.08
Standard Oil of Cal.	4.78	4.67	0.11	47.97	46.75
Atlantic Richfield	4.38	4.89	− 0.51	52.35	51.64
Phillips	3.92	4.07	− 0.15	56.27	55.71
Sun	3.67	3.84	− 0.17	59.94	59.55
Union	3.06	2.95	0.11	63.00	62.50
Continental	2.31	2.34	0.03	65.31	65.73
Cities Service	1.67	1.80	0.13	66.98	67.53
Marathon	1.52	1.58	− 0.06	68.50	69.11
Ashland	1.49	1.49	0.00	69.99	70.60
Clark	1.26	1.14	0.12	71.25	71.74
Standard Oil of Ohio	1.23	1.21	0.02	72.48	72.95
Hess	1.00	1.01	− 0.01	73.48	73.96
BP Oil	0.81	1.12	− 0.31	74.29	75.08
Tenneco	0.79	0.83	− 0.04	75.08	75.91

Source: McGraw-Hill Publications, *National Petroleum News*, Mid-May Issue, May 1973.

Table 2-22.   Concentration in Gasoline Sales: Based on Total U.S. Sales

		Market Share	
	1954	*1970*	*1973*
Top four	31.2%	30.7%	30.0%
Top eight	54.0	55.1	52.4
Top fifteen	72.5	74.2	N.A.

Table 2-25. Firms are ranked according to their relative importance as domestic crude producers, refiners, and marketers. With few exceptions, the top fifteen crude producers are among the top fifteen refiners and marketers. While some differences exist, the dominant feature is the similarity in relative importance at the different stages.

Table 2-23. Percentage of Gasoline Sales, Top Ten Marketers in Nine Regions—1973

	Percent of U.S.	New England	Mid Atlantic	South Atlantic	East North Central	West North Central	East South Central	West South Central	Mountain	Pacific
Texaco	7.97	10.46	8.2	83.	5.0	4.5	6.9	12.2		9.1
Exxon	7.63	7.02	12.7	13.2			8.9	14.0	3.2	
Shell	7.48	9.04	5.5	7.2	8.6		7.0	3.3	6.0	14.2
Amoco	6.88		5.7	6.7	11.2	15.9	3.1		5.9	
Gulf	6.75	7.83	8.1			2.7	9.9	9.3		4.5
Mobil	6.48	14.05	12.2	9.2	4.2	6.8		6.6		8.3
Standard Oil of Cal.	4.78								5.7	16.3
Atlantic Richfield	4.38	5.20	8.5							11.4
Phillips	3.92			3.5		7.0		4.7	7.1	5.1
Sun	3.67	4.07	8.1		5.0	2.8				
Union	3.06			3.9		3.5		4.6	6.2	9.6
Continental	2.31						4.1			
Cities Service	1.67									
Marathon	1.52				5.2					
Ashland	1.49						3.6			
Clark	1.26				5.4					
Standard Oil of Ohio	1.23				6.3					
Hess	1.00									
BP Oil	a¹	5.79								
Tenneco	a			2.8						
Kyso	a						9.6			
Skelly	a					3.9				
Shamrock	a							3.5		
Husky	a								3.3	
Chevron (west)	a								5.9	

aLess than 1 percent.

Source: NPN Factbook, MidMay Issue, May, 1974.

Table 2-24. Summary Table: Regional Concentration in
Gasoline Sales 1973

| | *Percentage of Sales Accounted for by* | |
Region	Top Four	Top Eight
New England	41.38%	63.46%
Middle Atlantic	41.6	74.85
South Atlantic	37.9	54.8
East North Central	31.5	50.9
West North Central	34.2	47.1
East South Central	35.4	53.1
West South Central	42.1	58.2
Mountain	25.2	43.3
Pacific	51.5	78.5

Source: NPN Factbook, Mid-May Issue, May, 1974.

Table 2-25. Relative Positions of the Leading Petroleum
Firms In Domestic Crude Production, Refining, and Gasoline
Sales

| | *Company Rank* | | |
	Crude *1970*	*Refining*[b] *1972*	*Gasoline Sales* *1973*
Exxon	1	1	2
Texaco	2	2	1
Gulf Oil	3	6	5
Shell Oil	4	3	3
Standard Oil of Calif.	5	4	7
Standard Oil of Ind.	6	5	a
Atlantic Richfield	7	8	8
Getty	8	a	a
Mobil	9	7	6
Union Oil of Calif.	10	11	11
Sun Oil	11	9	10
Marathon	12	a	14
Continental Oil	13	13	12
Phillips	14	12	9
Cities Service	15	15	13

[a]Not ranked in Top Fifteen.

[b]The tenth firm in refining is Standard Oil of Ohio.

Joint Venture Activity

Joint ventures represent an important element of market structure in the oil and natural gas industry. In general terms, a joint venture results in

> ... the creation of a new business entity by two or more corporate partners. The new entity may operate in the same geographic or product market as one or more of the creating firms, or it may operate in a totally different market.[21]

In the case of energy, joint venture activity occurs in the following areas: (1) joint lease acquisition; (2) joint ownership of pipelines; (3) joint ownership and production from oil and gas leases; and (4) international joint ventures.

A joint venture arrangement can be viewed as a quasi-merger. In certain instances, clear economic justification exists for creating joint activity, but it is also clear that joint activity creates, in a direct manner, an interdependency among the firms involved. Such an interdependency among sellers may lead to a loss of competition, but at the same time joint venture arrangements may result in greater efficiency. The greatest likelihood of anticompetitive results from joint venture activity occurs when the participants are horizontally related.

There are several economic justifications for the use of joint venture arrangements.[22] One, joint ventures may be created in order to share unusually large risks. In such a situation, large risks may prevent single firm entry but, by allowing risk to be spread over a wider base, joint ventures facilitate entry and encourage competition. Two, joint ventures are also economically justifiable in situations where, because of very important scale economies, separate operations would involve inefficient operations. Three, joint venture arrangements are justified in a situation where external benefits flowing from a given investment are distributed in an indiscriminate manner to firms in an industry. In this case, the incentive for a single firm to carry out the investment, often an investment in R&D, is reduced by an inability to capture the benefits. A joint venture allows more of the externalities to be captured.

Professor Stanley Boyle has conducted research in this area and the results are valuable in contributing to knowledge about the basic characteristics of joint ventures in the U.S.[23] The analysis is based upon some 276 joint ventures in 1965. Several findings that bear directly on the justifications commonly offered for joint activities emerge from the Boyle research.[24] One, the bulk of joint ventures, some 66 percent of the total studied, occur in the manufacturing sector. Two, the likelihood of and frequency of involvement in joint activity increases continually as firm size increases. In other words, small firms are not as heavily involved in such arrangements. This casts some doubts on the efficiency argument commonly offered as a justification for joint venture arrangements.

Three, some 44 percent of the parent-subsidiary combinations were found by Boyle to manufacture products in the same 4-digit SIC industry; and in those cases where a horizontal relationship was not present, some 85 percent of the parent-subsidiaries had a vertical relationship. In only 10 percent of the cases did the firms produce totally unrelated products.

Four, perhaps the most significant finding is that 55 percent of the parent companies were direct competitors in the sale of one or more products. While the extent of the competition between parents is not ascertainable, clearly the majority of joint ventures fall into the area with the greatest likelihood of anticompetitive effects, i.e., where parents are horizontally related. Five, in an analysis of joint ventures in interstate oil pipelines, Boyle found that over 85 of the joint subsidiaries were relatively small operations, thus weakening the justification that joint ventures are necessary to enter activities requiring very large capital outlays.

Professor Boyle concludes by noting that the arguments typically offered by firms in support of the joint venture forms of organization assume that

> ... the participating firms are either technically or financially unable to engage in the desired activity above, a line of reasoning that implicitly assumes, of course, that the parent companies are relatively small. In fact, however, such situations are rarely found.[25]

Thus, Boyle's evidence casts doubt on the argument that joint ventures are heavily utilized by small firms, but there can be no doubt that such arrangements do create an interdependency among sellers.

The petroleum and natural gas industries are the most joint venture intensive of all manufacturing. The extent of joint bidding activity for and joint ownership of federal offshore leases is indicated in Tables 2-26 and 2-27. The evidence presented by John W. Wilson[26] and contained in Table 2-26 lists the major bidding partnerships involved in lease sales in 1970-1972. Exxon is the only major firm not involved in joint bidding; in several cases, such as Getty, Phillips, and Union Oil, firms engaged in joint bidding did not submit any independent bids during this period. The degree of interdependency among the participants is greater than the composition of any single joint arrangement would imply, because most of the major firms participate in a multiple number of joint bidding groups.

The ownership pattern of federal offshore leases presented in Table 2-27 reveals a similar situation. For example, Continental has 119 leases, one of which is owned independently, and is engaged in some 300 joint ventures.

Wilson concluded that the close relationships inherent in joint activities creates a community of interest among sellers. And, as a result, the precondition for free markets that sellers be structurally and behaviorally independent of each other does not exist in the petroleum industry.[27] However, no evidence in support of this assertion was presented.

Table 2-26. Joint Bidding In Federal Offshore Lease Sales (1970-1972)

Company	No. of Independent Bids	Bidding Partners and No. of Joint Bids with Each	
Amerada-Hess	0	Signal	50
		La. Land	51
		Marathon	51
		Texas Eastern	16
Amoco	6	Texas Eastern	117
		Union	96
		CNG	79
		Transco	15
		Shell	14
Atlantic-Richfield	12	Cities	106
		Getty	73
		Continental	114
Chevron	79	Mobil	25
		Murphy	17
		Gen. American	17
		Pennzoil	12
		Pelto	13
		Superior	9
		Gulf	7
		Burmah	4
		Mesa	4
Cities Service	7	Atlantic	106
		Getty	100
		Continental	163
		Tenneco	3
Continental	27	Atlantic	114
		Cities	163
		Getty	102
		Tenneco	5
Exxon	80	—	
Getty	0	Atlantic	73
		Cities	100
		Continental	102
		Placid	4
		Superior	2
Gulf	17	Mobil	17

		Pennzoil	8
		S.O. Cal. (Chevron)	7
Marathon	24	Signal	65
		La. Land	69
		Amerada	51
		Texas Eastern	29
Mobil	8	Pennzoil	30
		S.O. Cal. (Chevron)	25
		Mesa	16
		Burmah	13
		Gulf	17
		Ashland	2
Phillips	0	Skelly (Getty)	69
		Allied Chem.	66
		Amer. Petrofina	34
Shell	59	Transco	47
		CNG	15
		S.O. Ind. (Amoco)	14
		Fla. Gas	17
Sun	115	Pennzoil	2
Texaco	15	Tenneco	32
Union	0	Amoco	96
		Texas Eastern	96
		Texas Gas	48
		Fla. Gas	5

Source: U.S. Senate, Subcommittee on Antitrust and Monopoly, "Testimony of John W. Wilson," Washington, June 27, 1973.

A more complete record of joint venture activity among the largest petroleum firms has been constructed by Prof. Mead. As indicated in Table 2-28, joint ventures are common in (1) pipeline operations and other transportation facilities, (2) refining, (3) exploration, and (4) oil and gas properties. Excluded are such joint activities as: (1) bids for oil and gas leases, (2) nonoperating concessions, and (3) producing properties operated under state unitization laws. Information was obtained from public sources and is probably incomplete. The results indicate a large number of joint ventures among the horizontally-related firms. For example, Mobil was involved in 349 joint ventures, and 300 of the total were with other large oil firms. Most of the joint ventures were among the largest companies. The frequency of joint ventures drops substantially for smaller firms.

Table 2-27.   Joint Ownership of Federal Offshore Producing Leases

Company	No. of Leases	Independently Owned	Major Partners and Number of Joint Ventures	
Amerada-Hess	15	0	Marathon	13
			Signal	14
			La. Land	14
				41
Atlantic-Richfield	94	3	Cities	85
			Getty	83
			Continental	87
			Tenneco	4
			S.O. Cal. (Chevron)	2
			El Paso	2
				263
Cities Service	101	1	Atlantic	85
			Getty	93
			Continental	91
			Mobil	2
			Tenneco	7
			S.O. Cal. (Chevron)	2
				280
Continental	119	1	Atlantic	87
			Cities	91
			Getty	87
			Mobil	19
			Tenneco	8
			S.O. Cal. (Chevron)	2
			Superior	2
			Transocean	2
			Southern Natural	2
				300
Getty	100	2	Atlantic	83
			Cities	93
			Continental	87
			Mobil	8
			Tenneco	4
			S.O. Cal. (Chevron)	3
			Phillips	3
			Superior	2
			Transocean	2
			So. Natural	2

			Allied Chemical	3
				290
Gulf	51	34	Mobil	7
			S.O. N.J. (Exxon)	6
			Phillips	4
			Kerr-McGee	2
				19
Marathon	18	0	Amerada	13
			Signal	13
			La. Land	13
			Union	5
			Sun	3
				47
Mobil	52	6	Continental	19
			Cities	8
			Getty	8
			Gulf	7
			S.O. Cal. (Chevron)	5
			S.O. N.J. (Exxon)	4
			S.O. Ind. (Amoco)	4
			Pennzoil	2
				57
Phillips	16	3	Kerr-McGee	7
			Gulf	4
			Getty	3
			S.O. Ind. (Amoco)	3
			Sun	3
			So. Natural	2
			Allied Chemical	3
				25
Shell	68	64	S.O. Cal. (Chevron)	2
				2
Chevron	105	86	Mobil	5
			Getty	3
			Atlantic	2
			Cities	2
			Continental	2
				14
Amoco	60	3	Texaco	29
			Union	12
			S. Natural	8
			Mobil	4

Table 2-27. (cont.)

Company	No. of Leases	Independently Owned	Major Partners[a] and Number of Joint Ventures	
			Kerr-McGee	4
			Superior	4
			Tenneco	3
			Phillips	3
			Pennzoil	4
			Texas Eastern	2
				73
Exxon	52	43	Gulf	6
			Mobil	4
				10
Sun	19	0	Burmah	11
			Murphy	10
			Kerr-McGee	4
			Union	3
			Phillips	3
			Marathon	3
			Cabot	3
			Diamond Shamrock	3
			Anadarko	3
				43
Texaco	55	16	S.O. Ind. (Amoco)	29
			Tenneco	9
				38
Union Oil	37	18	S.O. Ind. (Amoco)	12
			Marathon	5
			Superior	4
			Sun	3
			Texas Eastern	2
				26

[a]Major Partners.

Source: U.S. Senate, Subcommittee on Antitrust and Monopoly, "Testimony of John W. Wilson," Washington, June 27, 1973.

The interdependency created by extensive use of joint venture activity in the domestic oil and gas industry is duplicated in the foreign activities of the firms. Public information on this subject is not complete but some indication of the situation is available from data published in the *Oil and Gas Journal* and presented in Table 2-29. The data are limited to exploration and development activities of the firms.

The extensive use of joint ventures by petroleum firms is well documented. Such activity creates a direct interdependency among sellers and may reduce competition, but at the same time joint ventures may encourage competition by easing the entry of small firms that would not be able to enter individually. Evidence of the net effect is difficult to obtain, but the sales of oil and gas leases offers some insight into this question.

In the case of a lease sale, if four firms were to bid jointly and if each firm were able to bid independently, it would appear that three potential bidders have been eliminated. This suggests an anticompetitive effect. Another scenario, with competitive effects, is equally possible. One effect of the four firms joining in a series of bids is to multiply the number of tracts in which any single firm has an interest, thus reducing the risk of great loss borne by each participating firm. If this risk reduction causes the four member combine to bid more than four times as frequently as they would have bid separately, then, even in where the four firms were large, joint bidding may be competitive. In addition, the situation where a combination of small firms bids jointly but could not bid individually because of the capital requirements clearly increases competition.

The only behavioral evidence concerning the competitive impact of joint ventures concerns joint bidding for oil and gas leases in Alaska and the Gulf of Mexico. Professor Mead[28] examined data from Alaskan lease sales between 1959 and 1966 in order to identify the impact of joint bidding on bidding activities of the participants against each other and against nonparticipants in the joint activity. If joint bidding does not restrain competition, the frequency of bidding activities by a partner against other partners and nonparticipants should be similar in subsequent lease sales. Mead's results indicate that in a two-year period following a break up of a joint bidding agreement, the bidding patterns indicate a clear tendency of restrained bidding among former partners.[29] Over a longer time period, the results indicate that the cooperation among partners tends to dissipate rather quickly. A similar analysis for lease sales in the Gulf of Mexico yields the same results: a clear, but limited, propensity to avoid competition among the participants in previous joint bidding arrangements.[30] These results suggest that joint arrangements among horizontally-related partners may lead to temporary reductions in competition.

Some insight into the process of formulating a joint bid is provided in a deposition of Otto Miller, Chairman of the Board of Directors, Standard Oil Company of California.[31] In preparation of a joint bid by Standard Oil of

Table 2-28.   Frequency of Joint Ventures Among the 32 Largest Oil Companies[a]

	Standard (N.J.)	Mobil	Texaco	Gulf	Royal Dutch Shell Group	Standard (Ind.)	Standard (Cal.)	Phillips	Continental	Sinclair	Cities Service	Sun Oil
Standard (N.J.)	6											
Mobil	62											
Texaco	24	31										
Gulf	3	11	14									
Royal Dutch Shell[b]	59	51	37	19								
Standard (Ind.)	12	10	15	8	16							
Standard (Calif.)	21	21	50	8	22	2						
Phillips	8	10	9	5	9	6	4					
Continental	13	11	11	9	21	10	5	8				
Sinclair	5	11	9	7	6	10	1	7	13			
Cities Service	16	7	18	5	8	8	1	5	24	14		
Sun Oil	2	6	5	5	4	6		4	5	3	5	
Tidewater	11	6	8	5	7	5	4	3	4	4	12	6
Atlantic Refining	9	8	10	6	8	9	5	13	8	7	8	11
Sure	2	3	7	7	4	7	1	2	5	10	8	8
Signal	3	6	5	2	5	1	7	9	2	1	1	
Union	4	3	5	5	8	5	5	4	8	8	5	5
Standard (Ohio)	2	4	1	3	2	2	1	1		2	3	4
Marathon	9	8	6	2	7	5	1	2	19	6	6	2
Sunray DX	4	7	6	5	11	5	4	24	10	2	9	3
Ashland	2	4	3	1	6	1	4	8		1		
Hess												
Richfield	3	4	3	5	5	2	8	1	4	2	12	1
Skelly	1	1	2	1	1	2		3	5	5	4	1
Kerr-McGee	1	2		1	2		1	5	1		2	
Amerada	4	2		2	6	4		1	8	1		1
Amer. Petrofina	4	4	4	1	4	4	3	6	3		1	1
Murphy	3	2	1		4	1		1	4	2	1	3
Kern Co. Land												
Commonwealth			1		3							
Superior	1	4		1	4	6	2	1	2	3	4	6
Clark		1	1		1	1	1		1		1	1
Other oil companies	51	49	28	27	51	21	22	21	40	15	16	13
Total 32 largest	302	300	286	152	340	163	182	159	214	140	191	97
Total incl. other	353	349	314	179	391	184	204	180	254	155	207	110

[a]Sales ranking for 1964, *Fortune*, August 1965, pp. 24-40.

[b]The asset size rank for the Royal Dutch Shell Group is based on the position of the U.S. company, Shell Oil.

Tidewater	Atlantic Refining	Pure	Signal	Union	Standard (Ohio)	Marathon	Sunray DX	Ashland	Hess	Richfield	Skelly	Kerr-McGee	Amerada	American Petrofina	Murphy	Kern Co. Land	Commonwealth	Superior	Clark	Other
7																				
1	3																			
2	4	2																		
6	8	3	2																	
2	2	5	2	1																
3	7	7	2	6	1															
9	1	1	7	4	1	3														
2	2		6		2			6												
					1															
6	2	1	7	4	1	3	3	1												
4	1	3	1	3		1	7			1										
5	2		1	2		1	5	2		2	2									
2	2		1	3		8	1						1							
1	1							1	1		1									
1	5	1	1	4	1	2	2						1	1						
5	7		1	4		3	4	3		2	2	5	2	1	2					
	1		1			1	2											1		
18	9	2	4	9	6	6	7	6	2	8	6	3	4	8	10		1	6	1	
131	166	92	81	116	42	125	157	54	1	83	52	45	50	42	45	0	4	75	14	470
149	175	94	85	125	48	131	164	60	3	91	58	48	54	49	55	0	5	81	15	–

Source: Walter Mead, "The Structure of the Buyer Market for Oil Shale Resources," *Natural Resources Journal,* October 1968, p. 619, from data in *Moody's Industrials; Oil and Gas Journal; Wall Street Journal*; annual reports of each company, Federal Trade Commission, Staff Report to U.S. Senate Subcommittee on Monopoly of the Select Committee on Small Business, *The International Petroleum Cartel*, 82nd Cong., 2nd Sess. (1952).

Table 2-29.   Reported Foreign Joint Ventures in Exploration and Development (1960-1970)[a]

	1	2	3	4	5	6	7	8	9	10	11	12	13	14	15	16	17	18
1 Amerada																		
2 Atlantic	1																	
3 Cities	2	31																
4 Continental	15	9	11															
5 Getty	–	6	2	2														
6 Gulf	–	1	–	1	–													
7 Marathon	21	18	13	29	1	–												
8 Mobil	2	10	–	1	–	9	–											
9 Phillips	–	15	–	3	–	2	3	2										
10 Stand. of Cal. (Chevron)	–	1	–	–	–	9	–	14	3									
11 Stand. of Ind. (Amoco)	7	10	5	–	–	7	–	10	2	–								
12 Stand. of N.J. (Exxon)	–	5	1	1	5	7	1	33	5	8	13							
13 Stand. of Ohio (Sohio)	–	2	3	4	–	12	–	15	1	4	–	10						
14 Shell	4	3	–	6	–	15	4	23	2	17	2	63	66					
15 Sun	8	36	4	4	12	2	4	5	24	–	9	4	1	1				
16 Texaco	2	9	5	3	–	28	2	15	2	59	4	10	6	22	6			
17 Union	5	17	2	6	2	1	2	2	1	–	9	2	–	2	19	1		

[a]As reported in The Oil and Gas Journal.

Source: U.S. Senate, Subcommittee on Antitrust and Monopoly, "Testimony of John W. Wilson," Washington, June 27, 1973.

California, Exxon, and ARCO, Miller testified that in meetings where representatives of the three firms were present, "The only information that was exchanged was 'How much do you want to bid on this parcel?' And then some discussions on the geological aspects of it. That was all."[32] In the process of arriving at a consensus on the joint bid, Miller stated:

> You just put the number on the table that on this parcel we want to bid $32 million. They come in and say, 'No, we want to bid higher.' This one comes in and says, 'No, we want to bid 60.' And then we would look at the geology jointly. Not all of it. And the experts would talk to the likelihood of finding $20 million or $40 million or $250 million or a billion. . . . And if we decided that we wanted to bid 32, we tried to bring them down. If they are below us, bring them up. And sometimes they bring us up. And we try and reach a concurrence. If we can't reach a concurrence, then the one that wants to bid the high number is free to come out and bid completely by himself. That's part of the procedure. And that's all we discuss.[33]

If the firms were unable to reach agreement on a single bid to be submitted by the combine, then each firm is free to bid separately. But Miller testified that while the firm wanting to bid the largest amount could separately bid that amount or more, firms that wished to bid lesser amounts were not permitted to bid more than the amounts that they exposed in the prebid meeting.[34] When asked if there was a written agreement constraining subsequent bidding, Miller replied, "It's an understanding. I'm quite sure it's written."[35]

The Miller testimony indicates two points of competitive importance. First, of necessity, potential joint bidders must exchange vital bidding, market, and geological appraisal information in the process of attempting to reach agreement on the bid amount. Second, in the event of no agreement, firms wishing to bid lesser amounts are constrained from subsequently bidding (on their own or perhaps in other joint bidding combines) amounts in excess of their previously exposed bid. This means that the firm proposing the highest bid is aware of the maximum bid of his potential partners. If the firm exposing the highest bid can rely on there being no other bidders beyond his potential partners, then this information is capable of reducing competitive bids. If other bidders must be reckoned with, then the information would be of little value. The evidence of past bidding shows that for OCS lease sales since 1954, the average number of bidders per tract bid on has been 3.64. In recent sales, the average has been rising. The five large sales from 1970 through 1974 have averaged from 3.53 to 8.21 bidders per sale. This evidence suggests that the prebid negotiations have only a minor effect on competition.

The record of participation in oil and gas lease sales by joint bidding provides some insight into the competitive significance of joint bidding. The evidence, collected by Susan M. Wilcox and presented in Table 2-30 indicates

Table 2-30. Assets and Utilization of Joint Bidding by Firm
Size

Firm	Percent Joint Participation[a]	Current Assets[b] (million $)	Number Tracts Involved
Shell	10	1234	276
Sun	12	560	82
Exxon	35	6791	199
St. California	49	2010	276
Superior	52	64	44
Gulf	57	2652	164
Texaco	58	2820	137
Phillips	60	903	73
Tenneco	64	1065	44
Union	78	615	152
Mobil	83	2205	138
Arco	83	1082	206
Continental	84	978	164
St. Indiana	86	1466	187
Signal	87	617	31
Kerr-McGee	90	270	51
La. Land & Expl.	91	39	35
Skelly	94	169	62
Amerada	94	470	33
Marathon	96	366	26
Burmah	96	156	26
Allied Chemical	97	561	32
Pennzoil	98	257	47
Cities Service	99	584	150
Getty	100	450	117
Texas E. Transmission	100	162	41
Cabot	100	139	39
American Petrofina	100	87	21
Pennzoil Offshore	100	30	33
Mesa	100	24	27
Felmont	100	6	37
(Placid)	(52)	N.A.	(33)

aWeighted by bidding shares for joint bids.

bCurrent Assets from Moody's (1972).

Source: Susan M. Wilcox, Entry and Joint Venture Bidding in the Offshore Petroleum Industry, unpublished Ph.D. dissertation, University of California, Santa Barbara, 1974, p. 75.

that the degree of participation in joint bidding is inversely related to firm size. In contrast to evidence presented earlier, it appears that smaller firms are more frequent users of joint ventures in bidding for oil and gas leases than are large firms.[36] This suggests that joint bidding may be an important factor in enabling smaller firms to challenge large firms successfully. Further support for this position arises from the fact that joint bidding is increasing relative to solo bidding, and the big eight oil companies are winning fewer leases over time.[37]

In conclusion, the evidence indicates that crude oil and natural gas production commonly involve joint venture arrangements. Such activity creates a direct interdependency among participants, and as a result the concentration ratios presented earlier cannot be interpreted in the usual manner. The market has more interdependency than we would infer from the low concentration levels. It is however a leap of faith to conclude that the prevalence of joint ventures in oil and natural gas results in monopolization. There is no evidence from economic research to demonstrate such a conclusion. The review of bidding for oil and gas leases suggests that joint bidding by small firms promotes competition. At the same time, a case can be made for banning very large firms from joint bidding with other large firms because such firms are capable of solo bidding. Such a ban has recently been proposed by the Secretary of the Interior.

Natural Gas

Competition and the marketplace have, at present, only a limited role in the production of natural gas in the U.S. The FPC is responsible for setting prices at the wellhead for gas destined for interstate sales.

The absence of competition at the wellhead introduces a certain degree of ambiguity into an analysis of market structure in natural gas production. As noted by Professor Clark Hawkins, structural analysis is generally conducted within a context of possible antitrust actions.[38] Such actions are obviously less applicable to the case of a regulated industry. In addition, the fact that natural gas and crude oil often exist together in the same geologic formation makes them truly joint products and makes structural remedies such as dissolution inappropriate. In spite of these factors, however, recent discussion of plans to deregulate natural gas production indicates a need for an assessment of the present structure as a means of evaluating the prospects for effective competition if deregulation were to occur.

The natural gas industry includes three basic functions: (1) the search for and production of natural gas; (2) the transportation of gas from the field to the distributors; and (3) the local distribution of gas. Firms involved in producing gas are generally not involved in the other stages. This lack of vertical integration is in sharp contrast to the situation in crude oil. As indicated in Table 2-31, the majority of the 3,750 gas-producing firms in 1971 are firms with annual production of less than 100,000 mcf. As the size of annual production increases, the number of firms in the larger-size classes diminishes drastically.

Table 2-31. Number of Producers of Natural Gas Distributed by Size Groups: 1965-1971

Size Interval	*Number of Producers*					
	1965		*1970*		*1971*	
	No.	*%*	*No.*	*%*	*No.*	*%*
Over 500 million	N.A.	N.A.	9	0.2	9	0.2
100 million to 500 million	25[a]	0.7	17[c]	0.5	16[c]	0.4
50 million to 100 million	13[a]	0.4	13	0.3	15	0.4
25 million to 50 million	23	0.7	20[d]	0.5	20	0.5
10 million to 25 million	25	0.7	40	1.1	45[d]	1.2
5 million to 10 million	58	1.7	55	1.5	50[d]	1.3
2 million to 5 million	118	3.5	147	3.9	146	3.9
1 million to 2 million	150	4.4	146	3.9	143	3.8
500,000 to 1 million	233	6.9	246	6.6	228	6.1
250,000 to 500,000	302	8.9	267	7.1	279	7.4
100,000 to 250,000	336	9.9	386	10.3	399	10.6
Under 100,000	2100[b]	62.1	2400[b]	64.1	2400[b]	64.0
Total	3383	100.0	2746	100.0	3750	100.0

[a]Includes two foreign suppliers.

[b]Estimate.

[c]Includes three foreign suppliers.

[d]Includes one foreign supplier.

Source: Federal Power Commission, *Sales of Natural Gas to Interstate Pipeline Companies* (Washington: Government Printing Office, Various Years), Table H.

For example, there are only nine firms with annual production in excess of 500 million mcf. While small producers are more numerous, large firms clearly dominate natural gas production in the U.S. According to the evidence collected by the FPC and reported in Table 2-32, 73 percent of total production is attributable to the 25 largest firms. Firms in the less than 100,000 size class account for a majority of firms but produce only slightly more than 2 percent of total output. The existence of a large number of relatively small firms strongly suggests that entry barriers are not prohibitively high.

The concentration measures reported in Tables 2-33 through 2-36 are based on FPC data and represent interstate sales of natural gas.[39] A word of caution is necessary about the FPC gas data. As indicated, intrastate gas sales are not included. Perhaps more important is the fact that the sales attributed to a producer by the FPC do not necessarily indicate the amount of gas actually produced by that firm. In the case of gas produced by one firm in association with others, the FPC allocates the entire production of the group to the first producer named on the production certificate. Another complicating factor is

Table 2-32. Sales of Natural Gas to Interstate Pipelines
Arranged by Size of Producers: 1961-1971

Size Interval (thousands Cu. ft.)	Sales (mcf) Percent of Total			
	1961	1965	1970	1971
Over 500 million			46.7	46.1
100 million to 500 million	53.0[a]	65.3[a]	26.5	26.9
50 million to 100 million	13.6	8.4	5.9	6.6
25 million to 50 million	8.5	7.5	5.2	4.9
10 million to 25 million	6.6	3.6	4.1	4.4
5 million to 10 million	4.5	3.9	2.7	2.3
2 million to 5 million	4.6	3.6	3.2	3.1
1 million to 2 million	2.8	2.0	1.4	1.3
500,000 to 1 million	1.9	1.6	1.2	1.1
250,000 to 500,000	1.0	1.0	0.7	0.7
100,000 to 250,000	0.5	0.6	0.4	0.5
Less than 100,000	3.0	2.5	2.0	2.1
Total sales	8,339,355	10,315,400	14,440,737	14,597,827

[a]Represents share of production in the over 100 million size interval.

Source: Federal Power Commission, *Sales by Producers of Natural Gas to Interstate Pipeline Companies* (Washington: Government Printing Office, Various Years), Table H.

that producers may sell gas they purchased from others. There is no way to determine if these factors impart a serious and systematic bias to the FPC data.

The twenty largest sellers of natural gas and the market share of each for 1955, 1961, 1965, and 1970 are reported in Tables 2-33 through 2-36. It is interesting to note the general stability in the makeup of the dominant suppliers over the 1955-1970 period. The identity of the top eight producers is the same for 1961 and 1970.

The level of concentration at the four, eight, and twenty-firm levels is summarized in Table 2-33. For benchmark purposes, estimates of natural gas concentration in 1955, computed by Cookenboo, are included in Table 2-37. The four firm and eight-firm concentration levels are 25.3 percent and 42.7 percent respectively in 1970. The corresponding twenty-firm measure is 67.3 percent. The resulting concentration measures are not sufficiently high to represent a clear danger to competition. The data indicates rising concentration during the 1955-1971 period; the four-firm figure increased some 10 percent during the period. A similar movement occurred at the eight-firm level. These figures do not indicate an anticompetitive market structure, but some concern is obviously generated by the existence of a moderate trend towards higher

Table 2-33.   **Natural Gas Sales by Producers to Interstate Pipelines, 1955**

Rank	Producer	Volume (mcf)	Percent of Total	Cumulative Percentage
1	Phillips	511,113,955	8.74	8.74
2	Standard (Indiana)	324,296,232	5.54	14.28
3	Humble	268,144,970	4.58	18.86
4	Union Producing	231,320,866	3.95	22.81
5	Cities Service	214,508,599	3.67	26.48
6	Shell	190,547,320	3.26	29.74
7	Magnolia	163,906,059	2.80	32.54
8	Chicago Corp.	150,237,142	2.57	35.11
9	Gulf	147,643,569	2.52	37.63
10	Texas Co.	131,512,727	2.25	39.88
11	Atlantic	124,790,611	2.13	42.01
12	Shamrock	89,595,951	1.53	43.54
13	Pure	87,971,532	1.50	45.04
14	Sun	82,351,101	1.41	46.45
15	Ohio	78,834,690	1.35	47.80
16	Tidewater	77,796,361	1.33	49.13
17	Sunray	72,341,319	1.24	50.37
18	Skelly	69,544,744	1.19	51.56
19	Southern Production	64,553,341	1.10	52.66
20	Superior	62,184,420	1.06	53.72
	Total: Top 20	3,143,195,509		
	Total: Other	2,707,714,263		
	Grand Total	5,850,909,772		

Source: Leslie Cookenboo, Jr., "Competition in the Field Market for Natural Gas," *The Rice Institute Pamphlet*, Monograph in Economics, vol. 44, no. 4 (January 1958), p. 48.

concentration levels. The interpretation of these figures is not without some ambiguity, however. Ideally, a measure of concentration should be based, not on past or current sales, but on the ownership pattern of natural gas reserves. While sales data is indicative of past events, reserve ownership patterns are clearly relevant to future events. Companies generally do not reveal such figures. In most cases, however, there is a strong positive correlation between a company's past sales and reserve holdings.

The concentration measures reported above are based on annual volumes of flowing gas and may be somewhat deficient as a guide to assessing competition in free markets. Competition occurs at the margin as the result of

Table 2-34. Natural Gas Sales by Producers to Interstate
Pipelines, 1961

Rank	Producer	Volume (mcf)	Percent of Total	Cumulative Percentage
1	Phillips	661,328,762	7.9	7.9
2	Pan American Petroleum (Stand. Ind.)	461,448,960	5.5	13.4
3	Exxon	358,481,858	4.3	17.7
4	Shell Oil	318,808,523	3.8	21.5
5	Texaco	273,957,034	3.3	24.8
6	Socony Mobil Oil	224,113,059	2.7	27.5
7	Gulf Oil	223,711,587	2.7	30.2
8	Atlantic Refining	181,846,018	2.2	32.4
9	Superior Oil	178,985,618	2.1	34.5
10	Sinclair Oil & Gas	164,629,233	2.0	36.5
11	Union Producing	161,347,284	1.9	38.4
12	Union Oil of Calif.	161,016,019	1.9	40.3
13	Sun Oil	158,337,777	1.9	42.1
14	Cities Service	154,909,902	1.9	44.0
15	Champlin Oil & Refining	145,238,628	1.7	45.7
16	Continental Oil	140,002,636	1.7	47.4
17	Union Texas Nat'l Gas	133,723,696	1.6	49.0
18	Tidewater Oil	112,442,443	1.3	50.3
19	Calif. Co. Div. of Chevron	105,877,983	1.3	51.6
20	Warren Petroleum	102,617,430	1.2	52.8
	Total: Top 20	4,419,858,680		
	Total: Other	3,919,497,320		
	Grand Total	8,339,356,000		

Source: FPC, *Sales by Producers of Natural Gas to Interstate Pipeline Companies*, 1962.

interactions between buyers and sellers in a market. Much of the flowing gas is supplied under long-term, vintage contracts and has no impact on current gas supplies available to various buyers. Competition and prices are affected by currently available supplies, not by gas flowing to meet past commitments.

Some methodological improvement may possibly be obtained by basing concentration measures on new gas commitments. The concentration figures presented in Table 2-38 represent dominant firm control of new gas commitments for each year during the period 1965-1970. The resulting figures are substantially higher than the figures presented above. However the trend for

Table 2-35.   Natural Gas Sales by Producers to Interstate
Pipelines, 1965

Rank	Producer	Volume (mcf)	Percent of Total	Cumulative Percentage
1	Exxon	656,856,220	6.4	6.4
2	Pan American Petro-leum (Stand. Ind.)	631,271,629	6.1	12.5
3	Phillips	611,540,672	5.9	18.4
4	Shell Oil	493,906,694	4.8	23.2
5	Gulf Oil	425,514,121	4.1	27.3
6	Socony Mobil Oil	413,090,344	4.0	31.3
7	Texaco	332,439,459	3.2	34.5
8	Continental Oil	331,244,000	3.2	37.7
9	Union Oil of California	283,387,275	2.7	40.4
10	Sinclair Oil & Gas	266,739,907	2.6	43.1
11	Atlantic Refining	227,188,368	2.2	45.3
12	Superior Oil	220,212,942	2.1	47.4
13	Sun Oil	211,992,960	2.1	49.5
14	Union Producing	171,000,689	1.7	51.2
15	Cities Service Oil	169,854,414	1.6	52.8
16	Alberta & Southern Gas (Canadian)	152,298,574	1.5	54.3
17	Westcoast Transmission (Canadian)	143,991,086	1.4	55.7
18	Champlin Petroleum	143,320,208	1.4	57.1
19	Skelly Oil	143,090,778	1.4	58.5
20	Union Texas Petro. Div. Allied	136,423,849	1.3	59.8
	Total: Top 20	6,168,609,200		
	Total: Other	4,146,790,800		
	Grand Total	10,315,400,000		

Source: FPC, *Sales by Producers of Natural Gas to Interstate Pipeline Companies*, 1966.

1965 through 1970 suggests declining concentration. Whether these figures more accurately reflect actual concentration is not clear. The use of regional markets is in itself almost certain to yield higher concentration ratios. In addition, during this period a substantial amount of new gas commitments involved the intrastate market and thus are not reflected in the data contained in Table 2-39. Because of these problems, doubt exists as to the meaning of the figures presented by Dr. Wilson. However, Wilson's point that the use of total annual volume of flowing gas, much of which represents vintage contract commitments, is not the most

Table 2-36. Natural Gas Sales by Producers to Interstate
Pipelines, 1970

Rank	Producer	Volume (mcf)	Percent of Total	Cumulative Percentage
1	Exxon	1,300,642,683	9.0	9.0
2	Gulf Oil	813,738,549	5.6	14.6
3	Shell Oil	785,667,041	5.4	20.0
4	Pan American Petroleum (Stand. Ind.)	767,430,589	5.3	25.3
5	Phillips	707,235,036	4.9	30.2
6	Mobil Oil	650,890,489	4.5	34.7
7	Texaco	607,433,789	4.2	38.9
8	Atlantic Richfield	561,540,880	3.9	42.8
9	Union Oil of Calif.	548,896,648	3.8	46.6
10	Continental Oil	461,297,727	3.2	49.8
11	Calif. Co. Div. of Chevron	367,213,888	2.5	52.3
12	Sun Oil	361,622,934	2.5	54.8
13	Alberta & Southern Gas (Canadian)	304,529,422	2.1	56.9
14	Tenneco	252,971,722	1.8	58.7
15	Cities Service	243,511,899	1.7	60.4
16	Superior Oil	240,211,285	1.7	62.1
17	Westcoast Transmission (Canadian)	223,257,230	1.5	63.6
18	Trans-Canada P.L. Ltd.	199,655,647	1.4	65.0
19	Pennzoil Producing	184,440,676	1.3	66.3
20	Getty Oil	173,480,911	1.2	67.5
	Total: Top 20	9,502,706,323		
	Total: Other	4,938,030,621		
	Grand Total	14,440,736,944		

Source: FPC, *Sales by Producers of Natural Gas to Interstate Pipeline Companies*, 1970.

preferred approach in an analytical sense is well taken. In an unregulated market, perhaps the most relevant concentration measure would be based upon uncommitted gas reserves. Because such data is lacking, seller concentration cannot presently be examined from this viewpoint.[40]

Future levels of seller concentration will depend on whether gas production is deregulated and whether offshore production is increased. It may be possible to obtain some insight into the impact of deregulation on market structure by comparing the structures of the interstate and intrastate natural gas

Table 2-37. Seller Concentration: U.S. Natural Gas Sales by
Producer to Interstate Pipelines: 1955-1970

Companies	*1955*[a]	*Concentration Ratio (%)*		*1970*	*Percent Increase 1955-1971*
		1961	*1965*		
Top four	23	21.7	23.2	25.3	10.0
Top eight	35	32.5	37.7	42.8	22.3
Top twenty	54	53.1	59.8	67.5	25.0

[a]Leslie Cookenboo, Jr., "Competition in the Field Market for Natural Gas," *The Rice Institute Pamphlet*, Monograph in Economics, Vol. 44, No. 4 (January 1958), p. 48.

markets. The latter market is not regulated. Comparisons are limited because of a lack of data on intrastate sales. In a 1958 study, Cookenboo computed concentration measures for both the interstate market and Texas markets. His results indicate very little difference between the two markets.[41] In a recent study, Clark Hawkins notes that interstate sales from Texas producers are more concentrated than the national market and, by inference, suggests that the Texas market may also be more concentrated than the national market.[42] However, the type of data necessary to investigate such an inference is not available.

If future production of natural gas results in increased drilling in deep water offshore areas, the high capital costs associated with this type of drilling may make it more difficult for small producers to compete. The main effect of this will be felt directly on entry conditions; an event that will likely cause small firms to place greater reliance on the use of joint ventures. Thus, if offshore production accounts for a larger share of total production, large firms have an advantage and concentration can be expected to rise, perhaps significantly. An offset to this will be the use of joint ventures by small producers.

The buying side of the natural gas market is more concentrated than the selling side. As indicated in Table 2-40, in 1968 the top four and top eight purchasers of natural gas accounted for 38.5 percent and 58.6 percent, respectively, of total gas purchases by interstate pipelines. The same figure for the top twenty is 92.5 percent. Since 1961, there has, however, been a clear downward trend in buyer concentration.

The seller concentration data reported here is consistent with results reported by others. Some difficulty occurs in interpreting concentration measures because of the role of regulation in natural gas price formation and the methodological controversy over the proper base for calculating concentration, i.e., flowing gas or uncommitted gas reserves. Cookenboo examined the structure of the industry in the mid-1950 period in order to assess the state of competition.[43] He was able to analyze both the interstate and intrastate market. Concentration figures based on new contract sales in the early 1950s were very similar to concentration levels calculated from all sales during this period and led

Table 2-38. Concentration of New (non *et al.*) Gas Contract
Volumes Sold to Interstate Pipelines 1965-1970[a]

	4 Largest	8 Largest
South Louisiana		
1965	52.1%	68.9%
1966	63.8	82.6
1967	76.9	86.2
1968	51.6	68.9
1969	41.6	60.9
1970	41.1	68.1
Permian Basin		
1965	67.0	88.8
1966	89.0	94.3
1967	61.7	76.0
1968	76.5	84.7
1969	52.4	81.1
1970	66.8	82.7
Texas Gulf Coast		
1965	86.9	92.3
1966	95.9	98.9
1967	71.3	89.5
1968	83.7	98.6
1969	43.9	71.9
1970	64.9	74.2

[a]Ratios calculated on the basis of natural gas volumes sold to interstate pipelines under non *et al.* vintage year contracts in the first full year of sales following the vintage contract year. Thus, the 1970 concentration ratios are based on actual annual volumes in 1970 from new non *et al.* contracts culminated during 1969. The ratios are calculated by dividing the non *et al.* volumes accounted for by the 4 and 8 largest producers, respectively, in each year by total non *et al.* volumes. The implicit assumption in this calculation is that these producers shared proportionally in volumes sold under *et al.* contracts. Non *et al.* contracts generally accounted for over 75 percent of the total annual volume. To the extent that the largest firms had either a disproportionally large share of the *et al.* volumes, or a disproportionally small share, the actual concentration ratios would be either slightly higher or lower. In any event, the ratios reported here tend to underestimate actual concentration to some extent as they do not take account of all corporate affiliations (e.g., Getty, Skelly, and Tidewater are treated independently although all were controlled by Getty).

Source: U.S. Senate, Subcommittee on Antitrust and Monopoly "Testimony of John W. Wilson," Washington: June 27, 1973, pp. 13-14.

Cookenboo to conclude that field markets were characterized by low concentration and effective competition.

Professor Neuner conducted a similar analysis for the 1945-53 period, but placed somewhat greater emphasis on new contracts. Neuner concluded:

**Table 2-39. Seller Concentration: U.S. Natural Gas Sales
by Producer to Interstate Pipelines: 1955-1970**

Companies	Concentration Ratio (%)				Percent Increase 1955-1971
	1955[a]	1961	1965	1970	
Top Four	23	21.7	23.2	25.3	10.0
Top Eight	35	32.5	37.7	42.8	22.3
Top Twenty	54	53.1	59.8	67.5	25.0

[a]Leslie Cookenboo, Jr., "Competition in the Field Market for Natural Gas," *The Rice Institute Pamphlet*, Monograph in Economics, Vol. 44, No. 4 (January 1958), p. 48.

Table 2-40. Natural Gas Purchases by Interstate Pipelines[a]

	1961	1965	1968[a]
Top four	45.2%	40.5%	38.5%
Top eight	63.5	59.1	58.6
Top twenty	93.5	90.8	92.5

[a]Purchases by dominant interstate pipelines as a percent of total purchases by all interstate pipelines.

Source: Federal Power Commission, *Sales by Producers of Natural Gas to Interstate Pipelines Companies* (Washington: Various Years).

It is evident that the level of concentration in natural gas production is not excessive by comparison with manufacturing industry. The structural condition of natural gas fuel markets, by these tests, does little to support a claim of gas produces monopoly.[44]

The question of competition in interstate gas markets has been addressed by Prof. Paul MacAvoy in an innovative manner.[45] MacAvoy used a multiple regression technique with price as the dependent variable and various independent variables that, theoretically, are determinants of gas prices. In this way, MacAvoy analyzed differences in pricing patterns as a means of assessing competition. He concludes that "gas markets are diverse in structure and behavior, and were generally competitive or were changing from monopsony toward competition."[46]

Thus, there has been a general conclusion emerging from studies concerned with the structural setting of natural gas production. The studies, based on a variety of time periods and methodology, strongly suggest that gas markets are competitive to a significant degree. The concentration measures reported here are consistent with these findings. However, simply observing relatively low seller concentration levels is not sufficient to allow any sweeping generalizations as to the general competitiveness of market structure. The key structural feature is the degree of interdependence, and concentration measures—while allowing some inference—do not always reveal the total extent of seller interdependency. Other factors, such as joint venture arrangements, directly create interdependency but are not reflected in levels of seller concentration.

Coal

Coal, during the years immediately following World War II, was the single most important source of energy in the United States. In 1947, coal consumption accounted for 46.3 percent of total energy consumption, while oil and natural gas accounted for 33.7 percent and 15.5 percent respectively. Since then, however, there have been several basic changes in coal consumption patterns. Among the major changes are (1) the elimination of railroad demand for coal due to the introduction of the diesel locomotive; (2) the virtual elimination of coal from the heating market; (3) the increased demand for coal from the electric utility sector; and (4) strict environmental controls on sulfur emissions from electric utilities.

The survival of coal producers has been and continues to be directly related to their ability to meet the energy needs of electric utilities. It is important to note that in spite of an increase in the absolute amount of coal consumed by electric utilities, coal's relative position in the utility sector has been eroded due to competition from other fuels. As will be discussed below, the fact that utilities have become the sole user of coal has had a major impact on market structure. In the future, the development of a coal-based synthetic fuel process will provide an additional source of demand.

The twenty largest coal producing groups and their respective market shares are identified for various years in Tables 2-41 through 2-44. Producing groups represent companies under a common control or the equivalent of an independent decision-making unit. The economist, concerned with the impact of structure or competition, is interested in the number of such decision-making units in a market. At least one study failed to note this distinction and, according to Prof. Moyer, resulted in serious error.[47]

Included among the major coal producers are several companies, such as U.S. Steel, who are captive producers of coal. Captive production refers to coal production used solely by the company and not sold in the market. Captive production in 1971 amounted to 73.9 thousand tons or about 13.3 percent of total coal production.[48] Captive production is mainly due to steel and public utilities and is included in total production because, while captive producers are not established sellers in the coal market, they can be considered to be potential entrants into the commercial market, and thus they exert an influence on the market. In certain instances, captive producers have, in response to relative prices, supplied coal to the commercial market.

Levels of seller concentration in coal production are summarized in Table 2-45. In 1972, the top four and eight producers accounted for 30.4 percent and 40.4 percent of total coal production respectively. The share of the top twenty is 55.1 percent in 1972. The 1972 figures are a result of a constant long-term upward trend in concentration levels during the 1955-72 period. For

Table 2-41. Coal Production by 20 Largest U.S. Operating Groups: 1955[a]

Rank	Group	Tonnage (bituminous and lignite) (tons)	Percent of Total	Cumulative Percentage
1	Pittsburgh Consolidation Group	28,001,014	6.0%	6.0
2	U.S. Steel (C)	25,159,319	5.4	11.4
3	Peabody-Sinclair-Southern	13,663,920[b]	2.9	14.3
4	Island Creek-Pond Creek	10,247,134	2.2	16.5
5	Bethlehem Steel (C)	9,886,274	2.1	18.6
6	Eastern Gas & Fuel	9,307,068	2.0	20.6
7	Pittston Group	8,058,130	1.7	22.3
8	Pocahontas Fuel	7,968,603[c]	1.7	24.0
9	Truax-Traer	7,661,498	1.6	25.6
10	Freeman	6,550,000[d]	1.4	27.0
11	Republic Steel (C)	5,832,513	1.3	28.3
12	Sinclair-Southern Group	5,660,837[e]	1.2	29.5
13	North America Group	5,330,895	1.1	30.6
14	Jones & Laughlin Steel (C)	5,313,434	1.1	31.7
15	Berwind-White	4,895,840	1.1	32.8
16	Ayrshire Collieries Corp.	4,871,068	1.0	33.8
17	Rochester & Pittsburgh	4,602,489	1.0	34.8
18	Old Pen	4,238,921	0.9	35.7
19	West Kentucky	4,093,715	0.9	36.6
20	Omar Mining	3,967,994	0.9	37.5
	Total: Top 20	175,310,696		
	Total: Others	289,322,712[f]		
	Total U.S. Production	464,633,408		

(C) Captive Production.

[a]Keystone Coal Buyers Manual, *Coal Production in the United States for the Year 1955,* (New York: McGraw-Hill, Inc., 1956), pp. 3-10.

[b]Includes Sinclair-Southern figures after merger, July 11, 1955.

[c]Includes American Coal Co. of Allegheny Co. after merger, July 1, 1955.

[d]Includes Chicago, Wilmington & Franklin Coal Co., whose 1954 production was 3,274,337.

[e]Production only before merger, July 11, 1955.

[f]Bureau of Mines, *Minerals Yearbook 1955* (Washington: Government Printing Office, Vol. 11, 1958).

example, during this period four-firm concentration increased by 84 percent. Similar, but somewhat smaller increases were recorded at the eight and twenty-firm levels. While 1972 levels are best described as moderate, the upward trend raises a major competitive concern because, assuming the trend continues

Table 2-42. Coal Production by 20 Largest U.S. Operating Groups: 1965[a]

Rank	Group	Tonnage (bituminous and lignite) (tons)	Percent of Total	Cumulative Percentage
1	Peabody Coal Co.	49,050,412	9.6%	9.6
2	Consolidation Coal Co.	48,643,000	9.5	19.1
3	Island Creek Coal Co.	20,584,301	4.0	23.1
4	U.S. Steel Corp. (C)	17,991,000	3.5	26.6
5	The Pittston Co.	13,611,466	2.7	29.3
6	General Dynamics Corp.[b]	12,667,669	2.5	31.8
7	Bethlehem Mines Corp. (B)	11,951,504	2.3	34.1
8	Eastern Associated Coal Corp.	11,602,624	2.3	36.4
9	Ayrshire Colieries Corp.	8,313,496	1.6	38.0
10	Pittsburgh & Midway Coal Co.	8,186,248	1.6	39.6
11	Old Ben Coal Corp.	6,261,160	1.2	40.8
12	North America Coal Corp.	5,982,960	1.2	42.0
13	Westmoreland Coal Co.	5,847,966	1.1	43.1
14	Wending Gulf Coals	5,280,000	1.0	44.1
15	Southwestern Illinois	5,130,198	1.0	45.1
16	Republic Steel Corp. (C)	4,839,610	0.9	46.0
17	Bell & Zoller Coal Co.	4,468,374	0.9	46.9
18	Valley Camp Coal Co.	4,270,018	0.8	47.7
19	Rochester & Pittsburgh Coal Co.	4,252,232	0.8	48.5
20	Central Ohio Coal Co.	3,801,561	0.7	49.2
	Total: Top 20	252,735,841		
	Total: Others	259,352,422		
	Total U.S. Production	512,088,263[c]		

(C) Captive Production .

[a]Keystone Coal Buyers Manual, *Coal Production in the United States for the Year 1965* (New York: McGraw-Hill, Inc. 1966), pp. 3-10.

[b]Includes Freeman Coal Mining Corp. and United Electric Coal Companies.

[c]Bureau of Mines, *Minerals Yearbook 1965* (Washington: Government Printing Office, Vol. 11, 1967).

at the same pace, concentration at the four-firm level would soon approach the critical range of 50 to 55 percent.[49]

The movement towards higher concentration can be examined more fully by observing the movement in the market share of individual firms. In 1955, Pittsburgh Consolidation was the largest producer with only 6 percent of total coal production, but by 1965 Peabody Coal had become the largest supplier with a market share of 9.6 percent. In 1972, Peabody, now owned by Kennecott Copper, had increased its relative importance to the point where it

Table 2-43. Coal Production by 20 Largest U.S. Operating Groups: 1970[a]

Rank	Group	Tonnage (bituminous and lignite) (tons)	Percent of Total	Cumulative Percentage
1	Peabody Coal Co.	67,850,339	11.3%	11.3
2	Consolidation Coal Co. (O)	64,062,000	10.6	21.9
3	Island Creek Coal Co. (O)	29,722,041	4.9	26.8
4	Pittston Group	20,540,379	3.4	30.2
5	U.S. Steel (C)	19,631,000	3.3	33.5
6	Bethlehem Mines (C)	14,604,998	2.4	35.9
7	Eastern Associated Coal Corp.	14,538,990	2.4	38.3
8	Ayrshire Collieries Corp.	14,426,777	2.4	40.7
9	General Dynamics	14,092,297	2.3	43.0
10	Old Ben Coal Corp. (O)	11,687,056	1.9	44.9
11	Westmoreland Coal Co.	11,346,979	1.9	46.8
12	North American Coal Corp.	9,674,136	1.6	48.4
13	Pittsburgh & Midway Coal Co. (O)	7,837,511	1.3	49.7
14	Utah Construction	6,070,950	1.0	50.7
15	Southwestern Illinois	5,715,067	1.0	51.7
16	Valley Camp Coal Co.	5,533,699	0.9	52.6
17	Central Ohio Coal Co.	5,525,948	0.9	53.5
18	Republic Steel Corp. (C)	4,950,159	0.8	54.3
19	Bell & Zoller Coal Co.	4,700,255	0.8	55.1
20	Rochester & Pittsburgh Coal Co.	4,707,327	0.8	55.9
	Total: Top 20	337,227,908		
	Total: Others	265,704,092		
	Total U.S. Production	602,932,000		

(C) Captive Production
(O) Oil Ownership
[a]Keystone Coal Industry Manual, *U.S. Coal Production by Company ... 1970* (New York: McGraw-Hill, Inc., 1971), pp. 9-11.

accounted for 12.2 percent of total coal production. There has been some shifting of positions and movement among the top eight producers, but by 1970 the disparity of market share among the top four is somewhat greater than was the case in 1955.

A striking feature of the data presented in Tables 2-41 through 2-44 is that the market shares of the individual firms drop quite rapidly as relative size decreases. In 1970, for instance, the market share of firms ranking below the tenth position are less than 2 percent. For firms ranking lower than 16th,

Table 2-44. Coal Production by 20 Largest U.S. Operating
Groups: 1972[a]

Rank	Group	Tonnage (bituminous and lignite) (tons)	Percent of Total	Cumulative Percentage
1	Peabody Coal Co.	71,595,310	12.1%	12.1
2	Consolidation Coal (O)	64,942,000	11.0	23.1
3	Island Creek Coal (O)	22,605,114	3.8	26.9
4	Pittston Coal	20,639,020	3.5	30.4
5	Amax	16,380,303	2.8	33.2
6	U.S. Steel (C)	16,254,400	2.8	36.0
7	Bethlehem Mines (C)	13,335,245	2.3	38.3
8	Eastern Associated Coal Corp.	12,528,429	2.1	40.4
9	North American Coal Co.	11,991,004	2.0	42.4
10	Old Ben Coal Corp. (O)	11,235,910	1.9	44.3
11	General Dynamics	9,951,263	1.7	46.0
12	Westmoreland Coal Co.	9,063,919	1.5	47.5
13	Pittsburgh & Midway Coal Co. (O)	7,458,791	1.3	48.8
14	Utah International	6,898,262	1.2	50.0
15	American Electric Power	6,329,389	1.1	51.1
16	Western Energy Co.	5,500,700	0.9	52.0
17	Rochester & Pittsburgh	5,137,438	0.9	52.9
18	Valley Camp Coal	4,777,674	0.8	53.7
19	Zeigler Coal Co.[b]	4,201,164	0.7	54.4
20	Midland Coal	3,899,478	0.7	55.1
	Total: Top 20	320,428,813	55.1	
	Total: Others	269,571,187	44.9	
	Total U.S. Production	590,000,000		

(C) Captive Production.
(O) Oil Ownership.

[a]Keystone Coal Industry Manual, *U.S. Coal Production by Company ... 1972* (New York: McGraw-Hill, Inc., 1973).

[b]Owned by Houston Natural Gas.

individual market shares are less than one percent. It should be noted, however, that these rankings are on a national basis and the importance of a given firm may vary from region to region.

The increasing relative importance of the largest coal producers is reflected in the data presented in Tables 2-46 and 2-47. In terms of numbers, the number of groups in the largest size class, annual production in excess of 3,000,000 tons, declined from 31 in 1955 to 24 in 1970, but the share of total

Table 2-45. Seller Concentration: U.S. Coal Production 1955-1970ᵃ

Companies	1955	1965	1970	1972	Increase 1955-1972
Top four	16.5%	26.6%	30.2%	30.4%	84.2%
Top eight 24.0	24.0	36.4	40.7	40.4	68.3
Top twenty	37.5	49.5	55.9	55.1	46.9

ᵃKeystone Coal Buyers Manual, *Coal Production in the United States for the Years 1955-70* (New York: McGraw-Hill, Inc., 1955-70).

Table 2-46. The Number of Coal Operating Groups Distributed by Sizeᵃ

Size Classᵇ	Number of Groups			
	1955	1960	1965	1970
3,000,000 and Over	31	25	24	24
2,000,000 to 2,999,999	11	11	15	15
1,000,000 to 1,999,999	37	30	32	30
700,000 to 999,999	26	18	26	20
500,000 to 699,999	26	16	38	37
400,000 to 499,999	31	20	24	33
300,000 to 399,999	41	41	40	59
200,000 to 299,999	64	78	85	90
100,000 to 199,999	171	176	202	224
Less than 100,000ᶜ	4691ᵈ	6994ᵈ	6433ᵈ	3386ᵈ

ᵃKeystone Coal Buyers Manual, *Coal Production in the United States for the Years 1955-70* (New York: McGraw-Hill, Inc., 1955-70).

ᵇTons of bituminous and lignite.

ᶜEstimate.

ᵈNumber of groups not available. Figures represent companies.

output accounted for by the largest size group increased from 45.3 percent to 59.0 percent. At the same time, a dramatic decline occurred in the output attributable to the smallest size class, less than 100,000 tons of annual output. In 1955, this group accounted for some 15.4 percent of total production, but by 1970 the figure was only 6.2 percent.

The increasing importance of very large firms in the coal industry, and, as a result, a trend at the national level, towards higher concentration is confirmed by Prof. Reed Moyer in an analysis of midwestern coal markets.[50] Moyer's study covered the 1934-1962 period and focused upon coal production

Table 2-47. Coal Production Accounted for by Various Size
Classes[a]

Size Class[b]	Percentage of Total Production			
	1955	*1960*	*1965*	*1970*
3,000,000 and Over	45.3	48.8	51.9	59.0
2,000,000 to 2,999,999	5.8	6.5	7.1	6.0
1,000,000 to 1,999,999	11.3	10.0	9.2	7.4
700,000 to 999,999	4.6	3.8	4.1	2.7
500,000 to 699,999	3.2	2.3	4.3	3.7
400,000 to 499,999	3.0	2.2	2.1	2.5
300,000 to 399,999	3.0	3.5	2.7	3.4
200,000 to 299,999	3.4	4.5	4.0	3.7
100,000 to 199,999	5.0	6.1	5.5	5.4
Less than 100,000[c]	15.4	12.5	9.4	6.2

[a]Keystone Coal Industry Manual, *United States Coal Production by Company* (New York: McGraw-Hill, Inc.), various years.
[b]Tons of bituminous and lignite production.
[c]Estimate.

in a market encompassing much of Illinois along with parts of Indiana and Kentucky.

Moyer's findings in regard to concentration are presented in Table 2-48. Two points emerge from the results: (1) concentration in 1962 was significantly higher in the midwestern region than at the national level, and (2) during the period 1934-1962, concentration increased steadily. In a final appraisal of competition in the region, Moyer concludes:

> The industry can be accurately labeled an "oligopoly with a competitive fringe." The trend towards few sellers stems partly from attrition among the fringe which depends strongly on declining consumer market segments. But much of the increased concentration derives from a conscious policy of growth through mergers conducted by most of the remaining large firms.[51]

While a significant number of mergers occurred, Moyer did not find evidence to indicate that the merger activity was due to a desire to reduce competition; instead, the evidence suggests that the rising concentration represents an adjustment to changes in coal consumption patterns.[52] He states:

> The industry's salvation depends on its participation in the expanding electric utility market, the successful operator must gear himself

Table 2-48. Concentration of Midwestern Coal Production
Among Top Four, Eight, and Twenty Company-Groups 1934-1962
(tonnages in thousands)

	Four Largest	Eight Largest	Twenty Largest
1962	54.6%	74.2%	N.A.
1960	52.3	69.7	89.2%
1954	25.2	41.4	70.0
1949	26.8	39.3	61.4
1944	25.2	36.3	57.9
1938	22.6	33.8	55.9
1934	19.7	30.4	50.5
Increase 1934-1962	177.2%	144.1%	76.6%[a]

[a]Increase for 1960-1934.

Source: Reed Moyer, *Competition in the Midwestern Coal Industry* (Cambridge: Harvard University Press, 1964), p. 68.

>to meet its demands. And its demands call for large size, for large reserves strategically located to satisfy the needs of utilities operating in different market areas, and for large, capital intensive mining operations to guarantee low cost operations which are large enough to permit taking advantage of cost savings modes of transportation.[53]

Thus, the rising concentration observed in the midwestern coal market during the period 1934-1962 appears to represent a basic readjustment by coal producers to changes in coal use patterns, namely the demand characteristics of electric utilities.

The rising concentration reported here most likely represents a continuing adjustment by coal producers to the expanding needs of their sole remaining customer, namely electric utilities. It is clear that very large coal producers have clear advantages over smaller rivals in meeting utility demands.

One factor conferring a cost advantage on large producers is the introduction of unit trains, an innovation resulting in substantial scale economies due to transportation savings.[54] A unit train, composed entirely of coal hauling cars at the mine, travels intact between the mine and the generating plant, in this way the costs and time delays associated with the use of freight classification yards are eliminated. Unit trains are only economical when large tonnages are involved. Small producers cannot use this technique. Even combining individual production is not sufficient, because this defeats the purpose of the unit train technique. In short, it is the largest producers who can best take advantage of this innovation. In addition to conferring a cost advantage to coal supplied by large producers, the unit train has led to an expansion in the geographic areas that can be served from a given mine location.[55]

Another factor responsible for the growth of large coal producers is the combination of large increases in the size of electric generating plants and the increased use of long term contracts between utilities and coal suppliers.[56] Contracts between utilities and suppliers commonly run up to 25-30 years. In the case of a mine mouth generating plant, a supply contract must cover the entire fuel needs of the plant for its useful life, because once it is located at a mine mouth, there are few viable alternative fuel sources. Clearly, in order to compete for long-term contracts a firm must control huge coal reserves. Smaller companies are left to compete for spot market sales. Spot sales are not likely to account for a large proportion of coal sales to utilities and the continued viability of small producers is questionable.[i]

The upward trend is likely to be further accelerated by the development of coal gasification plants. These plants consume huge amounts of coal, and small firms with limited coal reserves are not able to participate in this type of development for the very same reasons that they are at a disadvantage when competing for utility sales.

In 1969, the federal government passed legislation requiring significantly stricter safety requirements for coal mining. The added cost of compliance, it is argued, falls most heavily upon the smallest mines. These mines, already marginal operations in many cases, are not likely to remain economic. During the 1968-1970 period, the number of mines producing less than 25,000 tons annually declined from 2,715 to 1,968, a 28 percent decline in two years. This decline accelerates the trend towards higher concentration. The coal demands of utilities, the decline of the spot market, and the new safety requirements combine to confer significant advantages on large mines and, as a result, suggest a continuing upward trend in concentration levels.

There is another factor, however, that might offset at least some of the upward pressure on concentration, and that is how the federal government disposes of the vast coal reserves on the public lands in the midwest. For instance, the results of a study of ownership patterns in five Southwestern states, presented in Table 2-49 indicates that the federal government controls 45 percent of the total acreage.[57] In the three-state region of Montana, Wyoming, and North Dakota there are some 39,496 acres of strip reserves of which the federal government controls some 5,434 million acres. The coal companies controlling most of the reserves in this area account for about one-third of national coal output or, in other words, two-thirds of the U.S. coal producers are not represented among the firms who control the reserves in this area. Thus,

---

[i]Small companies operating in the spot market play a vital role. Reduction in the number of small firms might lead to greater price volatility and cause the average spot market price to exceed the average long term contract by an even greater amount.

**Table 2-49.   Estimated Ownership of Coal Lands in Five
Southwestern States[a]**

State	U.S. Gov't	State Gov't	Indian Tribes	West. RR's	Private	Total
			Ownership (in Percentages)			
Wyoming	48	12	–	20	20	100
Utah	67	15	1	–	18	100
Colorado	36	14	6	20	24	100
Arizona	–	–	100	–	–	100
New Mexico	34	18	40	5	3	100
Total	45	13	13	12	17	100

[a]Provisional estimates based on incomplete data.

Source: *Southwest Energy Study*, report prepared for the Secretary of the Interior, November 1972, pp. 3-4.

expansion of output from this area would tend to reduce concentration at the national level.

Impact of Air Pollution Standards on Market Structure.   Air pollution standards have reduced the competitiveness of coal in many utility markets, and effective means of controlling stack emissions are not currently available. Because high sulfur coal exists more heavily in the East and low sulfur coal is more abundant in the West, Western coal production is likely to account for a larger share of U.S. production in the future. The distribution of coal reserves, by sulfur content, is indicated in Table 2-50. The ability of Western coal to compete successfully in areas outside its traditional market territory depends, of course, on movements in relative fuel prices. The development of coal gasification plants would also reinforce a movement to Western coal, not because of its low sulfur content, but because of lower mining costs.

An increase in the relative importance of Western coal production contains certain implications for market structure. Coal reserve ownership patterns differ significantly in the Western area. In the West, the federal government controls a large percentage of the coal lands. In five Southwestern states, according to the evidence presented in Table 2-51, this amounts to 45 percent of the coal lands.

In addition, the private companies with large holdings in the West differ from the dominant coal producers in the East. As indicated in Table 2-52, oil companies have significant reserve holdings in this area. On the assumption that air pollution standards will lead to greater coal production from Western fields, the structure of the coal market could undergo substantial change. The extent of the change will depend on the relative growth rates of the Western coal firms and how the federal government grants access to the public lands.[58]

Table 2-50. Estimated U.S. Coal Reserves By Sulphur Content
(millions of tons)[a]

| | United States | | East of Mississippi River | |
Sulphur Content (%)	Tons	%	Tons	%
−0.7	720,060	46	50,062	11
0.7 - 1.0	303,573	19	45,219	9
1.0 - 3.0	238,374	15	177,281	37
+3.0	314,159	20	206,495	43
Total	1,576,166	100	479,057	100

[a]As of January 1, 1965.

Source: *U.S. Energy Outlook*, National Petroleum Council, December, 1972, p. 160.

Table 2-51. Estimated Ownership of Coal Lands in Five
Southwestern States[a]

| | Ownership (in Percentages) | | | | | |
State	U.S. Gov't	State Gov't	Indian Tribes	West. RR's	Private	Total
Wyoming	48	12	–	20	20	100
Utah	67	15	1	–	18	100
Colorado	36	14	6	20	24	100
Arizona	–	–	100	–	–	100
New Mexico	34	18	40	5	3	100
Total	45	13	13	12	17	100

[a]Provisional estimates based on incomplete data.

Source: *Southwest Energy Study*, report prepared for the Secretary of the Interior, November 1972, pp. 3-4.

Oil-Coal Mergers. In recent years, a number of mergers between coal producers and firms in other industries have occurred. The major interfuel acquisitions of coal companies by noncoal firms are listed in Table 2-53. In 1970, coal companies owned by major oil firms and ranking among the top twenty coal groups accounted for 18.7 percent of total coal tonnage in 1970. At the same time, steel firms accounted for 6.5 percent of the total.

It is not surprising that the trend towards more interfuel mergers has aroused considerable comment and controversy concerning the competitive impact. Some see the oil company entry as an attempt to monopolize the production of energy. Oil interests, on the other hand, defend their entry on the grounds that the creation of energy companies leads to savings and is perfectly

Table 2-52.   Estimated Strip Coal Reserve Ownership in Montana, Wyoming and North Dakota[a]

Company	Estimated Reserve (millions of tons)
Burlington Northern RR	2423
Montana Power*	195
Montana-Dakota Utilities	273
Peter Kiewit*-Pacific Power & Light*	1189
Star Drilling Co. (optioned to Peoples Gas Co.)	1000
Reynolds Aluminum	1000
H.F.C. Oil Co.	810
Carter Oil-U.S. Steel*	465
Carter Oil-Amax Coal Co.*	1295
Carter Oil-Consolidation Coal Co.*	1253
North American Coal Co.*-Consolidation Coal Co.*	380
North American Coal Co.*	855
Consolidation Coal Co.*	300
Westmoreland Coal Co.*-Amax Coal Co.*	1500
Gulf Oil Co.*	800
Peabody Coal Co.*-Montana Power	1440
Peabody Coal Co.*	5207
Federal, State, Burlington-Northern	677
Federal Government	5434
Carter, Amax*, Kerr-McGee Oil, Sun Oil, Mobil Oil, Atlantic-Richfield Oil, Peabody*, Federal	18000
Total	39496

*Coal-producing companies in 1972.

[a]It is not known whether tonnages for two or more companies are jointly owned or represent individual ownerships grouped together. The 18000 million ton figures, however, presumably represents individual ownerships.

Source: Coal mining company's confidential field report.

consistent with effective competition. Others view the interfuel mergers as having been mainly the result of special tax law provisions that—up until their change in 1969—made coal mines more profitable to a purchaser than to a firm that has owned the operation for a long time. As a result, outside interests have acquired many of the major coal groups. It is necessary to point out that while oil interests have attained a solid position among major coal-producing groups, a variety of other industrial groups are also represented. Mergers have greatly reduced the relative importance of independent coal groups, but this does not necessarily mean that competition has simultaneously been eroded.

Of the mines acquired, a substantial proportion, perhaps over 50

**Table 2-53. Principal Acquisitions of Coal Companies by
Oil Companies and Other Large Industrial Companies**

	Company Acquired	*Date of Acquisition*
Oil companies:		
Gulf Oil	Pittsburgh & Midway Coal	Late 1963
Continental Oil	Consolidation Coal	Sept. 15, 1966
Occidental Petroleum	Island Creek	Jan. 29, 1968
Do	Maust Properties	Aug. 8, 1969
Standard Oil of Ohio	Old Ben Coal	August 1968
Do	Enos Coal	Do
Eastern Gas & Fuel	Joanne Coal	June 18, 1969
Do	Ranger Fuel (30% interest)	January 1970
Do	Sterling Smokeless	Apr. 15, 1970
Other large industrials:		
General Dynamics	Freeman Coal	Dec. 31, 1959
Do	United Electrical Coal	Dec. 31, 1959[a]
Kennecott-Copper	Peabody Coal	Mar. 29, 1968
Wheeling-Pittsburgh Steel	Omar Mining	December 1968
American Metal Climax	Ayrshire Collieries	Oct. 31, 1969
Gulf Resources & Chemicals	C & K Coal Co.	Jan. 6, 1970
Alco Standard	Barnes and Tucker	July 1970
Acquiring company:		
Peabody Coal[c]	Midland Electric Coal	August 1963
Pittston Co.	Eastern Coal	October 1969
Westmoreland Coal	Winding Gulf	November 1968
Do	Imperial Smokeless	Jan. 31, 1969

[a]Acquired 25-percent interest in 1959, then increased to 37-percent on July 1, 1962, to 52.9-percent as of Dec. 31, 1963, and to 100-percent in November 1966.

[b]The following important acquisitions of coal companies were also made by several of the largest independent coal companies during the past decade.

[c]Later acquired by Kennecott, as noted above.

Source: U.S. House of Representatives, "Concentration by Competing Raw Fuel Industries in the Energy Market and Its Impact on Small Business," (Washington: Government Printing Office, July 12-22, 1971), p. 46.

percent, have been acquired by oil firms. This raises the question as to whether such mergers lead to results in cost savings or to gains in economic efficiency. At least one petroleum executive argues that oil's entry into coal is designed to ensure a secure base of supply for their R&D efforts in coal gasification and liquefaction. Howard Hardesty, Senior Vice President, Continental Oil Company, whose merger with Consolidation Coal led to demands for antitrust action, states:

> It is my firm belief that the Association of Continental and Consol has resulted in increased supplies of coal at fair and reasonable prices and that the blending of talents and research facilities of the two companies has made and will make a substantial contribution towards meeting the nation's escalating demands for fuel.[59]

Commenting on the interface between Continental's and Consol's research and engineering facilities, Hardesty states: "Our combined research efforts reach across the horizon and involve pollution-free synthetic fuel, mine health and safety, environment, efficiency, and transportation."[60]

Unlike Continental Oil, Exxon has entered coal by acquiring reserves, but has not engaged in any significant production. Carl G. Herrington, an Exxon Vice President, states:

> We concluded that coal mining and the marketing of coal as a utility fuel offered an attractive long term investment opportunity which draws upon Humble's experience in exploring for minerals and its established management and technical resources.[61]

Exxon's entry decision also contained a long-run element. Herrington adds:

> Humble recognized concurrently that coal at some future date could become a suitable raw material to supplement crude oil and natural gas as an economically attractive source of hydrocarbons.[62]

Mergers occur for a variety of reasons. In any case, if one assumes profit maximization as a goal, two conditions are necessary for a merger to take place: (1) the capitalized value of the acquired firm, when integrated into the acquiring firm's operations, must exceed the capitalized value of the acquired firm operating independently; and (2) the price paid by the acquiring firm must not exceed the cost of achieving the same result in some alternative way, e.g., internal expansion.

Several factors can cause a firm to be worth more after acquisition than it was prior to acquisition. Some examples include (1) the creation of monopoly power from the merger, (2) significant cost reductions via scale economies or complementarities in production or distribution, (3) pecuniary scale economies in sales promotion, and (4) market imperfections that prevent the price of the acquired firm from being bid up to its true value. The argument offered by oil firms that coal ownership is essential to their R&D efforts in coal conversion is not totally convincing. It is true that patent protection in R&D in this area is not perfect, because much of it is supported by public funds, and as a result a coal firm must be able to quickly enter into gasification or liquefaction in order to capture the benefits of the R&D effort. But this can be accomplished without coal ownership before coal conversion technology becomes commercial. Oil firms could simply buy coal in the open market when conversion efforts become commercial. Thus, coal ownership does not appear to be necessary for an oil firm to engage in R&D activity concerned with coal conversion.

If, however, coal were priced above marginal cost, a firm expecting to use large amounts of coal in the future would have an incentive to enter coal mining, via exploration for coal reserves, in order to obtain coal at marginal cost. Such a situation (coal prices set above marginal cost) would also attract other entrants to the industry. There is little evidence to indicate that coal prices are presently, or are likely to be, in excess of marginal costs. Rate of return data indicates that independent coal firms are not at a disadvantage relative to coal producers associated with oil interests. Independent coal firms earned an average 14.5 percent rate of return in 1970, and Continental Oil's Consolidation Coal Co. earned only 10 percent.[63]

Another possible explanation for oil's dramatic entry into coal is that oil firms can supply large amounts of new capital to coal mining. This, however, assumes that the capital market is rationing capital to coal operations. In the past there have been capital shortages, but there is little evidence to indicate that a capital shortage had a relatively greater impact on coal than in other industries. If one considers recently announced intentions of the Pittston Company to construct a $250 to 300 million oil refinery, the argument that independent coal firms are unable to obtain capital appears to be unfounded.

The available evidence provides little support to the contention that the creation of energy companies has resulted in a monopolization of coal and uranium resources. The impact of the mergers on market structure has not been great. Oil interests have become relatively more important in the coal and uranium areas, but it cannot be concluded that such entry result in domination. It should be emphasized, however, that the use of production data undoubtedly understates the relative importance of oil interests in other fuels. A more precise view of oil involvement could be obtained if the ownership pattern of fuel reserves were known.

Uranium

The production and milling of uranium ore is one part of the nuclear fuel cycle. Other stages are: (1) conversion of uranium ore concentrate (U_3U_8) to uranium hexafluoride (UF_6); (2) uranium enrichment; (3) conversion of enriched material into uranium dioxide (UO_2) pellets; (4) manufacture of zinconium tubing to hold the fuel pellets; (5) fuel fabrication; and (6) fuel reprocessing.

The Atomic Energy Commission, heavily involved at each point of the fuel cycle, has a legislative mandate to foster competition in the nuclear industry. The AEC act, in its first section states:

the development, use, and control of atomic energy shall be directed so as to promote world peace, improve the general welfare, increase the standard of living, and strengthen free competition in private enterprise.[64]

In addition, section 105 of the AEC act is devoted solely to antitrust provisions.

The major uranium mining and milling companies in 1967 and 1970 are listed in Tables 2-54 and 2-55. Market shares are based on milling plant capacity, because actual amounts of uranium milled are not public information.[j]

Table 2-54.   Uranium Mining and Milling Companies: 1967

	Approximate Milling Capacity[a] (tons per day)	Share of Total	Cumulative Percentage
Kerr-McGee	6000	23.9%	23.9%
United Nuclear-Homestake-Partners[b]	3500	13.9	37.8
The Anaconda Company	3000	12.0	49.8
Union Carbide	1800	7.2	57.0
United Nuclear[c,e]	1725	6.9	63.9
Atlas Corporation	1500	6.0	69.9
Western Nuclear, Inc.	1200	4.8	74.7
Susquehanna-Western Inc.	1000	4.0	78.7
Petrotomics Co.	1000	4.0	82.7
Utah Construction & Mining	980	3.9	86.6
Federal-American Partners	900	3.6	90.2
Mines Development, Inc.	650	2.6	92.8
American Metal Climax	500	2.0	94.8
Foote Mineral[c,d,e]	500	2.0	96.8
Dawn Mining[c]	440	1.8	98.6
Cotter	400	1.6	100.0%
Total	25095		

[a]AEC computation of capacity from the most recently published sources.

[b]This mill, 70 percent owned by United Nuclear Corp., processes ore of both Homestake and UNC.

[c]Inactive.

[d]Ceased operation June 30, 1968.

[e]Probably will not be reopened.

Source: A.D. Little Report, p. 162.

---

[j]This does not represent a serious weakness because, in 1971, the sixteen largest milling firms, in terms of capacity, also control over 80 percent of domestic uranium reserves.

Table 2-55. Uranium Mining and Milling Companies: 1971

	Nominal Capacity Tons Ore per Day	Share of Total	Cumulative Percentage
Kerr-McGee	7000	20.6%	20.6%
Union Carbide	5000	14.7	35.3
United Nuclear-Homestake	3500	10.3	45.6
The Anaconda Company	3000	8.8	54.4
Utah Construction & Mining	2400	7.1	61.5
Standard Oil (N.J.)	2000	5.9	67.4
Susquehanna-Western Inc.	2000	5.9	73.3
Continental Oil-Pioneer Nuclear	1750	5.2	78.5
Atlas Corp.	1500	4.4	82.9
Petrotomics Co.	1500	4.4	87.3
Western Nuclear	1200	3.5	90.8
Federal-American Partners	950	2.8	93.6
Mines Development, Inc.	650	1.9	95.5
Dawn Mining	500	1.5	97.0
Rio Algan Mines Ltd.	500	1.5	98.5
Cotter Corp.	450	1.3	99.8
Total	33900		

Source: AEC, *The Nuclear Industry* (Washington: Government Printing Office, 1971), Wash 1174-1171, p. 20.

Concentration measures, reported in Table 2-56, indicate high and stable levels of concentration during the period 1967-1971. The four-firm and eight-firm figures in 1971 are 54.4 percent and 78.5 percent respectively. In fact, there are only sixteen major milling companies in 1971. Such concentration levels are inconsistent with vigorous competition.

The market structure of uranium production is heavily influenced by a severe problem of overcapacity in mining and milling. Capacity exceeds demand for uranium today, but a tight supply situation could exist in the 1979-1980 period. In 1970, the AEC completed its uranium procurement program. As a result, uranium producers were placed in a position of total dependence on commercial customers for their uranium sales. The excess capacity results from the failure of nuclear power plants to become operational at the rate expected by the industry and the AEC. The problem of excess capacity is likely to become more severe, at least during the short run.[65] In some respects, the uranium market represents an "infant industry" situation.

Thus, in light of the problem of excess capacity due to a slowly

Table 2-56.   Concentration Levels in Uranium Milling

	Percentage of Capacity	
	1967	1971
Top four	57.0%	54.4%
Top eight	78.7	78.5
Top sixteen	100.0	99.8

developing nuclear segment of the electric utility sector, high concentration levels at the milling and mining stages are likely to persist and may even increase if firms decide to exit from this stage.[k] It is interesting to note that an AEC official reports that in spite of a serious oversupply, uranium prices in 1971 have stabilized to some degree.[66] Such stabilization during a period of severe excess capacity suggests that competition is lacking and also reflects (1) an AEC ban on imports of U_3O_8 and (2) AEC stockpiles of surplus uranium.

In recent years, there has been a substantial amount of oil company entry into uranium production and fabrication. This is not a surprising event in light of the prospects for nuclear power and the fact that oil firms already possess much of the geological skills necessary for uranium exploration and production. As to the competitive implications, they do not turn on the simple fact of oil company entry, but rather on the extent of the involvement.

Information released by the AEC provides a guide to the relative importance of oil interests in the uranium area. In presenting the information, the AEC, in order to protect individual company data, have combined the firms into three groups on the basis of the extent of the firms involvement in other fuels. As indicated in Table 2-57, group A companies represent firms in which oil is the major source of revenue, group B companies have oil as a subordinate product, and group C companies are primarily engaged in natural gas and/or coal production. The AEC does not provide any additional information concerning the classification procedure. The data reflects the situation as of December 31, 1970.

The extent of uranium involvement by these various groups is indicated in Table 2-58. In 1970, total drilling, in thousands of feet, for uranium is reported as 23,500 feet, of which some 55.3 percent is accounted for by group A companies, 12.8 percent by group B firms, and 10.2 percent is attributed to group C firms. In terms of reserve estimates (at $8), oil companies control 43.5 percent of the total. The relative importance of companies engaged in fossil fuel production is shown by the figures contained in the summation column (A + B + C). With the exception of land control, more than 60 percent of the

---

[k]In August 1971, Atlantic Richfield announced plans to discontinue its commercial fuel activities. Babock and Wilcox has assumed some of the ARCO fuel fabrication facilities.

Table 2-57. Oil Company Involvement in Uranium: Dec. 31, 1970

	Group A: Oil Companies	Group B: Oil as Subordinate Source of Revenue	Group C: Natural Gas or Coal Production Involved	A+B	A+B+C	Industry Total
Land control for uranium purposes (thousands acres)	22.9%	4.6%	4.2%	27.5%	31.7%	24,000
Surface drilling in 1970 (thousands feet)	55.3	12.8	10.2	68.1	78.3	23,500
$8 ore reserves estimate Jan. 1, 1971 (tons U_3O_8)	43.5	6.9	15.0	50.4	65.4	246,000
Production capability (tons U_3O_8 in concentrates per year): Current	31.4	17.9	20.0	49.3	69.3	14,000
1973 announced plans	33.3	14.4	16.6	47.7	64.4	18,000
Uranium delivery contracts–1971 forward (tons U_3O_8)	47.1	4.4	9.6	51.5	61.1	81,300

Source: U.S. House of Representatives, Subcommittee on Special Small Business Problems, "Concentration by Competing Raw Fuel Industries in the Energy Market and Its Impact on Small Business," July 1971, p. 255.

Table 2-58. Oil Companies and Other Fossil-Fuel-Related Companies Having Uranium Interest

Group A.	Oil Companies: Atlantic Richfield Cities Service Continental Oil Getty Gulf Inexco Kerr-McGee Mobil Marathon Pennzoil Phillips Standard Oil (Ohio) Standard Oil (N.J.) Standard Oil (Ind.) Sun Tenneco Union Oil of California
Group B.	Oil as Subordinate Source of Revenue: Atlas Corp. Denison Mines Earth Resources King Resources Newmount Mining New Park Santa Fe Ind. Union Pacific Union Carbide
Group C.	Natural Gas or Coal Production Involved: American Metal Climax Consolidated Oil & Gas Colorado Interstate Federal Resources Houston Natural Gas Pioneer Natural Gas Homestake Utah Construction & Mining Continental Oil Denison Gulf Union-Pacific Kerr-McGee

Source: U.S. House of Representatives, Subcommittee on Special Small Business Problems, "Concentration by Competing Raw Fuel Industries in the Energy Market and Its Impact on Small Business," July 1971, p. 255.

activity in each category is accounted for by fossil fuel companies, ranging from a high of 78.3 percent in the case of surface drilling to a low of 61.1 percent in terms of uranium delivery contracts—1971 forward.

Evidence of a trend towards greater oil company involvement in uranium can be seen from data presented in Tables 2-59 and 2-60. The AEC data simply divides companies into two categories: oil and others.

Table 2-59. Oil Company Involvement in Uranium: Drilling and Reserve Control, 1966-1971

	1966		1971	
Number of Companies Drilling for Uranium	*Number of Co.'s*	*Percent of Total*	*Number of Co.'s*	*Percent of Total*
Oil	3	4%	17	16%
Other	65	96%	89	84%
	68		106	
	1966		1971	
Control of U.S. Uranium Reserves	*Tons (000)*	*Percent of Total*	*Tons (000)*	*Percent of Total*
Oil	44	31%	128	48%
Other	97	69%	138	52%
	141		266	

Source: Hearings and AEC.

Table 2-60. Oil Company Involvement in Uranium: Production Capacity and Drilling Footage, 1966-1971

	1966		1971	
Production Capacity	*Tons*	*Percent of Total*	*Tons*	*Percent of Total*
Oil	2.7	26%	4.4	28%
Other	7.6	74%	11.1	72%
	10.3		15.5	
	1966		1971	
Drilling Activity	*Feet (10⁶)*	*Percent of Total*	*Feet (10⁶)*	*Percent of Total*
Oil	1.4	33%	13.3	56%
Other	2.8	67%	10.2	44%
	4.2		23.5	

Source: U.S. House of Representatives, Subcommittee on Special Small Business Problems," Concentration by Competing Raw Fuel Industries in the Energy Market and Its Impact on Small Business," July 1971.

The importance of oil firms in terms of the number of companies drilling for uranium and control of domestic reserves in 1966 and 1971 is indicated in Table 2-59. In 1966, oil firms represented only 4 percent of the number of drilling companies but by 1971 this figure had increased to 16 percent. The most important indicator is perhaps the control of reserves. In this case, the amount of reserves controlled by oil firms rose from 31 percent in 1966 to 48 percent in 1971.

Table 2-60 indicates increasing oil company importance in terms of production capacity and drilling. As an indicator of the future relative importance of oil companies, drilling statistics are perhaps the most important. In 1968, oil companies accounted for 33 percent of total surface drilling but by 1971 the corresponding figure was 56 percent. Assuming a relatively stable relationship between drilling and eventual additions to reserves, these figures indicate that, in the future, oil companies will control an increasing portion of domestic reserves. The trend, even over the relatively short period of 1966-1971, clearly indicates a greater oil company involvement in uranium.

Uranium Enrichment: A Special Problem.[1] Policy makers are faced with the need for a crucial decision in regards to the question of competition in uranium enriching services. The government currently has a monopoly in providing enriching services but the AEC announced in December 1972 that new capacity, needed by 1982-1982, should be constructed by private firms. The initial step of allowing private firms to enter has been taken but the AEC has not indicated the future role, if any, of the three enriching plants currently owned by the federal government.

While the AEC decision implies the creation of a competitive market for enrichment services, the ability to achieve this depends on the type of technology to be used in the enrichment process.[67] Uranium enrichment involves the separation of the isotopes U_{235} from U_{238}. The U.S. has the only substantial capacity in the free world to carry out this activity. In allowing private firms to enter, one of the crucial decisions will be the type of technology to be used. U.S. plants presently use a gaseous diffusion technology but there is also a possibility of using centrifuges. Centrigue technology has two important advantages over the gaseous diffusion process. One, it uses significantly less inputs of electricity. In the past, the AEC gaseous plants, located in the TVA service area, have been able to obtain electricity at less than 4 mills per kwh but, even then, electricity costs have accounted for about 50 percent of total cost and about 75 percent of out of pocket costs. In the future, electricity certainly will not be available at such low rates.

Two, centrifuge plants are much less subject to scale economies than are gaseous diffusion plants. The Tripartite Centrifuge Project, composed of a consortium of Germany, Netherlands, and Britain, plans a centrifuge plant of

---

[1]This section relies on a study provided by Prof. Thomas G. Moore. See Appendix E.

200-300 metric ton capacity.[68] The minimum efficient size plant for a gaseous diffusion plant however is about 8000 metric tons. The impact on capital costs is huge. Reynolds Metals proposed to the AEC in 1972 to construct a 8750 metric ton plant for about $2.2 billion while the plant planned by the Tripartite group is forecast for no more than $60 million for a 300 metric ton unit.

The attractiveness of centrifuge technology is illustrated by company response to an AEC invitation to request access to classified enrichment technology; seven firms responded and six have received permission to conduct R&D in centrifuge technology. Only one is engaged in R&D work on gaseous diffusion.

Demand forecasts indicate that by 1982 excess demand will create a need for one diffusion plant or many centrifuge plants. As demand increases, the need for more plants will also increase. Whether the enrichment services are provided in a competitive or oligopolistic market depends heavily upon the choice of technology. If centrifuge is adopted, there is a reasonable probability of a large number of relatively small firms and easy entry conditions. On the other hand, an oligopolistic market will probably result if the gaseous process were to be adopted.

Another crucial factor influencing entry is the willingness of the AEC to permit much wider access and company use of restricted data. Some members of the industry indicate that, while the AEC has invited private firms to enter this area, the AEC would retain almost dictatorial power over such things as prices to be charged for separative work, the firm's organizational structure, and even over the industry's structure. If this were to occur, a competitive market in enrichment is not likely to result.

Thus, the ability to successfully transform a present government monopoly in uranium enrichment services to a competitive market hinges on (1) the type of technology selected and (2) the willingness of the AEC to allow firms to act independently of AEC control.

Concentration: An Energy Market

In the following section, concentration measures for a combined fuels, or an energy market, are presented. The methodology involves the conversion of each firm's fuel production into BTU units and computing concentration in terms of the extent of dominant firm control of total BTU output. Unfortunately this procedure is limited to fossil fuels because of an inability to obtain uranium production data by individual firm.[m] The resulting figures may be understated because uranium production is much more concentrated than is the case with fossil fuels.[69] Energy market concentration measures are determined by (1) the level of concentration in each fuel, (2) the extent to which the same firms dominate production in the different fossil fuels,[n] and

---

[m]Concentration measures reported for uranium were based upon milling capacity and not actual production.

[n]As the extent of dominant firm overlap increases, energy market concentration will be greater.

(3) changes in the relative importance of individual fossil fuels in the total BTU output.

The twenty-five leading fossil fuel firms, defined in terms of total BTU production, for 1955 and 1970 are listed in Tables 2-61 and 2-62, respectively. Concentration levels for the energy market are summarized in Table 2-63. In 1970, the top four firms controlled 21.17 percent of total fossil fuel production and the eight largest firms accounted for 34.98 percent of the total.

Table 2-61.   Leading Fossil Fuel Producers: 1955

	Fossil Fuel BTUs	Share of Total	Cumulative
Standard Oil of New Jersey	1,107,945,075	3.48	3.48
Standard Oil Ind.	817,887,360	2.57	6.05
Texas	808,451,955	2.54	8.59
Phillips	779,154,900	2.44	11.03
Shell	749,492,145	2.35	13.38
Pittsburgh-Consolidated	733,627,352	2.30	15.68
U.S. Steel	659,154,157	2.07	17.75
Gulf	615,041,505	1.93	19.68
Socony Mobil	460,230,000	1.44	21.12
Cities Service	412,111,815	1.29	22.41
Peabody-Sinclair-Southern	357,994,704	1.12	23.53
Atlantic	301,244,685	0.95	24.48
Sun Oil	289,335,285	0.91	25.39
Ohio Oil	275,690,190	0.87	26.26
Island Creek-Pond Creek	268,474,910	0.84	27.10
Continental	268,134,000	0.84	27.94
Bethlehem	259,020,378	0.81	28.75
Tidewater	258,607,860	0.81	29.56
Eastern Gas and Fuel	243,845,181	0.77	30.33
Union Producing	239,417,235	0.75	31.08
Sinclair	238,119,000	0.75	31.83
Sunray-Mid Continent	234,952,935	0.74	32.57
Pittson	211,123,006	0.66	33.23
Pocohantas Fuel	208,777,398	0.66	33.89
Skelly	208,057,425	0.65	34.54
Total: Oil	13,620,807,000		42.73
Gas	6,055,690,815		19.00
Coal	12,173,395,289		38.20
Total	31,869,570,416		

The following conversion factors were used: (a) coal, 13,100 BTU per lb; (b) crude oil, 5,800,000 BTU per barrel; (c) natural gas, 1,035 BTU per cubic foot. Bureau of Mines, *Minerals Yearbook.*

Table 2-62. Leading Fossil Fuel Producers: 1970

	Fossil Fuel BTUs (000,000 BTU)	Share of Total	Cumulative
Standard Oil of New Jersey	3,460,098,852	6.94	6.94
Continental Oil	2,424,245,214	4.87	11.81
Texaco	2,423,035,578	4.86	16.67
Gulf	2,244,152,490	4.50	21.17
Shell	1,958,694,450	3.93	25.10
Standard Oil of Indiana	1,789,659,303	3.59	28.69
Peabody	1,703,043,508	3.42	32.11
Atlantic Richfield	1,431,581,274	2.87	34.98
Mobil	1,414,895,865	2.84	37.82
Union Oil of California	1,106,406,321	2.22	40.04
Phillips	999,599,226	2.01	42.05
Standard Oil of California	995,358,903	2.00	44.05
Getty	934,254,363	1.88	45.93
Standard Industries	897,170,694	1.80	47.73
Sun Oil	815,640,186	1.64	49.37
Island Creek	746,023,229	1.50	50.87
Pittston	515,563,512	1.03	51.90
Cities Service	504,624,498	1.01	52.91
Marathon	502,357,830	1.01	53.92
Tenneco	427,835,073	0.86	54.78
Standard Oil of Ohio	407,669,351	0.82	55.60
California Co. Div. Chevron	380,066,490	0.76	56.36
Bethlehem Steel	365,854,449	0.73	57.09
Eastern Associates	364,854,649	0.73	57.82
Ayrshire	362,112,102	0.72	58.54
Total BTU: Gas	14,946,161,760		30.0
Oil	19,743,446,850		39.6
Coal	15,133,593,200		30.4
Grand Total	49,823,201,810		

The following conversion factors were used: (a) coal, 12,290 BTU per lb; (b) crude oil, 5,613,290 BTU per barrel; (c) natural gas, 1,035 BTU per cubic foot. Bureau of Mines, *Minerals Yearbook.*

The 1970 figures are low but represent significant increases over the corresponding 1955 levels. At the four-firm level, the 1970 concentration measure represents a 91.9 percent increase from the 1955 level. The eight- and twenty-firm concentration levels in 1970 are approximately 75 percent greater than the 1955 levels.

Table 2-63.   Concentration Levels and Trends: An Energy
Market

	1955 %	1970 %	Increase %
Top four	11.03	21.17	91.9
Top eight	19.68	34.98	77.7
Top twenty	31.08	54.76	76.2

The oil-coal mergers have raised a substantial concern for competition in the production of fossil fuels. The impact of these mergers on energy market concentration can be easily determined. Among the top eight energy producers in 1970, two companies, Continental Oil and Gulf Oil, produced coal via subsidiaries. Continental Oil owns Consolidation Coal and Gulf operates Pittsburgh and Midway Coal. Both were acquired by merger. By subtracting 1970 coal production from the total BTU production of Continental and Gulf, it is possible to determine what concentration in the energy market would have been in the absence of these mergers. Continental Oil's total production would have been reduced by 1,607,956,200 million BTUs (64,062,000 tons of coal) and Continental would drop out of the top eight group. Eliminating the Pittsburgh and Midway production of 196,721,526 million BTU (7,837,511 tons) from Gulf's total leaves Gulf with a total BTU output of 2,047,430,964 million BTU. Recomputing the concentration ratios for the top four firms, Standard Oil of New Jersey, Texaco, Gulf, and Shell, yields a figure of 19.85 percent. Thus, if the two mergers had not occurred, 1970 energy market concentration would have been 19.85 percent rather than 21.2 percent at the four-firm level. At the eight-firm level, concentration, without the mergers would have been 33.7 percent rather than the 34.98 percent reported in Table 2-63. In addition, there would have been a change in the composition of the top eight firms, and the upward trend in concentration would be somewhat less. Thus, in terms of concentration levels, the structural indicator commonly used by economists to examine the state of competition in a market, the creation of energy companies has had only a limited impact.

The eight largest firms in each fossil fuel as well as in the energy market are indicated in Table 2-64. In terms of crude oil and natural gas, the dominant firm overlap is great. In the case of coal, none of the top eight crude producers appear among the top coal producers. However, Continental's ownership of Consolidation Coal, in combination with its production of oil and natural gas, is sufficient to make it the second largest firm in an energy market. As was emphasized in an earlier section, if coal reserve ownership data were available, oil company involvement in coal would be much greater.

The energy market concentration figures based on domestic fossil

Table 2-64. Leading Fossil Fuel Producers: 1970

		Product	
Crude Oil	*Natural Gas*	*Coal*	*Energy Market*
Standard Oil of N.J.	Standard Oil of N.J.	Peabody Coal (Kennecott Copper)	Standard Oil of N.J.
Texaco	Gulf Oil	Consolidation Coal (Continental Oil)	Continental Oil
Gulf Oil	Shell Oil	Island Creek Coal (Occidental Petroleum)	Texaco
Standard Oil Co.	Phillips Petro. Co.	U.S. Steel	Shell Standard Oil of Ind.
Standard Oil Industries	Mobil Oil	Bethlehem	Peabody
Atlantic Richfield	Texaco	Eastern Associated Coal	Atlantic Richfield
Getty Oil	Atlantic Richfield	Ayrshire Collieries Corp.	

fuel production are low and would imply a structural setting in which interfuel competition can be vigorous. The figures, however, are deceptive for several reasons. One, uranium production is not included, thus imparting a downward bias of some magnitude. Two, several oil firms are heavily involved in coal in the form of reserve ownership and not reflected in the data based on production. Three, crude oil production reflects domestic output and if imports of crude oil increase as expected, concentration of crude sales in the U.S. will rise because concentration in the world petroleum market is higher than in the domestic market. These factors all impart a downward bias to the figures but the extent of the bias, while not known, is probably not sufficient to cause the true level of concentration to reach a level where the market structure would, by itself, imply monopolization.

CONDITION OF ENTRY

Entry conditions are difficult to measure. The main sources include (1) scale economies, (2) absolute cost differences between established firms and potential entrants, (3) product differentiation advantages, (4) capital requirements and (5) certain legal restrictions on entry such as patents. In the energy industry, scale economies, absolute cost differences, and capital requirements appear to be the most significant sources of entry barriers.

A discussion of entry conditions in the case of energy may appear unimportant in light of the reported low levels of seller concentration. However, if entry barriers are high and potential competition weak, the preservation of competition among established sellers requires that concentration levels remain at a level consistent with vigorous competition. Thus, in the presence of high entry barriers, the upward trend in concentration reported would achieve greater significance.

Scale Economies

Crude Oil and Natural Gas. The natural size unit for crude oil production is the oil field. Economies result when oil fields are treated as single units. Economies of scale arising from multi field operations are difficult to identify.

Techniques designed to estimate the behavior of long run average costs have not attained a high degree of perfection in economic research. One technique commonly employed to determine the prevalence of scale economies is the survivor technique.[70] This procedure attempts to infer the shape of the long-run average cost function by determining the shares of industry output attributable to different sized firms in the industry. Plant or firm sizes which contribute an increasing share of total output are assumed to be efficient but size classes with declining shares are judged to be inefficient. In simple terms, firms which survive best are presumed to be operating at lower average costs. In this way, the survivor technique indicates the shape of the long run average cost curve and indicates the size group where average cost per unit is the lowest. The survivor technique is not without serious ambiguities. For example, the criteria for distinguishing between surviving and nonsurviving groups will necessarily be somewhat arbitrary. In addition, a changing distribution of relative importance of various size classes may reflect differences in efficiency in utilizing new technology rather than differences in production efficiency. Others have clearly spelled out inherent problems with the procedure[71] and, consequently, any interpretations based on the survivor technique must be put forth with caution.

Applying the survivor technique to crude oil and natural gas producers yields the results reported in Table 2-65.[o] Table 2-65 shows the percent of value added contributed by each size category of firms for 1958, 1963, and 1967. The results indicate that only the two largest size categories are increasing their share of total value added during the 1958-1967 period.[p] Even among the two largest classes, only the largest eight firms gained shares during each of the census periods.

Table 2-65. Selected Statistics on the Crude Petroleum and Natural Gas by Size Category of Company

Operating Companies Ranked by Value of Shipments & Receipts	Avg. Number of Barrels of Oil Shipped in 1,000 Barrels (1967)	Percentage of Total Value Added			Value Added per Manhour 1967	Value Added per Dollar Capital Expenditure 1967
		1958	1963	1967		
First 8 companies	206	43.2	46.1	54.0	$ 152	4.3
9-16 companies	65	15.6	13.9	17.4	90	4.9
17-24 companies	23	8.9	7.5	6.4	95	5.1
25-32 companies	10	3.2	3.5	3.4	116	3.0
33-68 companies	10	7.6	7.2	6.5	59	5.0
69-100 companies	2	2.8	2.9	2.0	41	5.9
All others Less than	0.1	18.7	19.0	10.3	27	3.6
Total Less than	2	100	100	100	84	4.3

Source: U.S. Bureau of Census, *Census of Mineral Industries: 1963*, Vol. 1, Summary and Industry Statistics (Washington: Government Printing Office, 1967), p. 13B98.

U.S. Bureau of Census, *Census of Mineral Industries: 1967*, Industry Series: Supplement M1667(1)-13A, "Crude Petroleum and Natural Gas," p. 4.

[o]The discussion of scale economics in oil and coal draws upon a study prepared by Prof. T.G. Moore. See Appendix B.

[p]The companies ranked 25-32 displayed a slight and insignificant increase in share of industry value added.

These results suggest that only the largest firms are clearly economic. This finding is consistent with the observation that as oil exploration spreads to offshore areas and the Arctic, risks and capital requirements increase greatly and the very large firms are better able to undertake this type of activity. Small firms face substantial problems in the future in their ability to conduct oil exploration in order to expand production. The survivor technique does not, however, indicate the extent to which smaller firms are at a disadvantage.

On the other hand, certain tax law characteristics clearly provide an attraction for venture capital to enter crude oil exploration and reduce entry barriers. The elimination of this provision, on the grounds of equity factors, would reduce the ability of small firms to remain competitive. Most of the existing tax provisions benefit major companies far more than they aid the smaller, independent firms. For example, the foreign tax credit is of no value to domestic companies and the percentage depletion allowance has the greatest value to the integrated firm.

While entry, or the possibility of entry, is a key factor in maintaining competition in any industry, it is especially important in the petroleum industry that easy entry exist at the point of access to new reserves of crude oil and natural gas.[72] If firms have relatively easy access to new reserves and can therefore engage in production, then any attempt by established sellers to exercise monopoly power and restrict production in downstream markets will be unsuccessful. Consequently, it is crucial to determine the condition of entry into the business of producing crude oil and natural gas.

Potential new reserves of oil and natural gas are owned by private landowners, state governments, and the federal government. It is generally believed that the most productive unexplored potential oil and gas lands are on the outer continental shelf controlled by the federal government. Access to government controlled lands, both state and federal, is by oral or sealed bidding auctions with leases going to the highest bidder.[q] From 1954 through March 28, 1974, the federal government conducted 33 separate oil and gas lease sales on the OCS and offered 3,651 tracts for lease and 2,057 were actually leased. Joint bids are accepted and the only meaningful barrier to entry into the auction market is the capital requirement in the form of a bonus bid plus subsequent exploration and drilling costs. The average bonus bids for these tracts was about $5 million, a modest amount relative to entry into many other industries.

The record of bidding for OCS leases, presented in Table 2-66, clearly indicates relatively free entry conditions. During this period, 132 separate firms participated as winning bidders in 33 OCS lease sales. Participation involved bidding alone or jointly. Table 2-66 indicates the record of entry by successful entrants. Most of the entry occurred subsequent to the twentieth sale, June 13, 1967. Some 79 percent of the entry has occurred in the last seven years.

---

[q] In addition, the federal government issues rights to explore for and produce from its onshore lands, not on a "known geological structure," on a lottery basis.

Table 2-66. Entry, Proportional Concentration of Lease Acquisition, And Average Number of Bidders for OCS Oil And Gas Sales, 1954-March 28, 1974

Sale Number	Number of Tracts Leased	Number of New Entrants as Winning Bidders	Concentration Ratio, Eight Largest Buyers*	Average Number of Bidders	First Sale in New Area
1	90	22	76	3.4	La.
2	19	1	100	4.7	Texas
3	94	6	54	3.7	
4	27	0	100	1.2	
5	23	0	100	1.0	Florida
6d	19	4	80	2.0	
7	99	5	58	2.7	
8	48	0	100	2.2	
9	206	8	56	2.5	
10	195	5	59	3.3	
11	10	0	100	1.0	
12d	9	0	100	1.8	
13	57	0	100	1.2	California
14d	23	0	84	3.0	
15	74	0	95	2.2	Ore.-Wash.
16	27	0	100	2.1	
17d	17	2	69	3.6	
18d	24	0	94	2.5	
19d	1	0	100	7.0	
20	158	10	54	4.3	
21	71	6	90	2.2	
22	110	17	62	3.9	
23d	16	1	68	1.8	
24d	20	0	87	1.5	
25d	16	1	80	3.6	
26d	19	0	62	2.8	
27	118	7	36	8.2	
28d	11	0	74	2.5	
29	62	8	70	4.4	
30	116	11	40	5.8	
31	100	14	41	5.3	
32	87	0	62	4.2	Miss.-Ala.-Fla.
33	91	4	42	3.5	

dDrainage sales.

*The firms identified as the eight largest buyers differ from sale to sale and are not the same firms as the big eight oil companies.

Source: Susan M. Wilcox, "Entry and Joint Venture Bidding in the Offshore Petroleum Industry," Unpublished Ph.D. dissertation, University of California, Santa Barbara, 1974, p. 66.

The relative ease of entry is reflected in the downward trend in the concentration ratios for successful bidders. In the case of small sales, there often are only eight or fewer successful bidders and consequently, concentration levels will be 100 percent.[r] The weighted concentration ratio for the first 19 sales is 85.5 percent but, for the remaining fourteen sales, the average declined to 62.0 percent.

The relative ease of entry is also confirmed by the trend in the average number of bidders per tract over the period. For sales 1-19, there were 2.7 bidders per tract, but for the remaining sales the average had increased to 3.9.

Thus, the presence of auction markets and relatively low capital requirements indicates no significant barriers to entry exist in the case of crude oil and gas production. Entry has occurred, 132 firms have been successful bidders, and serves as a major force preventing the exercise of any monopoly power in downstream markets by existing firms.

Petroleum Refining. There have been several attempts based on different techniques to estimate the importance of scale economies in petroleum refining. It is somewhat reassuring to note a general consistency in the results. Bain estimated the minimum efficient size refinery to be 120,000 barrels per day in 1951. This represents 1.75 percent of industry capacity and would cost between $225 and $250 million in 1951. Bain reports no evidence of multiplant economies.

Stigler was the first to apply the survivor technique and, in the case of petroleum refining, he reports that a plant size between 0.5 and 2.5 percent of industry capacity grew relatively rapidly between 1947 and 1954.[73] In addition, Stigler reports that the minimum efficient size, 0.5 percent of industry capacity, is the same for a plant and company. This suggests the absence of significant scale economies due to multiplant operations.

Rawleigh Warner, Jr., Chairman of Mobil Oil, in a speech before the Economics Club of Detroit in March 1973, stated the minimum efficient size of a new refinery as about 160,000 barrels per day and costing approximately $250 million. This would be equivalent to 1.2 percent of industry capacity. Professor Leonard Weiss, using the survivor technique for the 1958-1961 period, found two optimum size classes, one for small, relatively specialized refineries and another for large diversified refineries. The minimum efficient size for the latter was about 150,000 barrels per day. The Pittston Company, primarily a coal producer, has announced an intention to construct a 250,000 barrel per day refinery in Eastport, Maine. The estimated cost is $300-$350 million dollars including the construction of supertanker facilities.

Census data for 1963 and 1967 confirms the bimodal results

---

[r]Where the concentration ratio is less than 100 percent, there may be an upward bias in the concentration ratio due to the method of analysis. Where joint bidding occurs, each partner is counted as a winning firm. If two firms that are partners in a winning joint bid are both among the big eight winners, then the concentration ratio would be overstated.

reported by Weiss. During this period, refineries with less than 20 employees and refineries with 100-1000 employees have grown more rapidly than the industry. The growth of small refineries may have been stimulated due to a favorable allocation of crude oil import tickets.

In a study of scale economies in British industries, Pratten and Dean[74] find evidence of scale economies up to the largest refinery size then available in Great Britain, approximately 200,000 barrels per day.

On the basis of available evidence, scale economies do not appear to require the construction of relatively large refineries, in the sense that the minimum efficient size plant contains a significant proportion of industry capacity. Capital requirements are large but may not represent a significant barrier in light of the apparent ability of the Pittston Company, a newcomer to refining, to obtain the necessary capital to construct a refinery costing one-third of a billion dollars.

The main source of entry barriers recently have been due to 1) an inability to obtain adequate crude oil supplies because of restrictions on imported oil and 2) environmental factors. The recent liberalization of oil imports should substantially reduce the barriers associated with petroleum refining.

Uranium. The availability of unit cost data for uranium production is limited to an AEC report that assesses the costs of operating a production center of varying sizes.[75] A production center, the most common type of arrangement in the industry, is defined as a processing mill plus a mine supplying the mill with ore. The AEC study is based on an engineering approach to estimating cost behavior and is limited in that such an approach generally overlooks such factors as management efficiency, shipping and stockpiling costs, and the effects of variations in capacity utilization on unit costs. As a result, the engineering approach almost always indicates that the largest sizes are the most efficient.

The mining of uranium ores is accomplished by two methods: (1) surface or open pit mining and (2) underground mining. Among the factors determining the choice of mining method are the size, shape, grade, depth, and thickness of ore deposits. In the early years of production, 1948-1952, underground mining accounted for over 90 percent of ore production, but by 1971 underground operations accounted for some 47 percent and open pit mines for 52 percent of total production.[76]

The AEC cost estimates represent

an analysis of the economics of a future uranium venture starting with the decision to explore for uranium to the point of mining and milling uranium ore and producing uranium concentrate.[77]

Cost estimates are presented for both open pit and underground mining and represents costs as of January 1, 1972. Capital costs cover a range from land acquisition to mill construction and represent approximately 55 percent of total costs. Operating costs include mining, hauling, milling, and royalty costs.

The results indicate substantially lower unit costs for open pit mines and, in addition, the existence of significant scale economies. Assuming a common ore grade and production rate, production costs per pound of concentrate from open pit mines range from $0.95 to $3.70 less than the cost from underground mines.[78] Thus it is not surprising to find an increasing share of ore production coming from open pit mines.

The relationship between unit cost and mine size, as estimated by the AEC, is indicated in Table 2-67. For an open pit mine, unit costs range from $5.99 per pound of ore recovered for a 500 TC/D mine to $4.67 for a 5000 TC/D mine. Unit costs for an underground mine behave in the same manner.

If a 5000 TC/D mine is the minimum efficient size mine, this would represent some 6.4 percent of total industry capacity. Most of the currently existing operations are substantially smaller than this size.

The results suggest that the largest uranium producing units have substantially lower unit costs than do smaller rivals. The minimum efficient size operation, a mine with a 500 ton per day capacity, would contain 6.4 of total industry milling capacity. This suggests that scale economies are likely to lead to rather high entry barriers but, given the weaknesses of the engineering approach and an inability to determine whether scale economies continue beyond the 5000 ton/day mine size, conclusions as to the height of entry barriers are very tentative.

Coal. There is substantial evidence to suggest that large mines have significant advantages over small size operations. Evidence presented in Table 2-68 indicates that the proportion of total coal output accounted for by the fifty largest mines has increased consistently from 19.1 percent in 1955 to 24.0 percent in 1970.

Table 2-67.   Uranium Mining Costs

| | $/lb Recovered | |
Mine Size	*Open Pit*	*Underground*
500 TC/D	$ 5.99	$ 7.30
1000	5.32	6.26
2000	4.96	5.67
3000	4.79	5.40
5000	4.67	5.18

Source: Klemenic, pp. 2, 7.

Table 2-68. Relative Importance of the Fifty Largest Coal
Mines

	Total Coal Production (000 tons)	Fifty Largest Mines Percent of Total
1955	464,633	19.1%
1960	415,512	22.6
1965	512,088	23.4
1970	602,932	24.0

Source: Keystone Coal Industry Manual, *U.S. Coal Production by Company, 1971* (New York: McGraw-Hill, 1971), p. 5.

The survivor technique, when applied to coal, confirms the finding that the largest producers have cost advantages but some ambiguities do appear in the results. As indicated in Table 2-69, the largest size mines have increased in relative importance during this period. This group accounted for 49.3 percent of total coal production in 1960 but by 1970 this figure had increased to 59.6 percent. At the same time, the shares of total output accounted for by the two smallest groups declined during the 1960-1970 period and in addition, mines in the 200,000-500,000 category also declined in relative importance. Slight increases in relative importance appear in the 50-100 and 100-200 thousand categories. On the basis of this evidence, the largest mines possess substantial cost advantages over smaller rivals.

Relatively little attention has been given to the role of scale economies in coal mining. In a study of the midwestern coal market, Moyer concluded that in terms of output per man-day, underground mines smaller than 50,000 tons annually had lower output per man day than did larger mines. In

Table 2-69. Relative Change in Coal Output by Mine Size
Underground and Strip, 1960-1970

Size of Mine (000 tons)	Total Output 1960 (000 tons)	1970	Percent of Total 1969 %	1970 %
Less than 10,000	20,164	9,227	4.9	1.5
10,000-50,000	44,237	50,849	10.6	8.4
50,000-100,000	27,894	43,310	6.7	7.2
100,000-200,000	37,203	55,729	9.0	9.2
200,000-500,000	81,013	84,297	19.5	14.0
More than 500,000	204,999	359,516	49.3	59.6
Total	415,512	602,932		

Source: Bureau of Mines, *Minerals Yearbook* (Washington: Government Printing Office).

the case of strip mines, he reports that productivity increases as mine size increases. Whether the latter relationship reflects scale economies is debatable; higher labor productivity may result because large mines use more capital but their unit costs may not be less than smaller mines. Moyer did find however that larger mines, both strip and underground, on average, had better resource bases, i.e. thicker seams, and larger stripping ratios, than did smaller operations.

In a further attempt to study the labor productivity relationship, tons per man/day were regressed against several variables for underground and strip mines in Illinois in 1971.[s] The strip mine regression includes two independent variables; mine size and stripping ratio. In the case of underground mines, the variables are: mine size, seam thickness, seam depth, age of mine, and a dummy variable indicating whether the mine has a shaft or slope opening.

The strip mine results are:

$$Y = 44.717 + \underset{(2.62)^*}{0.00669X_1} - \underset{(2.12)^*}{895X_2}$$

$$R^2 = 0.376 \quad N = 25$$

where

Y = tons per man day

X_1 = mine size

X_2 = stripping ratio

*Significant at 0.05 level
(t value in parentheses)

Mine size and stripping ratio are both significant at the 0.05 level.

Regression results for underground mines are:

$$Y = 14.30 + \underset{(2.19)^*}{0.0027X_1} - \underset{(0.19)}{0.167X_2} - \underset{(0.86)}{0.0056X_3} +$$

$$\underset{(1.03)}{2.456X_4} - \underset{(0.14)}{0.116X_5}$$

$$R^2 = 0.579 \quad N = 23$$

where

Y = tons per man day

X_1 = size of mines

X_2 = seam thickness

X_3 = seam depth

*Significant at 0.05 level
(t value in parentheses)

[s]The following results are based upon a study prepared by Prof. Reed Moyer. See Appendix D.

X_4 = slope or shaft entry

X_5 = age of mines.

Mine size is the only statistically significant variable in the underground mine regression. Thus, mine size seems to be a major determinant of labor productivity.

Two studies, one published and the other unpublished, provide some light on the question of optimum mine size. The unpublished study indicates cost conditions in 1972 for a West Virginia mine operating a 72-inch seam for a coal gasification plant. The optimum size mine was reported as one with an annual output of 3.8 million tons.

The published Bureau of Mines study presents a cost analysis based on cost data developed by geographic area, rank of coal, and output capacity for twelve hypothetical mines. Seam thickness and depth and type of overburden are assumed as typical for the area in which the hypothetical mine would be located. Costs include (1) use of new equipment, (2) prevailing union wage scales and (3) the payment of all miscellaneous costs including royalties, contribution to miners' welfare fund and license and permit fees. Transportation costs are not included.

The results are reported in Table 2-70. Production cost per ton was found to be sensitive to variations in seam and overburden thickness. The results

Table 2-70. Bureau of Mines Cost Estimates: Surface Mining by Size of Mines

Production (mm ton/year)	Operating Costs	
	$/ton	$/per million BTU
Bituminous		
Eastern Province		
1	$4.15	15.7
2	3.06	11.6
Interior Province		
1	3.90	16.3
1 (2 seams)	2.98	12.4
3	2.58	10.8
Subbituminous		
Rocky Mountain and Northern Great Plains Region		
1	2.37	16.5
5	1.68	11.7

Source: See Appendix D.

indicate that as mine size increases from 1-3 and 3-5 million tons per year unit operating costs decline. When mining 6-10 foot thick seams, costs for the larger mines (3-5 million) average 30 percent below costs for 1 million ton per year mines. The cost, dollars per million BTU, decreases in the same proportion as reductions in the cost per ton.

These results suggest that 3-5 million tons per year mines have significant cost advantages over smaller operations. It is interesting to note that mines of this size would rank among the top fifty mines in 1971, at which time the mine ranked number fifty produced 1.5 million tons. These results indicate that scale economies do not constitute significant entry barriers because even a 5 million ton per year mine represents less than 1 percent of total 1971 production. The cost advantages suggested by the results are probably the explanation for the trend towards higher concentration levels.

The National Petroleum Council's energy study presents some evidence of capital costs. Their findings, summarized in Table 2-71, are estimates for 1970 and represent average original capital outlays for mines currently in operation. Capital costs per annual ton for underground mines is predicted to approach $10 by 1985. Increases are also indicated for strip mines and by 1985 capital costs of about $9 per ton are indicated. The NPC estimates have limited value because in the years beyond 1970 the figures represent average capital costs.

An unpublished Bureau of Mines report indicates that the optimum size underground mine has 3.8 million annual ton capacity. The capital cost of this size mine, including a cleaning plant and shaft or slope excavation costs, is approximately $36 million. Similar estimates have been prepared for strip mines by geographic area.[t] In the Eastern Province-Appalachian Region, a 3 million ton per year mine requires an outlay of $28 million and in the Interior Region an

Table 2-71. Estimated Capital Investment per Annual Ton of Production at U.S. Coal Mines

	Underground Mines				*Surface Mines*			
Operating year	1970	1975	1980	1985	1970	1975	1980	1985
Original capital Investment	$ 7.15	$ 8.46	$ 9.20	$ 9.84	$ 6.39	$ 7.33	$ 8.07	$ 8.78
Total capital investment over life of mine[a]	19.66	23.17	25.03	26.64	10.59	12.15	13.79	14.44

[a]Less salvage value. N.B.: 30 year life (constant 1970 dollars).
Source: National Petroleum Council, *U.S. Energy Outlook*, p. 145.

---

[t]The estimates do not include outlays for a cleaning plant but facilities for screening and crushing are included.

outlay of $24.9 million is required. In both areas, the capital costs for a 3 million ton per year mine are about twice the amount required for a 1 million ton per year mine.

Capital costs for a mine producing 5 million tons annually in the Rocky Mountain area range from $13.9 million to $28.7 million. These figures can only be considered as estimates. It is safe, however, to conclude that capital costs for a mine in the 3 million annual tonnage class will most likely range between $20-$26 million, the exact figure being heavily influenced by the quality of the natural resource base.

Outlays of this magnitude may be large in relation to past levels of capital expenditures made by coal firms when fragmented production was the general pattern. However, such outlays are clearly not beyond the financial capabilities of many corporations.

A question can be raised concerning the impact of coal gasification plants on future capital requirements for entry into coal production. El Paso Natural Gas Co. has published cost estimates for a mine designed to supply coal to a large gasification plant. Excluding leasehold costs, capital costs for a mine of 8.84 million tons are estimated to be about $45.7 million. When leasehold costs are included, which according to El Paso have been excessively high, outlays increase to $64.6 million. Such a mine would serve a gasification plant of 250 million cubic feet per day, which appears to be about the minimum efficient size.

Northern Natural Gas and Cities Service Co. have announced plans to evaluate the possible construction of a combined coal gasification and pipeline venture involving four gasification plants, each with a capacity of 250 million cfd. This would increase the coal needs and capital costs by a substantial amount.

The capital needs created by gasification plants are high when compared to the traditional capital outlays associated with coal production. The mine size required to supply such a facility will certainly exceed the 6.7 million tons produced in 1971 by the largest U.S. coal mine. This does not mean that entry barriers are high for all potential entrants, because even a mine producing 6.7 million tons contains only a very small percentage of total coal production for 1971. It does, however, suggest that future investment in coal production may flow from firms outside of coal.

It is extremely difficult to predict the future course of entry conditions in an industry. Usually the factors creating entry barriers are long lived and change only slowly. The exception to this would be institutional factors, such as patent laws, which affect entry. In the case of energy, events do not suggest that entry barriers are likely to reach a point where entry is difficult. They may increase to some extent in the future but the increase would appear to be limited. The consideration of a time horizon which encompasses the more radical sources of energy suggests that entry conditions may actually be lower.

For instance, a long run movement away from fossil fuels to something such as solar power creates a whole new set of potential entrants.

VERTICAL INTEGRATION

One of the most visible elements of petroleum production is the extent of vertical integration among major firms. Major oil firms dominate each production stage from exploration and production of crude oil to gasoline marketing. For example, the Department of Justice estimates that less than 10 percent of crude oil refining is attributable to nonintegrated refiners.[79] Professor Tom Moore reports that approximately 90 percent of the crude oil transported by pipeline flows through pipelines controlled by major marketing and refining companies.[80] At the same time however, there are independent refiners, independent marketers, as well as independent crude producers.

The degree of vertical integration refers to the extent to which a firm performs different successive stages in the production of a particular product. To the extent to which vertical integration is present, internal organization is substituted for market processes.

The treatment of vertical integration has presented substantial problems at both theoretical and policy-making levels. Economic theory has traditionally assumed zero costs of operating competitive markets and, as a result, vertical integration represents an anomaly.[81] At the policy level, primary concern has been with the possibility that integration represents a strategy designed to achieve anticompetitive results. The lack of a theoretical explanation of why firms integrate, except in cases where technological interdependencies exist between stages, results in a tendency to regard integration as having dubious social benefits.

It is now clear, however, that significant information and transaction costs are incurred in the use of markets to carry out transactions. Thanks to pioneering efforts by Prof. Williamson,[82] several sources of cost savings due to integration have been identified. Integration, resulting in cost savings, is likely in those instances where the use of independent firms involves frequent and difficult bargaining due to uncertainties and complex technologies. Integration would increase the degree of internal coordination and lead to increased efficiency. Integration is also likely where the costs of obtaining information are high and/or the costs of enforcing contracts are high. Thus, the prospects of cost savings via vertical integration are much more widespread than was originally believed to be the case.

The competitive implications of vertical integration are generally discussed in terms of the relationship between integration and entry barriers. It is sometimes argued that, if a firm is required to enter an industry at successive stages in order to gain access to raw materials, the amount of capital required for entry, relative to single stage entry, increases and entry barriers become higher and potential competition is reduced. The higher entry barriers only occur if capital market imperfections constrain the potential entrant's ability to raise the necessary funds. When potential entrants are large established firms in other industries, there seems to be little a priori reason to believe that they would have less access to funds, internal or external, than would established firms in the industry. Thus, the argument that vertical integration, via increased capital requirements for entry, has an anticompetitive effect is seriously weakened.

The type of cost advantages gained from vertical integration are not readily identifiable in the case of petroleum. The product is relatively simple and standardized and requires little highly specialized investment. In addition, the information costs of using markets to obtain crude oil, rather than to rely upon backward integration, do not appear excessive. Some incentive for large oil firms to integrate into oil exploration is created by the extremely high risks in such activities. The risk to a very large firm would be substantially less relative to the small firm conducting only limited exploration activities. On the other hand, provisions in the tax laws make it more profitable for individuals with marginal tax rates in excess of the maximum corporate rate to explore for oil. In this situation, oil exploration should be dominated by wealthy individuals in the high tax brackets and large oil companies.

Historically, major oil firms have been both buyers and sellers of crude oil. In order to meet the crude oil needs of their own refining operations, most companies have had to purchase oil in the market. Since 1969 domestic crude oil and NGL production has declined causing self sufficiency rating to decline in most cases. The composite index declined from 69.6 percent in 1969 to 67.4 percent in 1972.[u]

In the world oil market, Prof. Adelman points out that, among the eight largest crude producers, during the 1950-65 period a crude surplus was generally the rule.[83] It is interesting to note that while total crude production doubled in this period, the volume of crude not retained within the producer's integrated operations increased by less. In any case, there is a wide difference among the firms during this period in the extent of their surplus position. The fact that some firms are willing to purchase their crude oil inputs suggests that there is no inherent technical or cost necessity for vertical integration. The decision to produce or buy appears to depend on the particulars of the operation.

Professors Kahn and deChazeau argue that the highly integrated nature of petroleum production represents, in part, a desire of firms to escape the uncertainties and instabilities that result from a dependence on imperfect intermediate markets in petroleum.[84] For example, prior to prorationing, the discovery of a new oil field, such as the East Texas field, brought a vertical flood of new oil to the market causing drastic reductions in price. Given such a possibility, refiners increased the extent of their control of pipelines and

---

[u]See Appendix F by Lichtblau.

marketing in an attempt to block the new oil from forward markets. As a result, nonintegrated crude producers were forced to integrate into refining and marketing in order to have outlets for their oil. Kahn and deChazeau sum up the historical record of vertical integration by stating:

> All it really says is that companies have sought managerial control over their raw material supplies and product distribution because they wanted the greater assurances that financial control brings. Attempts to supply narrower and more precise interpretations invariably lack conviction.[85]

Thus, in the early period of the industry, vertical integration served as a means of introducing stability into the raw materials markets.

Vertical integration, per se, does not create monopoly power. However, in situations where integration encompasses a market where such power exists, vertical integration can serve a means by which this power can be used to affect events at other markets within the integrated operation. It has been alleged that because institutional features have made crude oil production relatively more profitable than other stages, integrated firms have been able to establish a vertical price squeeze on nonintegrated rivals.

Professor Alfred Kahn, a long time student of the economics of petroleum, argues that nonintegrated refiners have been subject to the type of price squeeze indicated above.[86] An independent refiner buys crude oil inputs in a market where price has been held above marginal costs because of government sponsored controls over supply and sells his output, gasoline in a relatively competitive market. The price at which crude oil can be obtained in the market is a crucial matter to the independent refiner but the price to the integrated firm represents only a transfer price. The high crude prices create a squeeze on the independents' margin but have no real effect on the operation of the integrated firm. The only effect in the integrated firm is that more of its profits will occur at the crude level than at other stages. In this situation, independent firms are likely to integrate backwards into crude production in order to (1) ensure adequate crude supplies and (2) eliminate their susceptibility to any price squeeze.

SUMMARY

The structural characteristics of a market, especially the level of seller concentration and entry conditions, provide a partial guide to evaluating the state of competition in a market. Defining the relevant market for such an analysis is not an easy task. Substantial variations in the degree of interfuel substitutability exist among the various end uses of energy. In the case of electric utilities, fossil fuels and nuclear energy are substitutable and thus indicate that an energy market exists. In many other end uses direct substitution is significantly less but indirect substitution, in the form of increased use of electricity, is possible. In terms of the geographic dimensions of the markets, uranium clearly trades in a national market; the market for crude oil tends to be national, if not international in scope. Natural gas also trades in a national market. In the case of coal, the results are not without some ambiguity, due to variations in sulfur content; what was once a regional market now appears to be a national market.

Levels of seller concentration, based on production figures, are not high in the case of the fossil fuels. The most significant findings is that a steady upward trend in concentration levels has been occurring since 1955. Concentration levels in uranium tend to be higher, but significant amounts of excess capacity exist and it is not certain if present concentration levels will persist. In addition, there is reason to believe that the upward trend will continue. The proper interpretation of observed concentration levels in the case of crude oil is complicated by the existence of large numbers of joint venture arrangements which create a direct interdependency among sellers.

Interfuel mergers have caught the public's attention in recent years, and have raised a concern that oil company entry represents an attempt to monopolize the industry. The analysis presented here indicates that, while concentration levels have been rising, oil company entry is not the major reason for the upward trend.

It cannot be concluded from the results of the structural analysis that effective competition exists in the production of fossil fuels and nuclear energy in the U.S., nor can it be concluded that competition is ineffective. Such conclusions would be premature and must await an analysis of market conduct and performance. It should also be emphasized that the structural analysis of energy markets is weakened by a necessity to rely upon production data rather than data which reveals the ownership pattern of reserves of the various fuels.

In conclusion, the structural analysis indicates: (1) concentration levels tend to be low but a strong upward trend is present and (2) oil companies have entered, to a significant degree, the coal and uranium areas.

NOTES

1. Interdependency occurs in a market where a few sellers exist. In this setting, sellers recognize that the actions, such as price changes, of any single seller will have an impact on rivals and thus likely cause a reaction by rivals. The opposite of interdependency, independence, occurs in markets populated by many, relatively small sellers.

   Interdependency has been described as " . . . significant for markets in which there are sellers or buyers (or both) who are sufficiently important to warrant the assumption that the effects of their individual actions or their competitors will not be disregarded." William Fellner, *Competition Among the Few* (New York: Augustus M. Kelley, 1965), p. 7.

2. The variety of ways for measuring concentration level is discussed in: U.S. Senate, Subcommittee in Antitrust and Monopoly, "Testimony of Willard F. Mueller," *Economic Concentration, Part I* (Washington: Government Printing Office, 1964), pp. 111-21. Alternative concentration indices such as the Herfiendahl index and Gini Coefficient are discussed in Douglas Needham, *Economic Analysis and Industrial Structure* (New York: Holt, Rinehart and Winston, Inc., 1969), pp. 83-96.

3. Several qualifications need to be mentioned. One, as stated, the relationship between concentration and allocative efficiency represents a central tendency subject to random deviation. Thus, the relationship should be observed in a group of industries with varying degrees of seller concentration but it doesn't necessarily hold for an individual industry. Two, economic theory is not able to specify the quantitative equivalent to "high" and "low" concentration levels. Three, the hypothesis is not able to state a priori whether the relationship is continuous or discrete.

   Discussion of the relationship between concentration levels and profitability can be found in: Joe S. Bain, *Industrial Organization* (New York: John S. Wiley and Sons, 1968) and F.M. Scherer, *Industrial Market Structure and Economic Performance* (Chicago: Rand McNally and Co., 1970).

4. A detailed, critical evaluation of past studies can be found in Norman R. Collins and Lee E. Preston, *Concentration and Price Cost Margins in Manufacturing Industries* (Berkeley: University of California Press, 1968) and Leonard Weiss, "Quantitative Studies of Industrial Organization," M.S. Intriligator, ed., *Frontiers of Quantitative Economics* (Amsterdam: North Holland Publishing Co., 1972).

   Doubt as to the proper interpretation of empirical results and their ability to support a structural approach to antitrust are expressed in: Yale Brozen, "The Antitrust Task Force Deconcentration Recommendation," *Journal of Law and Economics*, October 1970, Yale Brozen, "Bain's Concentration and Rates of Return Revisited" *Journal of Law and Economics*, October 1971, Stanley I. Ornstein, "Concentration and Profits," *Journal of Business*, October 1972, Harold Demsetz, *The Market Concentration Doctrine* (Washington: American Enterprise Institute, 1973).

5. Wm. J. Baumol, "Informed Judgement, Rigorous Theory, and Public Policy," *Southern Economic Journal*, January 1962, pp. 142-43. Confirmation of Baumol's position has been offered by Prof. H. Michael Mann, Director, Bureau of Economics, Federal Trade Commission. See: H. Michael Mann, "A Structuralist Direction for Antitrust: The View of a Policy Advisor," Speech before the Midwest Conference of the Continuing Legal Education Program of the Minnesota State Bar Association and University of Minnesota, Minneapolis, Minnesota, April 28, 1973.

6. P. Balestra, *The Demand for Natural Gas in the United States: A Dynamic*

Approach for the Residential and Commercial Market (Amsterdam: North-Holland Publishing Co., 1967).

7. As a result, attempts at establishing market boundaries regardless of methodology, always contain a subjective element in the analysis. For a discussion of the limitations associated with using estimates of cross elasticity coefficients to establish product market boundaries see: Federal Trade Commission, *Interfuel Substitutability in the Electric Utility Sector of the U.S. Economy* (Washington, D.C.: Government Printing Office, 1972), pp. 24-30 and Klaus Stegemann, "Is the Coefficient of Cross Elasticity An Appropriate Measure of Closeness of Substitutes?" (Queen's University: Discussion Paper No. 36), unpublished, undated. Empirical results are presented in: David Schwartzman, "The Cross Elasticity of Demand and Industry Boundaries: Coal, Oil, Gas, and Uranium," *The Antitrust Bulletin*, Fall 1973.

8. Report prepared for the Interdepartmental Energy Study by the Energy Study Group under the direction of Ali Bulent Cambel, *Energy R&D and National Progress* (Washington: Government Printing Office, 1964), p. xxv.

9. FTC, *op. cit.*

10. *Ibid.*, p. 2.

11. Thomas D. Duchesneau, *Interfuel Substitutability in the Electric Utility Sector of the U.S. Economy*, Economic Report to FTC, February 1972, p. 80.

12. Kenneth G. Elzinga and Thomas F. Hogarty, "The Problem of Geographic Market Delineation in Antimerger Suits," *The Antitrust Bulletin*, Spring 1973, pp. 73-5.

13. *Oil and Gas Journal*, September 30, 1968, p. 36.

14. Leslie Cookenboo, *Crude Oil Pipe Lines and Competition in the Oil Industry* (Cambridge: Harvard University Press, 1955), p. 74 and de Chazeau and Kahn, *op. cit.*, pp. 17-20.

15. The FTC survey results are reported in: Permanent Subcommittee on Investigations of the Committee on Government Operation, "Preliminary Federal Trade Commission Staff Report on Investigation of the Petroleum Industry," *Investigation of the Petroleum Industry* (Washington: Government Printing Office, July 12, 1973), pp. 5-7.

16. M.A. Adelman, *The World Oil Market* (Baltimore: The Johns Hopkins Press, 1972).

17. Adelman, *op. cit.*, p. 82.

18. For a discussion of the changing structure of international crude oil production and the role of OPEC see: Subcommittee on Antitrust and Monopoly, "Statement of Prof. Edith Penrose," *Government Intervention in the Oil Industry, Part I* (Washington: Government Printing Office, March-April 1969), pp. 156-181, 429-433.

19. *Oil and Gas Journal*, September 22, 1969.

20. *Business Week*, "All Set to Build Some Oil Refineries," May 12, 1973, pp. 48-9.

21. Michael Bergman, "The Corporate Joint Venture Under the Anti-trust Laws," *New York University Law Review*, 1962, 37, 712.

22. The following discussion draws upon work by Prof. Mead. See: Walter J. Mead, "The Competitive Significance of Joint Ventures," *The Antitrust Bulletin*, Fall 1967, pp. 819-48.

23. Stanley E. Boyle, "An Estimate of the Number and Size Distribution of Domestic Joint Subsidiaries," *Antitrust Law and Economics*, Spring 1968, pp. 81-92. Another attempt to document the use of joint arrangements is reported in: Martini and Berman, "Expansion Via Joint Subsidiaries," *Mergers and Acquisitions* (American Management Association, 1957).

24. *Ibid.*

25. *Ibid.*, p. 92.

26. U.S. Senate, Subcommittee on Antitrust and Monopoly, "Testimony of John W. Wilson," Washington, June 27, 1973.

27. *Ibid.*

28. Walter J. Mead, "The Competitive Significance of Joint Ventures," *The Antitrust Bulletin*, Fall 1967, pp. 819-48.

29. Of 16 former partners, only one bid more frequently against former partners than against non partners. *Ibid.*, p. 841.

30. In addition to a restraint in bidding competition, a possible adverse impact on pricing may result. For instance, results indicate that the high bid (Y) is positively related to the number of bidders (X);

$$\log\ Y\ =\ 0.48529\ +\ 1.28707 \log\ X$$
$$(0.05416)$$

The existence of joint bidding reduces the number of bidders and, as a result, lowers the high bid, *ibid.*, p. 844. Further evidence is presented in: Public Land Review Commission, *Study of the Outer Continental Shelf Lands of the United States* (Washington: No. PB188714, November 1969), pp. 513-21.

31. Walter Mead, *Competition in the Energy Industry* (Washington: Energy Policy Project, staff paper 1974), pp. 37-9.

32. Deposition of: Otto Miller, January 4, 1974, in the Superior Court of the State of California in and for the County of Sacramento, No. 241, 392, p. 48.

33. *Ibid.*, pp. 40-1.

34. *Ibid.*, p. 44.

35. *Ibid.*, p. 42.

36. For the 32 lease sales during the 1954-1973 period, 46 percent of all bids were joint bids and joint bidders won 44 percent of the tracts. Susan M. Wilcox, "Entry and Joint Venture Bidding in the Offshore Petroleum Industry," (Santa Barbara: University of California, unpublished Ph.D. dissertation, 1974), p. 77.

37. *Ibid.*

38. Clark A. Hawkins, "Structure of the Natural Gas Producing Industry" in

Keith Brown, ed., *Regulation of the Natural Gas Producing Industry* (Baltimore: The Johns Hopkins University Press, 1971), p. 138.

39. Federal Power Commission, *Sales by Producers of Natural Gas to Interstate Pipeline Companies* (Washington: U.S. Government Printing Office, 1971).

40. Dr. Wilson has also computed concentration in terms of uncommitted gas reserves as of December 31, 1971 and June 30, 1972. The results, generally the same at each point in time; indicate a four-firm and eight-firm concentration ratio of 51 percent and 73.9 percent, respectively. The proper interpretation of the figures is not clear. The existence of uncommitted gas reserves during a time of extreme gas shortages strongly suggests that the uncommitted gas is uneconomic, i.e. shut-in gas or reserves too small for commercial production. If so, the gas would not be a price determining factor in the market. Subcommittee on Antitrust and Monopoly, *op. cit.*, pp. 17-19.

41. Cookenboo, *op. cit.*, pp. 48-51.

42. Clark Hawkins, pp. 150-54.

43. Leslie Cookenboo, Jr., "Competition in the Field Market for Natural Gas," *The Rice Institute Pamphlet* (Monograph in Economics, Vol. 44, January 1958).

44. Edward J. Neuner, *The Natural Gas Industry* (University of Oklahoma Press, 1960), p. 245.

45. Paul W. MacAvoy, *Price Formation in Natural Gas Fields: A Study of Competition, Monopsony, and Regulation* (New Haven: Yale University Press, 1962).

46. *Ibid.*, p. 265.

47. Reed Moyer, *Competition in the Midwestern Coal Market* (Cambridge: Harvard University Press, 1964), pp. 67-8 and Hubert E. Risser, *The Economics of the Coal Industry* (Lawrence, Kansas: University of Kansas, 1958).

48. Keystone Coal Buyers Manual, *U.S. Coal Production by Company, 1972, op. cit.*, p. 5.

49. James Meehan and Thomas Duchesneau, "The Critical Level of Concentration: An Empirical Analysis," *The Journal of Industrial Economics*, November 1973.

50. Reed Moyer, *Competition in the Midwestern Coal Industry* (Cambridge: Harvard University Press, 1964).

51. *Ibid.*, p. 302.

52. For a list of the mergers see: *Ibid.*, pp. 76-7 and Moyer Appendix.

53. *Ibid.*

54. For an analysis of the negative impact of ICC regulation policy on the rate of innovation of unit coal trains see: Paul W. MacAvoy and James Sloss, *Regulation of Transport Innovation* (New York: Random House, 1967).

55. This is consistent with the evidence of wider trading areas for coal in recent years reported by Prof. Hogarty.

56. Reed Moyer, "Requirements Contracts and Energy Transportation Development in Coal," *The Southern Economic Journal*, April 1965.

57. *Southwest Energy Study*, Report for the Secretary of the Interior, November 1972, pp. 3-4.

58. For a complete discussion of the latter question see: Public Land Law Review Commission, *Study of Outer Continental Shelf Lands of the U.S.*, November 1969.

59. Subcommittee on Special Small Business Problems, *Concentration by Competing Raw Fuel Industries in the Energy Market and Its Impact on Small Business* (Washington, July 12-22, 1971), p. 98.

60. *Ibid.*, p. 101.

61. *Ibid.*, p. 129.

62. *Ibid.*

63. Subcommittee on Special Small Business Problems, *Concentration. . . .*, p. 104.

64. Arthur D. Little Co., *Competition in the Nuclear Power Industry: Report to the USAEC and the U.S. Department of Justice* (Washington: Government Printing Office, 1968), p. 3.

65. Some of the excess capacity may be absorbed by sales to foreign governments. For instance, a German group has been discussing the possibility of purchasing uranium enrichment services from Russia and the U.S. It was reported that the delegation was informed by U.S. authorities that the Russian enrichment fee would be more attractive than the U.S. AEC's. McGraw-Hill Publication, *Nucleonics Week*, Press Release, March 29, 1973.

66. Remarks by Ernest B. Tremmel, at the 1971 Annual Conference of The Atomic Industrial Forum, Miami Beach, Florida, October 20, 1971.

67. See: Joint Committee on Atomic Energy, "Statement of Thomas E. Kauper," Washington: August 1, 1973.

68. France has recently announced plans to construct a massive enrichment plant costing at least $2.5 billion. It will employ the gaseous diffusion technology. The French plant will have to compete with the Tripartite Centrifuge project.

   This addition of a large increase in enrichment capacity creates a possibility that, if the AEC continues with its expansion plans for new U.S. enrichment capacity, there may be an oversupply of enriched uranium by the 1980s. *Wall Street Journal*, Jan. 7, 1974.

69. Possible biases in this procedure are discussed in: Ralph Turvey and A.R. Nobay, "On Measuring Energy Consumption," *Economic Journal*, December 1965, pp. 787-93.

70. For a discussion of the technique see: George J. Stigler, "The Economies of Scale," *Journal of Law and Economics*, October 1958, pp. 54-71.

71. W.G. Shepherd, "What Does the Survivor Technique Show About Economies of Scale?", *Southern Economic Journal*, July 1967, pp. 113-22.

72. The analysis of bidding patterns for OCS leases as an indicator of entry conditions draws upon a paper by Prof. Walter Mead. See: Walter J.

Mead, "Competition in the Energy Industry," (Ford Foundation: Energy Policy Project, staff paper, May 3, 1974).

73. Stigler, *op. cit.*, pp. 65-72.

74. C. Pratten and R.M. Dean, *The Economics of Large-Scale Production in British Industry* (Cambridge University Press, 1965).

75. John Klemenic, "Examples of Overall Economics in a Future Cycle of Uranium Concentrate Production for Assumed Open Pit and Underground Mining Operations," (Grand Junction, Colorado: U.S. Atomic Energy Commission, 1972).

76. Statistical Data of Uranium Industry 1971.

77. Klemenic, p. 1.

78. Klemenic, *op. cit.*, p. 2, 7.

79. U.S. House of Representatives, *Hearings Before the Subcommittee on Communications and Power* (Washington: April 18, 1967).

80. Thomas G. Moore, "The Petroleum Industry," in Walter Adams, ed., *The Structure of American Industry* (New York: The MacMillan Co., 4th ed., 1971), p. 125.

81. Oliver E. Williamson, "The Vertical Integration of Production: Market Failure Considerations," *The American Economic Review*, May 1971, pp. 112-23.

82. *Ibid.*

83. Adelman, *op. cit.*, pp. 93-5.

84. Melvin G. deChazeau and Alfred E. Kahn, *Integration and Competition in the Petroleum Industry* (New Haven: Yale University Press, 1959), pp. 75-118.

85. *Ibid.*, p. 104.

86. U.S. Senate, Subcommittee on Antitrust and Monopoly, "Testimony of Alfred E. Kahn," *Economic Concentration Part 2*, March 18, 1965, pp. 591-622.

Chapter Three

Conduct

An analysis of competition cannot rest solely on a study of market structure. While structure may be a partial determinant of market conduct and performance, wide variations in conduct are consistent with a given structural setting. As a result, implications concerning the viability of competition derived from an analysis of market structure can only be confirmed by an analysis of market conduct and performance.

CRUDE OIL

In competitive markets, prices tend to reach the level at which quantity demanded equates quantity supplied; at the competitive equilibrium price, the opportunity cost of the last unit supplied just equals the value of the last unit demanded by consumers. Such an equilibrium represents an optimum allocation of resources. Competitive markets are a mechanism which, under definable conditions, continuously seek the optimum allocation of resources.

In the domestic crude oil industry, price policies are determined largely by government policies rather than by demand and supply forces. Government intervention has been systematic and comprehensive ranging from policies designed to stimulate production to policies which attempt to control output in order to support crude oil prices.[1] As a result, resource allocation decisions are mainly the result of government decision-making and the role of competition, in the sense of freely fluctuating prices equating quantity demanded to quantity supplied, is substantially reduced.

The domestic crude oil industry has historically suffered from a problem of excess production capacity. The existence of chronic excess capacity indicates that competition is not allowed to exercise its allocative function. The price mechanism has, in effect, been prevented from equating quantity demanded to quantity supplied.

The lack of competition in crude oil markets tends to be confirmed when fluctuations in crude oil prices are compared with price movements of various petroleum products. Professor Kahn notes that product prices tend to display the numerous fluctuations expected in markets where competitive forces are dominant, while crude oil prices display significant amounts of rigidity. Such rigidity suggests that market forces play only a limited role in crude markets.[2]

The following section deals with the various government policies designed to influence crude oil markets and notes their impact on industry price policies.

Percentage Depletion Allowances

Percentage depletion allowances, established in 1926, provide a tax subsidy designed to attract additional resources to oil exploration. Under the provisions of the law, a crude oil producer can subtract from its gross income, after payment of any royalties, an amount equal to 22 percent of its total income from crude production. The deduction is limited to 50 percent of the firm's net income. Where the full value of the allowance is taken, a firm's federal income tax rate is reduced by half. In addition, the law allows oil companies to expense certain exploration and drilling costs rather than depreciating them over a longer period. This also leads to a reduction of tax liability.

The effect of the depletion allowance is, at the time of initial implementation, to make investment in crude oil exploration and production relatively more attractive in comparison to investment in other industries without such provisions. This should lead to an increase in crude oil production and, assuming constant demand, lead to lower prices for crude oil. The depletion allowance can also make investment in crude production more profitable than investment in refining and marketing.

Prorationing

The potential price-depressing effects of depletion allowance were supplemented in 1930 by the discovery of the East Texas oil field. This discovery resulted in a large increase in crude oil supplies at a time when demand was declining and, as expected, oil prices fell dramatically and created a large amount of instability in crude oil markets. This led to a plea for government intervention to reduce instability and resulted in the implementation of prorationing.[3] Prorationing involves two aspects: (1) production controls designed to ensure the maximum recovery of oil (MER) and (2) controls which aim to restrict production levels to market demand. The latter is called market demand prorationing.

The combination of a price system with gain-motivated private conduct, while representing a powerful vehicle for achieving economic efficiency in resource use, does not always lead to optimum performance. In the case of crude oil production, unregulated competition is not capable of bringing about the optimal rate of production. The inability of competitive markets to create the proper incentives to attain optimum conservation levels arises in the case of crude oil from two factors: (1) the existence of two or more owners of operating interests on the surface overlying a single oil reservoir and (2) the fluid, hence migratory nature of reservoir contents. As a result, an economic incentive exists for each producer to produce more oil than is optimal from either the viewpoint of the operators as a group or society. The result of unregulated competition would be overinvestment and, ultimately, less than optimal recovery of oil.[a] Consequently, production controls are substituted for the forces of competition in order to assure an efficient recovery of oil resources. The level of production from each reservoir is limited to the maximum rate of oil production consistent with no significant loss of ultimate recovery. This is commonly referred to as a reservoir's MER and is based on the physical characteristics of the reservoir.

In addition to MER type of production controls, prorationing includes market demand prorationing which attempts to match oil production to estimated demand. Market demand prorationing is not required because of any technological factor, but represents a means of influencing crude oil production and prices. As Professor Mead states:

> While the need for MER-type prorationing is granted, the only justification for market demand prorationing is from the resource owner's view, who presumably would prefer high prices for his crude rather than low prices.[4]

Demand prorationing has been adopted in most major producing states and is administered by individual state regulatory bodies. In the process of matching supply to demand, explicit discussions of implications for crude oil prices do not occur but such implications do exist and cannot be denied. Professor Stephen L. McDonald states:

> The implicit assumption in the hearing is either the price will remain unchanged or that demand is so inelastic with respect to price in the short run that prospective price has negligible bearing on prospective purchases.[5]

Hence, it is not surprising to observe the rigidity of crude oil prices. Total allowable production in a state is ultimately translated into allowable production rates for individual wells. Historically, allowables have been substantially below the maximum efficient rate (MER) of production, or in other words, government imposed restrictions on output have been used to support crude oil prices at a level in excess of the market clearing level. Since 1972 however, in light of rising demand, allowables have been set at 100 percent of MER.

---

[a]High initial rate of production from new reservoirs would lead to a loss of otherwise recoverable oil.

It should be emphasized that MER prorationing is justified on the grounds of engineering efficiency in order to conserve oil, but that demand prorationing cannot be justified on the basis of conservation goals. Demand prorationing adds additional restrictions on supply and is a substitute for market forces. As a result, the price mechanism is prevented from clearing markets. Price, denied its allocation function, becomes simply a device for determining the incomes of buyers and sellers of crude oil.[b]

The pattern of production restrictions in the major demand-prorationing states during the 1948-67 period is indicated in Table 3-1. Declining demand factors indicate increases in the level of restriction on crude oil production. As shown in Table 3-1, market demand factors tended to decline up to the 1960-63 period. Professor McDonald notes that during periods of general decline in business activity, as was the case in 1948-49, 1953-54, and 1957-58, significantly lower demand factors were the case. In this way, state regulatory authorities were able to support crude prices against the downward pressures associated with periods of recession.[6]

States generally grant various exemptions from production controls. Such exemptions are clearly important when one considers that in 1963 some 43 percent and 58 percent of crude oil production in Texas and Oklahoma, respectively, came from exempt wells. Various types of exemptions exist, but the best known is perhaps the exemption of low capacity wells and fields, commonly referred to as stripper wells.[c] As a result, the data presented in Table 3-1 does not adequately indicate the extent to which productive capacity has been utilized. The ratio of crude oil output to productive capacity for five states is indicated in Table 3-2. The rates of capacity utilization for a given year are higher than the market demand factors reported in Table 3-1. This mainly reflects the extent of exemptions from demand prorationing provisions. The greatest production restraints occurred in the case of Texas followed by Louisiana. The figures indicate that output levels have been restricted in order to support higher crude oil prices.

The existence of substantial amounts of overinvestment and excess capacity in crude oil production is strong confirmation that demand prorationing is an economic tool used to support crude oil prices at a level in excess of the market clearing level. Some level of excess capacity may be socially desirable, but it would only be a coincidence if the amount of overinvestment existing at any given moment due to demand prorationing were consistent with society's goals.[7] In addition, it could be argued that alternatives may offer less costly means of assuring the availability of excess capacity.

---

[b]Crude oil prices in nonmarket demand states are imperfectly supported by production restrictions in market demand states.

[c]In terms of efficiency, the exemption of stripper wells promotes the production of oil from the least efficient wells and a reduction of output from more efficient wells.

Table 3-1. Annual Average of Monthly Market Demand Factors: Texas, Louisiana, New Mexico, and Oklahoma, 1948-1967 (percent)

Year	Texas	Louisiana	New Mexico[a]	Oklahoma
1948	100%	b	63%	c
1949	65	b	61	c
1950	63	b	69	c
1951	76	b	74	c
1952	71	b	68	c
1953	65	90%	63	c
1954	53	61	57	c
1955	53	48	57	60%
1956	52	42	56	53
1957	47	43	56	52
1958	33	33	49	45
1959	34	34	50	41
1960	28	34	49	35
1961	28	32	49	31
1962	27	32	50	35
1963	28	32	54	31
1964	28	32	54	28
1965	29	33	56	27
1966	34	35	65	38
1967	41	38[d]	74	50

[a]Southeast area only. Normal unit allowables converted to market demand factor based on 70 barrels per day as maximum normal unit allowable.

[b]No fixed allowable schedule.

[c]Comparable data not available.

[d]Not exactly comparable with preceding figures because of the introduction in this year of the new "intermediate zone" allowable schedule, which effectively increased the base on which the market demand factor applied.

Sources: Respective state conservation commissions: Stephen L. McDonald, *Petroleum Conservation in the United States: An Economic Analysis* (Baltimore: The Johns Hopkins Press, 1971), p. 164.

The price supporting effects of demand prorationing are indicated in Table 3-3. The evidence based on events in Texas, indicates that stable prices are associated with stable demand factors and, up to 1966, higher crude prices were supported by lower demand factors.[d] Demand factors fell dramatically in

[d]The effectiveness of state prorationing policies to control production was enhanced by the passage of the Connally Hot Oil Act in 1935. The act prohibits the interstate shipments of oil produced in violation of state prorationing laws.

Table 3-2.   Ratio of Crude Oil Output to Productive Capacity:
Texas, Louisiana, New Mexico, Oklahoma, and Kansas, 1954-1967
(percent)

Year	Texas	Louisiana	New Mexico	Oklahoma	Kansas
1954	71%	85%	91%	82%	92%
1955	72	87	99	83	88
1956	72	80	93	84	88
1957	70	69	93	83	88
1958	60	59	89	79	88
1959	63	63	92	79	90
1960	60	67	90	77	87
1961	60	65	88	79	88
1962	59	68	83	82	90
1963	62	69	84	84	89
1964	63	68	89	86	89
1965	64	66	92	87	91
1966	67	69	96	98	95
1967	70	69	98	101	97

Note: Capacity estimates as of 1 January of each year. Output is for the full year.
McDonald, p. 165.

Sources: Capacity—Productive Capacity Committee, Independent Petroleum Association of
America; Output—U.S. Bureau of Mines.

recession periods 1948-49, 1953-54, and 1957-58 while prices declined by small
amounts in the 1948-49 and 1957-58 recessions but actually increased during
the 1953-54 business decline. Such price behavior would not occur in a
competitive setting.

It should be emphasized that the evidence presented in Tables 3-1
through 3-3 is designed to indicate the historical relationship between demand
prorationing and capacity and prices. Since early 1972 individual states have
raised allowable production to the point where allowables are at 100 percent of
MER in most reservoirs. In this setting, state imposed restrictions no longer
influence the flow of crude oil, with the exception of MER restrictions, and the
market now warrants the previously state supported price level or a higher level.
However, current world prices, to which the U.S. price adjusts, reflect the
impact of the OPEC cartel rather than oil's true scarcity value. In light of the
changed situation, it is perhaps tempting to conclude that a discussion of
demand prorationing policies is of little value. Such a conclusion would be
incorrect for three major reasons. One, past policies illustrate the extent to
which industry price policies have been due to government regulation rather
than to market forces. Historically, competition has not played a major role in

Table 3-3.   Crude Oil Prices and Market Demand Factors in Texas: Annual Averages, 1948-1966

Year	Crude Oil Price (dollars per bbl)	Market Demand Factor (percent)
1948	2.61	100%
1949	2.59	65
1950	2.59	63
1951	2.58	76
1952	2.58	71
1953	2.73	65
1954	2.84	53
1955	2.84	53
1956	2.83	52
1957	3.11	47
1958	3.06	33
1959	2.98	34
1960	2.96	28
1961	2.97	28
1962	2.99	27
1963	2.97	28
1964	2.96	28
1965	2.96	29
1966	2.97	34

Sources: American Petroleum Institute (API), *Petroleum Facts and Figures*, biennial, 1959-67; and Texas Railroad Commission. McDonald, p. 189.

domestic crude oil markets. Two, the industry's current cost structure and market structure is, to an important degree, the result of policies such as prorationing. Three, is the federal-enabling legislation. The Connolly "Hot Oil" Act is still operative and at any future time when the oil industry is not satisfied with the level of crude oil prices, market demand factors may be lowered below their present 100 percent level.

Oil Import Quotas

The oil import quota program, instituted in 1957, represents another government program which has reduced competition in domestic crude oil markets.[8] Until recently, foreign oil delivered to U.S. ports was available at a price substantially below the price of domestic oil. Such a price differential could not persist if free trade were the case. In order to support the U.S. price, restrictions, initially of a voluntary nature but made mandatory in 1959, were placed upon the volume of imported crude oil. Professor Adelman notes that, while a mandatory program did not become operational until 1959, a review of U.S. oil import policy prior to 1959 indicates that imports have never been free since World War II.[9] Adelman indicates that since this period no company was free to consult its own interests and its own profit in determining the level of oil imports.[10]

Import restrictions had the effect of maintaining the price gap. Professor Dam describes the Mandatory Program as "the kind where the restricting government attempts to retain the scarcity values inherent in the quotas for its own nationals."[11] In simple terms, import quotas isolated, to a large degree, the domestic industry from the influence of foreign competition.

The import quota program was rationalized on national security grounds in the sense that such restrictions created an incentive to maintain a "strong" domestic oil industry and prevented the U.S. from becoming dependent on foreign sources of crude oil. The validity of the program and possible alternatives have been hotly debated in the literature[12] and in 1973 a fee system was substituted for the quota program. One of the effects of the program has been to raise entry barriers to the construction of new refinery capacity in the U.S. In recent years, little significant increase in refining capacity has occurred in the U.S.; construction has shifted to areas where foreign crude oil inputs could be obtained.

There is a general consensus that domestic oil production will continue to lag behind demand in the U.S. The gap between demand and domestic supply will have to be met by increasing imports. Crude oil imports in 1973 were approximately 1,172,409,000 barrels. The origin of imported oil in 1973 is indicated in Table 3-4. Canada provided some 31 percent of total imports and the Middle East accounted for 23 percent of the total. The share of imports coming from the Middle East increased until October 1973 when the five month embargo was instituted; the embargo was lifted in March of 1974 and imports are expected to increase.

Table 3-4. Origin of Crude Oil Imports: 1973[1]

Country	Imports (millions of barrels)	Percent of Total
Canada	364.3	31.1%
Middle East	280.7	23.9
Nigeria	168.3	14.4
Venezuela	116.1	9.9
Indonesia	73.4	6.3
Other	169.6	14.5
Total	1172.4	100.0

[1] Preliminary

Source: *Oil and Gas Journal*, January 28, 1974, p. 118.

The relative importance of imported oil in the domestic market in the future depends on the interaction of many complex forces including relative price movement, supply and demand elasticities, and the prevailing political climate. In addition, oil imports are sensitive to changes in the mix of total energy use in the U.S. The National Petroleum Council has prepared estimates of the relative role of oil imports to the year 1985. The estimates, based on different sets of assumptions, indicate that oil imports, defined as crude oil equivalent of the calculated BTU deficit, will increase sharply between 1970-1975 and account for between 42-51 percent of total oil supply.[13] In the period 1975-1985, estimates are sensitive to the assumptions made and the estimated importance of imports in 1985 range from 18 to 65 percent of total supply in 1985.[14] In any case, there appears to be little disagreement concerning the prediction of a rapid increase in oil imports in the period 1970-1975.

Foreign Tax Credit

Under U.S. income tax law, a credit against U.S. tax liability is allowed for foreign income taxes paid by U.S. companies. The credit for foreign taxes is designed to avoid international double taxation.[15] There is some controversy as to deciding what constitutes a tax payment, especially with regard to whether royalty payments represent a foreign tax or a normal business expense. Under current interpretations of the law, royalty payments paid to foreign governments are considered to represent tax payments.

The credit allows an offset for foreign taxes on income earned in foreign operations against potential U.S. tax on the same amount of income but not on income earned in domestic operations.[16] The foreign credit is limited to 48 percent of foreign income and can be applied on a country by country or on an overall basis.

Professor Mead notes that, in addition to the subsidy nature of the foreign tax credit, the credit provision can lead to a situation where it may be more profitable to invest in foreign countries rather than in the U.S.[17] This conclusion is supported by the Brannan Report.[18]

There has been a systematic intervention by the government into domestic crude oil markets. The intervention has ranged from the substitution of government decision-making for private decision making, as evidenced by demand prorationing, to the establishment of institutional features, such as the foreign tax credit, which causes a change in resource allocation. As a result, a rather comprehensive and sometimes conflicting system of subsidies has been created for the oil industry. The resulting pattern of investment and resource allocation is certain to be substantially different from the results which would occur under a free market setting. Economic theory would predict that demand prorationing policies, by supporting prices in excess of the market clearing level, would lead to a problem of overinvestment and excess capacity. In a similar manner, the import quota program promotes inefficiency in the use of domestic resources. Economic reasoning predicts that the net effect of such a subsidy system would be an overinvestment in oil production and a misallocation of scarce resources. In terms used by Professor Adelman, the government intervention program represents "a system of organized waste."[19] In the section of this book concerned with performance, evidence as to the efficiency of resource allocation will be presented. It suffices, at this point, to indicate that price and output policies on domestic crude oil mainly reflect a response to systematic government intervention.[e] Free market forces play a substantially reduced role.

Role of OPEC

This book, while primarily concerned with the domestic energy industries, recognizes the rising importance of the Organization of Petroleum Exporting Countries (OPEC). The OPEC group is oil rich and accounts for some 90 percent of oil reserves of oil producing countries. Among the member countries vast differences in oil reserves exist; Saudi Arabia has the greatest oil endowment and is clearly the dominant member of the OPEC group.

The OPEC group was formed in 1960 in response to the international oil companies' reduction in the posted prices of crude oil during the period 1958-1960, a period during which a world surplus of oil was the rule. Such an act reduced the earnings received by the producing countries. Rather than individual countries bargaining with companies over oil prices and production levels, OPEC was used as a means by which the various countries could bargain collectively. Such action represents the cartel nature of OPEC. Bargaining became more intense and OPEC began to make threats that, unless their demands for higher prices were met, oil production would be curtailed or halted entirely.

The direct impact of OPEC on the U.S. market is minimal, given our relatively small use of Arab oil. Imports from Arab countries reached a peak in October 1973 when they amounted to 8 percent of domestic consumption. All projections indicate that U.S. dependence on imports are likely to increase dramatically as domestic production capacity continues to fall, by increasing amounts, below U.S. demands. However, the impact is much greater when it is considered that prices of domestic production have increased to equate with world prices.

The creation of OPEC provides a vehicle by which the members can potentially make good on their talk that their oil reserves are exhaustible and should be conserved. This raises the issue of whether the OPEC cartel is an effective device for restricting output and elevating price.[20]

Economic reasoning points out that cartels, used to restrict output,

---

[e]It should be noted that the industry has generally supported the governmental policies.

contain a variety of opposing forces. On the one hand, strong economic incentives exist for members to adhere to any output restricting agreement but, at the same time, incentives exist for individual parties to any agreement to act in their own interests thus threatening the cohesion of the cartel. As an example, individual members of OPEC may have divergent ideas concerning the appropriate price level of crude oil and each country's share of total output, especially when substantial differences in production costs exist among countries. Consequently, it will be difficult to reach an agreement among cartel members as to a common price and output policy. In addition, given agreement on price and output, the closer the agreed upon price is to the monopoly level, the greater the incentive for any member to "cheat" or act independently of the group interest. The history of producer cartels indicates that the forces promoting instability often dominate but there is no assurance that this will always be the case.

Speculation has been intense as to the strengths of the OPEC cartel. In 1973 events in the world oil market have been largely due to OPEC's actions. OPEC has been successful in obtaining huge price increases since the 1971 Teheran Agreement. Perhaps even of greater significance has been the Arab states' decision, led by Saudi Arabia, to use oil as a political weapon. In response to the 1973 Arab-Israeli War, Arab countries announced an embargo on oil shipments to countries with a pro-Israeli foreign policy. This was to be accomplished by incremental cutbacks in production until a pro-Arab stance was adopted. Non-Arab states, while not involved in Arab efforts to use oil for political purposes, benefit nonetheless from the price rise resulting from production cutbacks. There is some indication that the destabilizing elements, present in any cartel, have surfaced in the action of the Arab states. The Arab total embargo of oil to the Netherlands and the U.S. did not fully prevent shipments of oil to these countries. Given the drastic price increases, there is a strong economic incentive for an individual country to sell oil to the U.S. and Holland regardless of the stated policy of a total embargo.

The question remains however as to whether OPEC will continue to be the main factor determining the state of the world oil market. In any case, OPEC has been successful in obtaining higher prices and the Arab producers have demonstrated a willingness to use oil as a political weapon. The higher oil prices, in addition to the increased volatility of Middle East oil supplies, will continue to be an important determinant of conditions in domestic energy markets. This suggests that host governments are becoming more powerful and international companies less powerful in world oil markets.

Nonmarket Transactions

Government intervention into domestic crude oil markets has acted to shield, to a substantial degree, private decision-making from the influence of free market forces. Oil companies themselves have demonstrated a desire to further reduce their reliance on markets in conducting transactions. Evidence of this rests in the degree of vertical integration, the rise of exchange agreements, and the existing of processing agreements. It should be pointed out that such features, while representing nonmarket transactions, do not necessarily indicate anticompetitive conduct.

Exchange Agreements.   The use of exchange agreements represents a reliance on barter as a substitute for direct market transactions. In geographic areas where a surplus of crude oil or refined products, usually gasoline, exists, oil companies have tended to enter into exchange agreements with other firms. As an illustration, company A with a gasoline surplus would agree to swap gasoline with firm B by agreeing to provide B's marketing operations in A's area with gasoline. In return, B would provide gasoline to A's retail outlets located near B's refinery.[f] In an FTC survey of independent refiners, most respondents were found to engage in exchange operations with both independents and majors. Among the respondents, 15 percent of their crude oil inputs and 51 percent of their gasoline output was accounted for by exchange agreements.[21]

Processing Agreements.   A processing agreement among firms occurs when a given firm contracts to refine the crude oil owned by another firm. The ownership of the crude oil doesn't change but the owner simply pays for the refining service. There is little information on the frequency of this practice in the industry but it is known that, in some cases, refineries which appear to be independents are under long-term contract to provide refinery service to major oil companies.

Vertical integration, as pointed out earlier, represents a decision by the firm to carry out transactions outside of an organized market. Vertical integration is a major feature of the oil industry. Such integration is not a determinant of competition at any given production stage, but can present competitive dangers when integration encompasses a productive stage at which market power exists. In this case, integration is the vehicle by which the market power existing at one stage can be used to transmit the power to other markets within the integrated structure. In the case of the oil industry, vertical integration encompasses crude oil production, which, relative to refining and marketing, may be more profitable because of the depletion allowance. As a result, companies tend to take their profits at the crude stage.

Crude Oil Self Sufficiency
and Prices

Imbalances in capacity between various production stages within the integrated firm's structure have been common. As an example, most of the major integrated oil firms have had to obtain part of their crude oil from outside

---

[f]Price controls may lead to barter if crude oil prices are less than their competitive value, then barter is a means of obtaining a competitive value.

sources. The opposite type of imbalance has occurred in the case of product flow between refining and marketing operations. This excess product flow has been an important source of supply to nonintegrated firms at the different production stages and according to de Chazeau and Kahn an important source of competition.

The degree of domestic crude oil self sufficiency for a group of integrated U.S. oil companies for the years 1956, 1966, and 1971 is indicated in Table 3-5. The figures indicate a clear trend towards greater self sufficiency during the period. As an example, during this period, self sufficiency for the top five firms increased from an average of 55.8 percent to 73.0 percent.

The degree of crude oil self sufficiency, in combination with vertical integration and the tax advantages resulting from percentage depletion allowance, has special importance in terms of crude oil price policies. Depletion allowance creates an incentive for resources to flow into crude oil production and, initially, tends to lower crude oil and product prices. The integrated firm is generally both a buyer and seller of crude oil; in the former role it has an

Table 3-5. Crude Oil Self Sufficiency for Fifteen U.S. Companies,[a] 1956, 1966, 1971

	Ratio of Not Crude & NGL Production to Refinery Runs		
	1971	*1966*	*1956*
Exxon	80.6%	71.3%	52.5%
Gulf	77.6	70.3	52.3
Texaco	80.3	77.6	62.0
Standard Oil of California	65.9	71.8	62.3
Mobil	60.4	49.6	50.0
Standard Oil of Indiana	58.6	55.7	46.0
Shell	65.1	64.7	63.1
Continental	85.0	90.8	116.2
Phillips	53.3	92.5	85.3
Marathon	90.0	80.8	257.1
Atlantic Richfield	62.8	57.4	48.6
Sun	72.6	51.3	46.9
Union	81.3	66.8	55.0
Cities Service	85.1	67.8	46.9
Standard Oil of Ohio[b]	8.9	20.5	25.2

[a]USA and Canadian production.

[b]With the Prudhoe Bay discovery and the exchange agreement between British Petroleum and Standard Oil of Ohio, the latter will be one of the most crude self-sufficient in the U.S.

Source: First National City Bank, *Energy Memo*, October 1967 and April 1973.

incentive for lower crude prices but in the latter role an incentive exists for higher prices. The net balance of these opposing interests depends on the degree of crude oil sufficiency and the extent to which higher crude prices are passed on to refined products.[22] De Chazeau and Kahn report that a company with a crude oil self sufficicncy in excess of 77 percent stands to gain from higher crude oil prices if none of the increase were passed on to product prices. The greater losses in refining operations would be more than offset by the lesser gains—lesser gains before tax but greater gains after tax due to depletion—in crude production operations. If some of the increase can be passed on, the critical level of self sufficiency, which, if exceeded, creates an incentive for higher crude prices, falls.

More recently, Kahn argues that changes in tax levels since his initial analysis have caused an increase in the critical level of crude oil self sufficiency to about the 80 percent level.[23] The most recent analysis is that presented by Prof. Richard Mancke in which an analysis of gains and losses to the integrated firm from higher crude prices is presented and it is concluded that the critical level is about 93 percent.[24] The 1971 self sufficiency ratios presented in Table 3-5 indicate that no listed firm has a ratio in excess of 90 percent.

The use of exchange and processing agreements and the extent of vertical integration in the domestic crude oil industry does not represent anticompetitive conduct per se, but does indicate the tendency towards mutual cooperation among competitors. Perhaps more importantly it represents a clear tendency to circumvent the use of markets in carrying out transactions. These factors, in combination with the government imposed restrictions on competition, indicate that firms have been able to some extent, to isolate themselves, from the influence of free market forces.

Elimination of the Mandatory
Oil Import Program

As pointed out in an earlier section of this book, a mandatory oil import program was instituted by executive action in 1959. Adopted on the basis of national security needs, the program provided protection to the U.S. petroleum industry so that higher cost U.S. production and refining capacity would be developed. Without such protection, market forces would have provided an economic incentive for U.S. companies to develop cheaper foreign sources of crude oil.

In 1972, crude oil production in most U.S. oil fields was set at 100 percent of MER. In order to meet the continuing short fall between demand and U.S. production, restrictions on imports were relaxed. In 1972, quotas were eliminated on heating oil between December, 1972 and April 30, 1973. Previously, quotas on imports of residual oil had been eliminated. Between 1969 and 1972, total oil imports rose by some 52 percent to 4.7 million barrels per day. During this period, foreign crude oil prices remained below domestic prices, thus making import tickets of substantial value. By 1973, foreign crude prices were higher than domestic prices.

The quota program, designed to stimulate domestic development has in effect resulted in an opposite set of incentives. Because crude oil was made artificially scarce in the U.S., causing domestic crude prices to be higher, companies faced a situation where the incentive was to explore and produce low-cost foreign crude and to sell refined products in the U.S. market. In recent years, there has been a decided movement in refinery construction away from the U.S. and to foreign lands. Thus, the quota program has contributed to the energy crisis.

The policy of restricting imports by the use of quotas was eliminated by executive proclamation on May 1, 1973. The action eliminated any quotas and suspended existing duties on crude oil and petroleum product imports. In its place, a system of license fees on imports of crude oil and petroleum products was established. Any company or individual can import crude oil or products by obtaining an import license and paying the fee level existing at that time.

The fee schedule sets a higher fee on imports of residual oil, distillates, and gasoline than on crude oil. This is designed to create an economic setting where the incentive is to build refineries in the U.S. rather than to refine products elsewhere and market the products in the U.S. The fee schedule is presented in Table 3-6. Fees are imposed as of May 1, 1973 and are increased in a series of steps up to their maximum level on November 1, 1973. In implementing the new program, oil import licenses in effect as of May 1, 1973 are to be considered as exempt from the fee schedule. In addition, certain imports will remain exempt for a period of time following the May 1 date. These exemptions will be gradually phased out by 1980. The extent of exemption from the new fee schedule is indicated in Table 3-7.

In order to stimulate the construction of new refining capacity, companies constructing new refineries or additions to existing capacity will be allowed to import crude oil, free of any fee, up to 75 percent of their additional input needs for their first five years of operation. This represents a substantial liberalization of the schedule for this particular activity.

Table 3-6. Fee Schedule (cents per barrel)

	May 1 1973	Nov. 1 1973	May 1 1974	Nov. 1 1974	May 1 1975	Nov. 1 1975
Crude	10 1/2	13	15 1/2	18	21	21
Motor gasoline	52	54 1/2	57	59 1/2	63	63
All other finished products and unfinished oils (except ethane, propane, and butanes	15	20	30	42	52	63

Source: Statement of William E. Simon, Deputy Secretary of the Treasury, Washington, D.C., April 18, 1973.

Table 3-7. Percentage of Initial Allocation Exempt from License Fees

After April 30	*Percentage*
1973	100
1974	90
1975	80
1976	65
1977	50
1978	35
1979	20
1980	0

Source: Statement of William E. Simon, Deputy Secretary of the Treasury, Washington, D.C., April 18, 1973.

In effect, what has occurred has been the elimination of the quota program, the adoption of a fee schedule, and for practical purposes, the suspension of the fee program for a 5-6 year period. Even as initially proposed, the fee limits are not prohibitively high. The elimination of the quota program certainly eliminates much of the restrictions on foreign oil entering U.S. markets and, more than anything else, recognizes that the quota program tended to tighten domestic petroleum supplies. The effects of the new program, while possibly having a long run impact on U.S. energy supplies, will have little short run impact.

U.S. Refining Capacity. During the 1960s there was little expansion in domestic refining capacity in recent years. Several factors are responsible for the situation but perhaps the most important is the past restriction of the flow of foreign oil to U.S. markets. Consequently, companies constructed new refineries in areas where low cost foreign crude oil could be obtained. The substitution of a fee system for the quota program attempted to create a setting in which companies had an incentive to expand domestic refinery capacity.

The new import policy appears to have been responsible for a rash of new refinery projects. Major refinery projects announced since May 1973 are listed in Tables 3-8 and 3-9. Capacity increases due to new construction range between 2.9 and 3.1 million barrels per day and another 1.7 million b/d capacity will result from expansion of existing facilities. The list of companies planning expansion includes majors and nonmajors, such as utilities and distributors. Much of the new capacity would not become operational until at least by 1976. However, substantial doubts have been raised by events in the Middle East and it is not clear as to how many of the projects listed in Tables 3-8 and 3-9 will be completed and operated. The Arab-Israeli conflict and the willingness, and

Table 3-8. Announced Plans to Construct New Refineries

Company	Location	Size, b/d	Type Refinery
Ashland Oil Co.	Ohio River Valley	100,000	Full range
Belcher Oil Co.	Manatee County, Fla.	200,000	Fuels
Charter, Florida Gas	Florida	150,000	Fuels/SNG
Crown Central	Baltimore	100,000	Fuels/SNG
Energy Co. of Alaska	North Pole, Alaska	15,000	Full range
Energy Corp. of Louisiana	Reserve, La.	200,000	Full range
Exxon	Northeast U.S.	150-200,000[b]	
Georgia Refining Co.[a]	Brunswick, Ga.	200,000	Fuels
Hampton Roads Energy Co.	Norfolk Va.	180,000	Fuels
JOC Oil	Burlington, N.J.	55,000	Fuels/SNG
New England Petroleum	Oswego, N.Y.	100,000	Fuels/SNG
No. Illinois Gas Co. Gibbs Oil Co.	Sanford, Me.	250,000	Fuels/SNG
Odessa Refining Co.	Mobile, Ala.	120,000	Fuels/SNG
Pacific Resources Inc.	Portland, Ore.	100,000	Fuels
Pacific Resources Inc. San Diego Gas & Elec.	Carlsbad, Calif.	100,000	Fuels/SNG
Pennzoil Co.	Pascagoula, Miss.	150-200,000	Fuels/SNG
Pioneer, Oklahoma Natural Gas, et al.	Houston area	100,000	Fuels
Pittston Co.	Eastport, Me.	250,000	
Refining Co. of La.	Coastal parish	100,000	Fuels
Shell Oil Co.	Logan Township, N.J.	150-200,000[b]	Full range
Steuart Petroleum Co.	St. Mary's County, Md.	100,000	Fuels/SNG
Total		2,870,000-3,070,000	

[a]Incorporated by Fuels Desulfurization Inc.
[b]Estimated.

apparent success, of OPEC countries to use oil as a political weapon have cast doubts as to whether Middle East crude oil can be considered as a dependable source of input to U.S. refineries. Major companies are likely to experience the least difficulty in obtaining long-term commitments, but substantial risk of interruption still exists. Nonmajor companies are likely to face substantially greater problems in obtaining Middle East crude, thus creating substantial difficulty in acquiring financing for construction. Evidence of a firm crude supply at a known price will be difficult to obtain.

Ashland Oil has attempted to ensure a continuity of supply. Ashland

Table 3-9. Announced Plans to Expand Existing Refinery Capacity

Company	Location	Size, b/d	Est. on Stream
Exxon	Baytown, Tex.	250,000	1976
	Baton Rouge, La.	100,000	1974-75
	Bayway, N.J.		
Socal	Richmond, Calif.	175,000	1976
	Perth Amboy, N.J.	75,000	1975
Texaco	Convent, La.	200,000	1977
	Anacortes, Wash.	15,000	1973
	Lawrenceville, La.	10,800	1973
Mobil	Paulsboro, N.J.	150,000	1977-78
ARCO	Philadelphia	25,000	1973
	Carson, Calif.	20,000	1973
	Houston	95,000	1976
Coastal States	Corpus Christi, Tex.	50,000	1973
Marathon	Robinson, Ill.	75,000[a]	1973
BP Oil Corp.	Marcus Hook, Pa.	45,000	1975
Clark Oil	Wood River, Ill.	42,000	1974
Amerada Hess	Corpus Christi, Tex.	47,000[a]	1974
American Petrofina	Port Arthur, Tex.	30,000[a]	1973
Amoco Oil	Whiting, Ind.	25,000	1974
Famariss Oil	Lovington, N.M.	25,000	1974
Southwestern Oil	Corpus Christi, Tex.	20,000	1973
		30,000	?
Howell Corp.	Corpus Christi, Tex.	12,000[a]	1973
		20,000	?
Navajo Refining	Artesia, N.M.	6,000	1974
Apco Oil	Arkansas City, Kan.	20,000	1974
Alabama Refining	Mobile	5,000	1974
Conoco	Wrenshall, Minn.	2,500	1973
Husky Oil	Salt Lake City	12,000	1974
Kerr-McGee	Wynnewood, Okla.	35,000	?
Hawaiian Indep.	Barber's Point	95,000[b]	1974
Somerset Refinery	Somerset, Ky.	2,000	1973
Total		1,714,300	

[a]Reactivation.

[b]Phased expansion.

Source: *The Oil and Gas Journal*, November 5, 1973, p. 20.

has announced plans to construct a new refinery of 100,000 b/d capacity and has proposed an arrangement with Iran. In effect a joint venture is proposed; Iran, in return for providing crude oil, would receive a share of Ashland's downstream operation. If accomplished, Iran's self-interest would lie in continuing a supply of crude to Ashland. The proposal is still being negotiated and given the range of investment opportunities available to countries such as Iran, there is no assurance that the deal will materialize.

Events in the domestic refining market have taken a strange twist. Initially, import quotas created disincentives for domestic refining investment, but the substitution of a fee system reversed this situation and resulted in plans to expand domestic capacity. Finally, expansion plans appear to be jeopardized, to a substantial degree, by the volatile crude oil supply situation in the Middle East.

COAL

The structure of domestic coal production has, in recent years, been characterized by rising levels of seller concentration and the acquisition of several large coal producers by oil and mining companies. Hence, it is not surprising that critics have expressed concern as to the effectiveness of present and future levels of competition in coal production. The following section evaluates the conduct of the coal industry in terms of pricing patterns, output policies, and the relationship between price and output policies. The output policies of acquired coal companies is analyzed and compared to those companies which were not acquired. Finally, the evidence offered by critics to support their charge of a conspiracy among coal producers is critically evaluated.

Output Behavior[g]

The output record during the period 1966-1972 for major coal producers involved in merger activity is summarized in Tables 3-10 and 3-11. As indicated in Table 3-10, by 1968, four coal-oil mergers had occurred: Consolidation-Continental Oil, Island Creek-Occidental Petroleum, Old Ben-Standard Oil of Ohio, and Pittsburgh and Midway-Gulf Oil.[h] In 1968, the four coal-oil companies produced 104.9 million tons of coal and by 1972 production stood at 106.4 million tons. Their share of total coal output declined slightly from 19.2 percent in 1968 to 17.9 percent in 1972. Their share of the total output attributable from the fifty largest coal companies also declined from 28 percent to 26.6 percent.

The output behavior of individual firms presents a mixed picture and

---

[g]See Appendix D by Moyer.

[h]Arch Mineral is unlike the four coal-oil companies. Initially owned by coal operators, ownership is now mainly by Ashland Oil and H.L. Hunt oil interests. Half of Arch's 1972 coal tonnage results from the acquisition of Southwestern Illinois Coal Co.

Table 3-10. Output Behavior of Major Acquired Coal Companies, 1966-1972 (million tons)

Coal Company	Parent Company	Year						
		1972	1971	1970	1969	1968	1967	1966
Peabody	Kennecott Copper	71.6	56.0	67.9	59.7	59.8[a]	59.4	54.0
Consolidation	Continental Oil	64.9	54.8	64.1	60.9	59.9	56.5	51.4[a]
Island Creek	Occidental Petroleum	22.6	22.9	29.7	30.4	25.9[a]	25.9	23.7
Amax	American Metal Climax	16.4	13.3	14.4	11.3[a]	9.3	8.6	8.5
Old Ben	St. Oil of Ohio	11.2	10.5	11.7	12.0	9.9[a]	10.3	9.9
Arch Mineral[b]	Ashland Oil-H.L. Hunt	11.2	2.1	—	—	—	—	—
Freeman-Unit. Electric	Gen'l Dynamics	10.0	11.5	14.1	14.0	13.0	14.1	13.6[a,c]
Pittsburgh & Midway	Gulf Oil	7.7	7.1	7.8	7.6	9.2	9.0	8.9[a,d]
Total U.S. Bituminous Coal tonnage		595.4	552.2	602.9	560.5	545.2	552.6	533.9
Tonnage of Top fifty coal producers		399.4	358.4	407.0	383.6	374.2	364.8	341.1

[a]Year when company was acquired.

[b]Much of Arch Mineral's increased output came in 1972 from its acquisition of Southwestern Illinois Coal Co.

[c]Year when General Dynamics achieved complete ownership. Large stock interest previously held.

[d]Control of Gulf Oil dates to 1963.

Source: See Appendix D by Moyer.

Table 3-11. Major Acquired Coal Companies' Output as
Percentage of Total Coal Output, 1966-1972

Coal Company	Parent Company	Year						
		1972	*1971*	*1970*	*1969*	*1968*	*1967*	*1966*
Peabody	Kennecott Copper	12.0	10.1	10.5	10.7	11.0	10.7	10.1
Consolidation	Continental Oil	10.9	9.9	10.6	10.9	11.0	10.2	9.6
Island Creek	Occidental Petroleum	3.8	4.1	4.9	5.4	4.8	4.7	4.4
Amax	American Metal Climax	2.8	2.4	2.4	2.0	1.7	1.6	1.6
Old Ben	Std. Oil of Ohio	1.9	1.9	1.9	2.1	1.8	1.9	1.9
Arch Mineral	Ashland Oil-H.L. Hunt	1.9	0.4	–	–	–	–	–
Freeman Unit. Electric	Gen'l Dynamics	1.7	2.1	2.3	2.5	2.4	2.6	2.5
Pittsburgh & Midway	Gulf Oil	1.3	1.3	1.3	1.4	1.7	1.6	1.7

Source: See Appendix D by Moyer.

the results provide no evidence of any coordinated output policy among the coal-oil companies. As indicated in Table 3-12, Consolidation Coal's market share increased subsequent to the acquisition and a slight increase occurred in the case of Old Ben. In the two remaining cases, Island Creek and Pittsburgh and Midway, market shares declined following their acquisition. Thus, output increased in two cases and declined in two cases during a time when total tonnage increased by a modest amount.

The failure of output behavior to yield even superficial evidence of a coordinated output policy subsequent to the acquisition is not surprising when one considers that concentration levels, while rising, are not presently at a level where successful coordination would be predicted. In 1972, the combined output of the four coal-oil firms represented less than 20 percent of total production. Consequently, an output restriction policy would have shifted market share to nonoil controlled firms and would have been ineffective as a means of elevating prices in the market.

It is sometimes contended that the acquisition of coal companies by large, noncoal interests is anticompetitive because the remaining independent companies are not able to compete successfully with the large coal-oil and coal-conglomerate firms. A comparison of output behavior between independent and nonindependent[i] coal companies doesn't offer any conclusive evidence to support the hypothesis that independents are unable to compete. Analyzing output data during the period 1968-1972, the market share for seven major

[i]Nonindependents represent coal-oil and coal-conglomerate companies.

Table 3-12. Index of Acquired Coal Companies' Output, 1966-1972 (1966 = 100)

Coal Company	Parent Company	Year						
		1972	1971	1970	1969	1968	1967	1966
Peabody	Kennecott Copper	133	104	118	110	110	110	100
Consolidation	Continental Oil	126	107	125	118	116	110	100
Island Creek	Occidental Petroleum	95	97	125	128	109	109	100
Amax	American Metal Climax	193	156	169	133	110	101	100
Old Ben	Std. Oil of Ohio	113	106	118	121	100	104	100
Arch Mineral	Ashland Oil- H.L. Hunt	N.A.	N.A.	N.A.	N.A.	N.A.	N.A.	N.A.
Freeman Unit. Electric	Gen'l Dynamics	74	85	104	103	96	104	100
Pittsburgh & Midway	Gulf Oil	86	80	88	85	103	101	100
Total U.S. bituminous coal output		112	103	113	105	102	104	100

N.A.—Not applicable.
See: Appendix D by Moyer.

nonindependent producers remained constant at 34.3 percent and for 11 independents with output in excess of 3 million tons annually and operating in both 1968 and 1972, market share declined from 15.6 percent to 13.8 percent during the period.[j] Moderately large independents have been fairly successful but the picture is mixed in the case of small independents. The decline of some small coal firms may reflect basic changes in mine health and safety standards as well as a decline in spot market sales to electric utilities. The following comment by an officer of a major independent company characterizes the situation:

> The independent coal company, they say is the model for the industry, the champion of free fuel competition, the first line of defense against the domination of fuel conglomerates. This is one part nostalgia and four parts baloney. . . . Today few companies have the 10 or 20 or 30 million dollars it costs to put in a new mine. . . . Where could we expect to raise 30 million dollars for a big new coal mine when that amount would at least equal our total investment in the mines we operate today?[25]

---

[j]Independents with annual sales between 1-3 million tons had a slight decline in shares from 4.5 percent to 4.4 percent.

In part, differences in management aggressiveness in shifting resources into western coal operations and into export sales and out of eastern coal markets explains the differences in performance among independent coal firms.

Mergers

The competitive impact of mergers involving coal companies is not easily ascertained. An assessment of coal-oil mergers rests, to a large degree, upon the extent of interfuel competition in various end uses. Coal-conglomerate mergers, which represent a change in ownership but no change in the number of sellers in the industry are even more difficult to evaluate.

Entry by noncoal firms into coal production has been by both merger and nonmerger means. It is extremely difficult to determine the motives behind the decision to enter coal production. Continental Oil, Standard Oil of Ohio, and Occidental Oil chose the merger route while other companies, the best example being Exxon, entered by acquiring coal reserves.

Mergers are not inherently anticompetitive; however in terms of competitive impact, *de novo* entry of oil firms into coal is generally preferred. *De novo* entry results in an increase in the number of independent decision-making units thus strengthening the forces of competition. Mergers between firms in the same market reduce the number of sellers and may be anticompetitive. In assessing mergers, the most crucial question is whether the noncoal companies would have entered on a *de novo* basis if entry by merger were denied. Some insight into this question is available from the Kennecott-Peabody Coal merger.

In 1968, Kennecott Copper acquired the Peabody Coal Company, an action which led to a Federal Trade Commission antitrust attack on the grounds that the merger reduced potential competition. The FTC argued that Kennecott was a potential entrant and, in the absence of the merger, would have entered on a *de novo* basis, which would have been procompetitive.

Testimony in the case indicated that Kennecott had acquired coal reserves in 1965 and subsequently attempted to acquire other reserves. Kennecott claimed the acquired reserves only represented a hedge to protect itself in its purchases of fuel for its western smelters. Given the difficulty of assigning motives, Kennecott's actions in acquiring reserves are consistent with an intention of entering *de novo* into coal production.

The FTC denied the merger on the grounds that Kennecott was a potential entrant to an industry with a discernible trend towards higher concentration.[k] The decision was upheld by the Supreme Court.

The Case for a Conspiracy[l]

Several organizations, reacting to the entry of noncoal companies

---

[k]As indicated earlier, the U.S. government owns some 45 percent of Southwestern U.S. coal reserves.

[l]A more complete discussion is presented in Appendix C by Prof. Moyer.

into the coal industry, have asserted charges of collusion and conspiracy to restrict coal supply in order to raise coal prices. The most vociferous of the proponents of this view include the TVA Public Power Association, the American Public Power Association, and the National Rural Electric Co Op Association.

Conspiracy charges rest upon price-output behavior in coal markets during the period 1969-1971. The supporting evidence was recently presented, in a joint statement, to the House Select Committee on Small Business.[26] The evidence, examined below is not compelling. The main points are:

    a) a 60 percent increase in coal prices between 1969 and December, 1970 in the face of an increase of only 4 percent in coal consumption by electric utilities during the period.

    b) the contrast between sharply rising prices in the period 1969-70 with the relative price stability of coal prices in the immediately preceding period.

    c) a marked slowdown in bidding actively for TVA and other public power company contracts.

    d) the failure of suppliers to deliver on long term contracts.

There is no question but that coal prices increased sharply during the period 1969-1971, but the 60 percent figure cited in the testimony is misleading and clearly unrepresentative of coal prices in general. The figure reflects on movements in the steam coal component of the Wholesale Price Index and is based on a small sample of major coal producers and is heavily influenced by events in spot markets. Spot market prices did increase over this period but the WPI may be an overstatement; the average realization on all coal shipments rose 25 percent from 1969 to 1970.

On the surface, the strongest evidence would appear to be the differences in price movements between various time periods. The APPA argues that from 1960 to 1965 coal prices were stable in spite of a 23 percent increase in total tonnage and a 40 percent increase in utility consumption. This price performance is in sharp contrast to the rapidly rising prices of the period 1969-1971.

The marked difference in price stability appears to be related to significantly different conditions during the two periods. During the period of price stability, expansion in production occurred at a time when excess capacity was the rule. Consequently, upward cost pressures due to greater output were reduced as excess capacity was reduced.

Perhaps of greater importance are the changes in labor costs and productivity during the periods. From 1960-1965 average hourly earnings in coal mining increased some 19 percent but productivity, in terms of output per man day, increased by 55 percent. However, from 1965-1971 earnings increased 40 percent but productivity rose by only 3 percent. During 1969-1971, the period in which rapid price increases occurred, output per man day declined by some 9 percent; in the case of underground mines the decline was 23 percent.

A closer inspection of the relationship between pricing behavior and productivity levels, wage levels, and capacity utilization before and after 1968 is presented in Table 3-13. Modified wage costs, reported in Table 3-13, represent a proxy for wage costs and are computed by converting tons per man day into output per hour and dividing the latter into average hourly earnings.[m]

As indicated in Table 3-13, during the period 1958-1963 coal prices, reported as average realization, steadily declined while total output and capacity utilization increased moderately. Most importantly, the declining prices were accompanied by declining wage costs, from $0.259 in 1958 to $0.200 in 1963. During the 1963-1968 span, coal output increased by 18.8 percent and the average price increased by only 6.4 percent. Wage costs were relatively stable during the period.

Events are quite different in the period 1968-1972. In this case, prices increased by some 64 percent while output rose by only 9.2 percent. It is this situation that led to conspiracy charges. But the evidence in Table 3-13 indicates that during this period wage costs increased by 67.8 percent, perhaps mainly due to productivity losses caused by enforcement of the 1969 Coal Mine Health and Safety legislation. This evidence places the rapid price increases in a very different perspective and strongly suggests that rapidly rising wage costs, rather than effective collusion, are the reason for the higher prices.

Sharply rising coal prices and differences in price behavior between time periods did occur but they provide little basis for believing that collusion was present. As discussed above, changes in other factors, such as excess capacity and earnings and productivity movements can be used to explain the observed price behavior.

There is no denying however that domestic coal supplies were very tight during the period 1969-1971. The key question is whether the tight supply resulted from a conspiracy of producers to limit output or whether other economic forces were the cause. Two factors appear responsible for the tight supply: one, a great expansion in export demand and two, an unfavorable climate for investment in new mines during the late 1950s to around 1965. Exports increased during 1968-1970 from 50.6 to 70.9 million tons, an increase that accounted for 40 percent of the total increase in coal tonnage during this period.

Shortages during 1969-1971 clearly reflect the lack of new mine development during the previous years. The time involved between the conception of a new mine and initial production can involve several years. Coal producers faced a stagnant market in the first half of the 1960s. Predictions were that nuclear reactors would soon displace coal from the utility market.

---

[m]Certain nonwage costs such as contributions to the union's welfare fund and social security payments are not included.

Table 3-13. Productivity, Wage Costs, Prices, Production, and Capacity Utilization, Bituminous Coal Industry, 1958-1972

Year	(1) Modified Wage Costs[a]	(2) Index of Wage Costs (1968 = 100)	(3) Average Realization	Realization Index (1968 = 100)	(5) Average Number Days Worked per Year[b]	(6) Annual Coal Output (MM tons)	(7) Output Index (1968 = 100)
1958	$0.259	130.2	$4.86	104.1	184	410.4	75.3
1959	0.255	128.1	4.77	102.1	188	412.0	75.6
1960	0.245	123.1	4.69	100.4	191	415.5	76.2
1961	0.225	113.1	4.58	98.1	193	403.0	73.9
1962	0.225	113.1	4.48	95.9	199	422.1	77.5
1963	0.200	100.5	4.39	94.0	205	458.9	84.2
1964	0.196	98.5	4.45	95.3	225	487.0	89.4
1965	0.199	100.0	4.44	95.1	219	512.1	93.9
1966	0.198	99.5	4.54	97.2	219	533.9	97.9
1967	0.196	98.5	4.62	98.9	219	552.6	101.4
1968	0.199	100.0	4.67	100.0	220	545.2	100.0
1969	0.213	107.0	4.99	106.9	226	560.5	102.8
1970	0.243	122.1	6.26	134.0	228	602.9	110.5
1971	0.273	137.2	7.07	151.4	210	552.2	101.3
1972	0.334	167.8	7.66	164.0	N.A.	595.4	109.2

N.A. Not available.

[a]Average hourly earnings/tons per man-day.

[b]Capacity is calculated by the Bureau of Mines to be 280 operating days a year, but this measure is of doubtful validity. It is probably overstated.

Source: *Bituminous Coal Annual, 1972*, National Coal Assoc., Bureau of Mines.

Expectations simply did not justify the investment outlays required for new mining capacity. However, when nuclear power failed to meet its timetable, utilities turned in part to coal as an alternative energy source and shortages inevitably were the result.

In addition, investment in new mines was held back by increasingly stringent rules on sulfur emissions in many metropolitan areas. Coal operators thus faced the risk of being unable to compete in the utility market. As a result, shortages were the result during 1969-1971. The pressures on coal supply, rather than being due to any conspiracy, appear to be the result of economic and institutional factors that were unfavorable to investment in new mines during the period preceding the 1969-1971 period. It should be pointed out that uncertainty as to whether the recently relaxed sulfur emission standards represent only a temporary lowering is likely to make coal operators unwilling to invest in mines in high sulfur fields, thus continuing the shortage of coal.

The TVA Experience. Electric utilities faced severe reductions in coal stockpiles during the period of tight supply. The TVA supply situation received the greatest publicity and was used by certain groups as evidence of anticompetitive action by coal-oil companies. TVA officials charged that companies reduced their bidding activity on TVA contracts during 1969-1971 whereas previously active bidding was the rule. The TVA chairman has stated that TVA's policy was to draw down stockpiles rather than to pay 27-28 cents per MBTU when it had previously paid 20 cents per MBTU.[27] As expected, TVA eventually yielded and paid the higher price. As to the conspiracy charge, Aubrey J. Wagner, TVA chairman, in response to a direct question by a Congressional committee as to evidence of a conspiracy, stated "I have not seen any evidence that I can put my hands on and say that is it."[28]

There is only limited information on the profitability of coal operations because most companies are controlled by parent companies and report earnings on a consolidated basis. Limited data indicates that substantial increases occurred from 1969 to 1970. As an example, Pittston's reported profits as a percentage of sales for 1969 and 1970 were 4.1 and 6.9 percent, respectively, and for Westmoreland Coal the corresponding figures are 1.5 and 5.2 percent.

The higher profits were offered by S. Robert Mitchell, in the APPA report, as proof of a conspiracy. The higher profits could reflect such an event or they could simply reflect the tight supply picture. In the earlier discussion, it was argued that the price behavior of the 1969-1971 period is entirely consistent with economic factors and offers little support to the conspiracy charge.

Perhaps the greatest price and output movements have taken place in the case of coal exports. Events in the export market during the period 1967-1972 are indicated by the data contained in Table 3-14. The export tonnage excludes sales to Canada and accounts for about 70 percent of total coal

Table 3-14. Export Coal Price-Output Behavior, 1967-1972[1]

Year	Export Tonnage (MM tons)	Exports as Percent of Total Production	Average FOB Mine Price, Export Shipments	Index of Export Prices (1967 = 100)	Average Nonexport Coal Price	Index of Nonexport Coal Price (1967 = 100)
1967	34.2	6.2	$ 5.92	100	$4.54	100
1968	33.9	6.2	6.06	102	4.58	101
1969	39.4	7.0	6.76	114	4.85	107
1970	52.3	8.7	9.67	168	5.93	131
1971	39.1	7.1	12.13	205	6.68	147
1972	37.8	6.4	13.28	224	7.28	160

[1] Excludes exports to Canada.
Sources: *Minerals Yearbook*, Volume I, U.S. Department of Interior. *World Coal Trade*, Coal Exporters Association of the United States Inc., U.S. Bureau of Mines.

exports and 83 percent of metallurgical coal exports. There are no published figures to indicate FOB mine price of overseas exports. The FOB figure presented in Table 3-14 has been estimated by taking the data on FOB prices in east coast ports and subtracting an estimate of the weighted average shipping and dumping charges per ton.

The most striking feature of the export market is the enormous price increase. During the period 1967-1972, while domestic coal prices increased by 60 percent, export prices increased by 124 percent! In 1972, the export price level was 82 percent above the average domestic coal price level. In 1960, a sample of District 8 mines reported that export prices *lagged* 11 percent behind the average price for all shipments and 19 percent behind domestic metallurgical shipments.

Much of the increase in export prices reflects an increase in world demand. Exports of metallurgical coal to Japan accounts for most of the increased U.S. exports. However, price behavior in this market is only partially explained by rising world demand. For instance, prices rose by 25.4 percent while exports fell by 25.2 percent in 1971. At the same time, in the domestic market a 12.6 percent price increase occurred while output declined 6.8 percent. In 1972, exports fell 3.3 percent while ore realization increased by 9.5 percent. During the entire 1970-1972 span domestic prices increased by $1.35 a ton as domestic production decreased by 1 percent; export prices increased $3.61 per ton while export tonnage decreased 28 percent.

Some of the price rise is certainly due to higher wage costs which increased about 46 percent during the period 1970-1972. However domestic production was subject to similar cost pressures yet domestic price increases were substantially smaller.

Further analysis of price behavior in export markets is hindered by a lack of data. No information is available concerning the prices of new export contracts relative to prices on existing contracts. One factor does stand out however, namely the structure of export markets. Concentration in the export market is significantly greater than in the case of domestic production. Twelve companies, members of the Coal Exporters Association of U.S., Inc., an affiliate of the National Coal Association, account for 75-80 percent of total exports.[n]

SUMMARY

Competition in the production of domestic crude oil markets has been systematically reduced by government intervention. Restrictions on oil imports and demand prorationing have had anticompetitive effects on conduct. These are the exact policies that a private monopoly would have pursued if it had the

---

[n]Pittston Company, a major coal exporter, had a profit increase of $9.5 to $23.5 million from 1969 to 1970 on nearly identical total sales volume in the two years but with an increase in export sales from 6.3 to 11.2 million tons.

necessary capability. An analysis of price and output patterns in crude oil indicates that prorationing, during periods of excess capacity, has been used as a means to support crude oil prices or, in other words, to prevent the price mechanism from carrying out its allocative function. The result has been an inefficient use of scarce resources in crude oil production in the U.S. The domestic crude oil market was further protected from the forces of competition as a result of the oil import quota program. Such restrictions were designed to protect the higher U.S. price in order to ensure a "strong" domestic oil industry. One effect of this intervention has been to create a shortage of domestic refining capacity. Other factors, such as the foreign tax credit, have caused investment in oil development to flow overseas.

Government intervention with anticompetitive consequences has characterized domestic crude oil markets. Price and output behavior indicates a noncompetitive pattern and is the result of policies such as demand prorationing.

Critics have been quick to assert that oil entry into coal represents an attempt to monopolize and that rising coal prices during 1969-1971 are the result of an oil company conspiracy to restrict output. A review of price and output patterns of coal firms fails to support such an assertion. No substantive evidence exists to indicate that oil-coal companies have restricted output in order to elevate price. The sharp price increase during the period 1969-1971 appears to be not due to oil company entry, but the result of a marked increase in wage costs and a sharp decline in productivity level. The tight coal supply situation of 1969-1971, often taken to be evidence of a conspiracy, appears to be due to a sudden rise in utility demand, an unfavorable climate for investment in new coal capacity in earlier years, and a sudden growth of coal exports. Thus, differences in conduct between the period 1969-1971 and earlier periods are real but were caused by differences in economic forces between the two periods rather than due to any conspiracy to restrict coal production.

NOTES

1. For a most comprehensive review and analysis of government policies in petroleum markets see: U.S. Senate, Subcommittee on Antitrust and Monopoly, *Government Intervention in the Market Place*, Vols. 1-5 (Washington: 1969).
2. U.S. Senate, Subcommittee on Antitrust and Monopoly, *Economic Concentration* (Washington, March 18, 1965), pp. 599-603.
3. An excellent analytical study of prorationing is available in: Stephen L. McDonald, *Petroleum Conservation in the United States: An Economic Analysis* (Baltimore: The Johns Hopkins Press, 1971).
4. *Op. cit., Government Intervention . . .* , p. 80.
5. McDonald, *op. cit.*, p. 57.
6. McDonald, *op. cit.*, p. 165.
7. McDonald, *op. cit.*, p. 188.

8. Analytical studies of the quota program include: James C. Burrows and Thomas A. Domencich, *An Analysis of the United States Oil Import Quota* (Lexington, Mass.: D.C. Heath and Co., 1970); Kenneth W. Dam, "Implementation of Import Quotas: The Case of Oil," *The Journal of Law and Economics*, April 1971; and Cabinet Task Force on Oil Import Control, *The Oil Import Question* (Washington, 1970).

9. Adelman, *op. cit.*, pp. 154-5.

10. *Ibid.*

11. Dam, *op. cit.*, p. 60.

12. Cabinet Task Force on Oil Import Control, *op. cit.*

13. National Petroleum Council, *U.S. Energy Outlook*, December 1972, p. 23.

14. *Ibid.* Two points are relevant. One, the forecast of a high level of imports is inconsistent with the Nixon administration goal of energy self sufficiency. Two, if the import price remains in the $9-10 per barrel range, U.S. demand for imports will be substantially reduced.

15. For a discussion of the tax and its impact see: Gerard M. Brannan, *The Role of Taxes and Subsidies on United States Energy Policy* (Washington: Energy Policy Project, 1974).

16. Brannan.

17. *Government Intervention . . . , op. cit.*, p. 85.

18. Brannan Report.

19. M. Adelman, "Efficiency of Resource Use in Crude Petroleum," *Southern Economic Journal*, January 1964, p. 105.

20. Conflicting views are expressed by: Adelman, *The World Oil Market* and James E. Akins, "This Time the Wolf Is Here," *Foreign Affairs*, April 1973.

21. Committee on Government Operations, *Preliminary Federal Trade Commission Staff Report on Investigation of the Petroleum Industry* (Washington: July 12, 1973), p. 6.

22. De Chazeau and Kahn, pp. 220-23.

23. *Government Intervention in the Market Place*, Vol. 1, p. 135.

24. Richard Mancke, "Petroleum Conspiracy: A Costly Myth," *Public Policy*, Winter 1974, pp. 12-13.

25. "Coal: Prospects and Problems," speech by Herbert S. Richey at the National Energy Forum, Washington, D.C., September 23, 1971.

26. Artificial Restraints on Basic Energy Sources" in *Concentration by Competing Raw Fuel Industries in the Energy Market and Its Impact on Small Business*, Hearings before the Subcommittee on Small Business Problems of the House Select Committee on Small Business, 92nd Congress, 1st Session, July 12-15, 20, 22, 1971, p. A445.

27. Hearings, *op. cit.*, Vol. 2, p. 11.

28. Hearings, *op. cit.*, Vol. 2, p. 19.

Chapter Four

Performance

Market performance is a multidimensional concept. Long-run rates of return and the extent of technical progress are among the more important aspects of performance.

PRICES AND PRODUCTIVITY TRENDS—OIL

Crude oil and gasoline prices during the 1950-1973 period are presented in Table 4-1. In terms of crude oil, many prices in 1973 are somewhat above their 1950 level but when the effect of inflation is accounted for, crude oil prices in 1973 are less than their 1950 level. During this period crude oil prices advanced at a rate less than the rate of inflation. Gasoline prices display a similar pattern. Actual prices increased over the 1950-1973 period but their rate of increase was less than the rate of inflation; consequently gasoline prices, in real terms, were lower in 1973 than they were in 1950. Since 1973 however, the real price of both crude oil and gasoline has increased significantly.

Productivity levels, defined as output per man hour, in petroleum refining for the 1939-1971 span are presented in Table 4-2. During the 1960-71 period, productivity advanced at an average annual rate of 5.8 percent.

The behavior of prices over long periods of time, while of obvious interest to consumers, does not bear directly on the question of how competitive an industry. Effective competition implies that prices will equate with opportunity costs; thus it is the relationship of price to cost that is directly relevant to the question of competition. An industry could display declining prices while at the same time have a widening price-cost gap, which in the long run would be indicative of ineffective competition.

PROFITABILITY

Evidence of long-run profit rates have traditionally been used as a means of assessing the extent to which resources have been efficiently allocated. The

Table 4-1. Gasoline and Crude Oil Prices, 1950-1973

	Crude Petroleum Price at Wells Oklahoma	Wholesale Price Index, BLS 1967 = 100	Deflated Prices Crude Oil	Average U.S. Gasoline Prices Excl. Tax	Consumer Price Index 1967 = 100	Deflated Price Gasoline
1974 March	6.33	154.5	4.10	0.36[a]	139.7	0.26
1973	3.87	135.5	2.86	0.28	133.1	0.19
1972	3.45	119.1	2.90	0.25	125.3	0.20
1971	3.41	113.9	2.99	0.25	121.3	0.21
1970	3.23	110.4	2.93	0.24	116.3	0.21
1969	3.18	106.5	2.99	0.23	109.8	0.22
1968	3.06	102.5	2.99	0.23	104.2	0.22
1967	3.02	100.0	3.02	0.22	100.0	0.22
1966	2.93	99.8	2.94	0.21	97.2	0.22
1965	2.92	96.6	3.02	0.20	94.5	0.22
1964	2.92	94.7	3.08	0.20	92.9	0.22
1963	2.93	94.5	3.10	0.20	91.7	0.22
1962	2.97	94.8	3.13	0.20	90.6	0.22
1961	2.97	94.5	3.14	0.21	89.6	0.23
1960	2.97	94.9	3.13	0.21	88.7	0.24
1959	2.97	94.8	3.13	0.21	87.3	0.24
1958	3.07	94.6	3.25	0.22	86.6	0.25
1957	3.05	93.3	3.27	0.21	84.3	0.25
1956	2.82	90.7	3.11	0.22	81.4	0.27
1955	2.82	87.8	3.21	0.21	80.2	0.26
1954	2.82	87.6	3.22	0.21	80.5	0.26
1953	2.72	87.4	3.11	0.20	80.1	0.25
1952	2.57	88.6	2.90	0.20	79.5	0.25
1951	2.57	91.1	2.82	0.21	77.8	0.27
1950	2.57	81.8	3.14	0.20	72.1	0.28

[a]January.

Source: *Survey of Current Business.*

existence of abnormally high long-run rates of return suggests inefficiency and little competition. On the other hand, normal profit performance suggests that effective competition prevails in a market.

The usefulness of profit data reported by modern corporations as a means of assessing the state of competition is limited. Increased diversification and the use of consolidated reporting makes it nearly impossible to measure performance in any given operation. For example, oil companies are highly diversified among various fuels and in addition have substantial petrochemical operations. The reported profits reflect a consolidation of their entire operations. The lack of meaningful product-line reporting is a major problem in such situations and acts as a constraint in formulating rational public policy.[1]

Oil

Profit data for twenty major oil companies is reported in Table 4-3. The figures indicate average profit rates during the 1967-1972 period and the year 1973. In addition, the average profit rate for all manufacturing during the period 1967-1971 is also reported. There is a substantial amount of variation in the reported figures, but on the average the top twenty firms earned at a rate comparable to the average rate for all manufacturing. Oil company profits in 1973 and the first quarter of 1974 increased at a very sharp rate. In terms of absolute profits earned, Gulf's profits in the third quarter 1973 were up some 91 percent from third quarter 1972 levels. Similar figures for Exxon and Mobil Oil are 81 and 64 percent, respectively. In nearly every case, profit rates in 1973 were substantially above the firm's average earnings rate for the 1967-1972 period.

In spite of 1973 results, oil company reported profitability has not been generally above average manufacturing profitability. This may seem to be a paradox in light of the importance of the special tax subsidies and special benefits associated with domestic crude oil production. Oil company executives often confront critics, who argue that the oil industry is the recipient of favored treatment, with the retort that oil company profits fail to reflect any special benefits on market power. Professor Kahn, in addressing this question, argues that the failure of oil company earnings to be high relative to other industries, which do not receive the special benefits of the oil industry, supports the contentions of those who argue that the special provisions such as demand prorationing and the depletion allowance have led to large inefficiencies and higher costs. The essence of resource allocation under competition is that resources flow to alternative investment projects in response to profit differentials. If an industry becomes more profitable relative to other industries, regardless of the reason, resources will flow into the more profitable industry. In the case of crude oil, special tax provisions increase relative profitability during the period following their implementation. Given the lack of high entry barriers, new investment entered crude production. The resulting excess capacity results

Table 4-2.   Output per Man Hour: Petroleum Refining
(1967 = 100)

Year	Output per Man-Hour[a] Production Workers
1939	29.9
1947	28.6
1948	b
1949	31.7
1950	36.6
1951	38.4
1952	39.9
1953	40.9
1954	43.1
1955	47.1
1956	49.1
1957	49.9
1958	52.5
1959	59.1
1960	62.1
1961	67.1
1962	73.5
1963	78.5
1964	83.0
1965	89.9
1966	97.1
1967	100.0
1968	103.7
1969	110.6
1970	110.2
1971[c]	114.9

[a]The output measures underlying the output per man-hour and output per employee indexes relate to the total production of the industry. They do not relate to the specific output of any single group of employees.

[b]Not available.

[c]Preliminary.

Source: Output based on data from the Bureau of Mines, U.S. Department of the Interior. and the Census, U.S. Department of Commerce. Employment and hours based on data from the Bureau of the Census, Department of Commerce, and the Bureau of Labor Statistics, U.S. Department of Labor.

in higher costs and, over time, costs rise to meet crude prices thus eliminating the higher profits initially earned during the period immediately following the implementation of the special benefits. The special benefits raise the expected earnings and cause investment to expand to the point where rates of return are

Table 4-3. Profits as a Percent of Invested Capital Twenty
Largest U.S. Petroleum Firms

Firm *(ranked according to 1971 sales)*	*Average Profit Rate (1967-1972)*	*Profit Rate (1973)*
1. Exxon	12.3%	19.1%
2. Mobil Oil	10.7	14.9
3. Texaco	13.8	16.2
4. Gulf Oil	10.4	14.4
5. Standard Oil of California	10.4	14.5
6. Standard Oil of Indiana	9.7	12.4
7. Shell Oil	10.6	10.7
8. Atlantic Richfield	8.4	8.7
9. Continental Oil	10.1	13.4
10. Tenneco	11.2	12.0
11. Occidental Petroleum	14.2	9.0
12. Phillips Petroleum	8.4	11.7
13. Union Oil of California	9.2	10.5
14. Sun Oil	9.5	11.9
15. Ashland Oil	11.6	15.5
16. Standard Oil of Ohio	8.2	7.9
17. Getty Oil	8.5	9.0
18. Marathon Oil	11.4	16.2
19. Clark Oil	15.2	27.9
20. Commonwealth Oil Refining	11.5	17.7
Average	10.8	13.7
Average: all manufacturing (1967-1971)[a]	10.8	

[a]Economic Report of the President, January 1973, p. 280.
Source: *Fortune*, May issues.

close to the level that would be earned on alternative uses of capital. The point
to be emphasized is that subsidies increase profits only at the time they are
established and in the case of crude oil have caused overcapacity and resource
misallocation. Thus, the oil industry defense of their preferred position cannot
be accepted; the existence of excess capacity and higher costs is clear evidence
that prices were not allowed to carry out their allocation function. The oil company contention that oil profit rates have not been excessive relative to returns
in manufacturing is appealing on the surface. However, given the presence of
excess capacity up to 1972 under the market demand prorationing system and
the fact that reported profits were not large is strong evidence that competition
was lacking in domestic crude oil markets. If competition had been strong, oil
company profits would have been much lower.

The array of government intervention into petroleum markets has created economic incentives that led to a waste of scarce resources. The effect of prorationing has been to create an incentive to (1) drill an excessive number of wells, (2) drill deeper wells, (3) to restrict production from low cost wells, and (4) to encourage production from high cost wells. The result has certainly been to increase the cost of domestic crude oil production. Inefficiency in domestic production has been effectively sheltered from the competitive pressures of imported oil. The mandatory import quota system was initiated in 1959 but some argue that the U.S. government has always frowned upon imports. This type of protection has allowed the inefficiencies to persist.[2]

Coal and Uranium

Profit data in the case of coal and uranium producers is extremely limited and thus of limited value in assessing competition. The main reason is that most of the major coal producers are controlled by noncoal companies and profits are reported on a consolidated basis, thus making it impossible to ascertain earnings from coal operations. A similar problem occurs in the case of uranium producers.

Profit data for certain independent producers is presented in Table 4-4. The average profit rate for the four companies over the period 1967-1972 is 9.8 percent. This is slightly below the average for all manufacturing. Distinctly lower earnings are the case among uranium producers, a situation which reflects the excess capacity during the period.

Table 4-4. Profits as a Percent of Total Capital Coal and Uranium Firms (average 1967-1972)

Coal	*Profit Rate (%)*
Pittston Co.	12.1
Eastern Gas and Fuel	8.7
North American	6.5
Utah International	11.9
Average	9.8
Uranium	
Kerr-McGee	8.6
Atlas Corporation	2.5[a]
United Nuclear	5.7[b]
Average	5.6

[a]1969-1972.
[b]1969-71.
Source: Value Line Investment Survey.

Profit data from coal and uranium operations is simply too limited to reach even a tentative judgment as to the effectiveness of competition in these markets. It is known that coal profits, which have been quite low for many firms during the 1960s, have improved considerably during the period 1969-1971. Given the structure and conduct of coal production, this is likely due to the growth of demand for coal from electric utilities. Profits have not increased to the level where monopoly power could be suspected to be the cause.

TECHNOLOGICAL PROGRESS

Technological progress is an important goal of the U.S. economy. Progressiveness covers a wide area encompassing R&D activity, invention and innovation. The following section, based upon a study by Professor Edwin Mansfield,[a] presents current knowledge concerning the relationship between firm size and invention, innovation, and diffusion of new ideas in the petroleum and bituminous coal industries. In addition, implications concerning merger and divestiture policy are discussed.

Firm Size and R&D Activity

Petroleum Refining. In absolute terms, the petroleum industry is among the largest spenders in research and development; in 1971, total R&D outlays were about $500 million, a figure which includes $128 million for chemical R&D conducted by oil companies.

The relationship between firm size and R&D expenditures is important. It is generally accepted that some minimum size is necessary if a firm is going to be able to support R&D activity; but the more important question is whether very large size yields proportionately greater R&D expenditures. Research results, based on nine major petroleum firms for the 1945-1959 period, indicate that, among the top firms, a one percent increase in firm sales yields an 0.86 percent increase in R&D expenditures.[3] Others have reported similar results. Thus, a firm's R&D intensity, in terms of R&D expenditures per unit of sales, does not appear to increase in proportion to increased size, once the threshold size is exceeded.

It is important to note that the R&D projects carried out by large petroleum firms tend to be generally safe from a technical view, have a high probability of success, and are expected to influence profits in five years or less. Only a small proportion of the outlays represent basic research. Enos presents evidence that laboratories operated by the leading oil firms were not responsible for the radical inventions discovered prior to World War II.[4] Such ideas as continuous processing, catalysis, and cracking by the application of heat and pressure are attributable to independent inventors. There is however some

[a]See Appendix G.

evidence to indicate that very large petroleum firms tend to do more basic, long term and technically risky R&D than do small firms. At the same time, differences between the R&D activity of very large and simply large petroleum firms are not great.

The lack of proportional increase in R&D expenditures with larger firm size does not rule out the possibility that, when R&D output is measured, the largest firms will be more productive. This possibility is examined by utilizing a list of major inventions in petroleum refining and petrochemicals to construct an index of inventive output for eight oil firms during 1946-1954. The index was regressed against the firm's size and the average of the firm's R&D expenditures in 1945 and 1950. The coefficients of both independent variables were not statistically significant, thus failing to support the hypothesis that large size leads to proportionately greater R&D output.

In another test, a subjective ranking of companies in terms of their inventive output per dollar of R&D was regressed against firm size and size of R&D expenditures.[5] Similar results were obtained, i.e., no evidence of higher R&D productivity in the largest firms and no evidence of scale economies in R&D among the large firms. Thus, on the basis of historical evidence, it can be stated that very large firm size does not appear to yield proportionately greater levels of R&D activity, in terms of either R&D expenditures or productivity.

There is some evidence to suggest that changes have occurred in the rate of increase in R&D expenditures and profitability of R&D investment. During the period 1963-1966, petroleum R&D rose by about 33 percent; between 1966-1969, the increase was 25 percent; but between 1969 and 1971, there was no increase in the level of R&D outlays.[6] One explanation for the change is that expected returns from R&D efforts have declined in recent years. R&D activity is a costly process and costs have increased by very large amounts in recent years. A review of six major cracking processes indicates that while costs have leaped upward, the time involved in developing the new processes has not changed on the average. Investment in R&D by a firm must obviously be based upon the expectation of profit. Enos has estimated returns from major cracking innovations.[7] The results, presented in Table 4-5, probably understate the actual return but indicate that the successful process innovations in refining were very profitable. It should be noted however, that estimated returns tended to decline over time.

A simple model, used to estimate marginal rates of return from all R&D, developed by Mansfield for five major oil companies in 1960 indicates (1) marginal rates of return were very high, averaging between 40-60 percent and (2) rates of return appear to be lower in the largest firms than in somewhat smaller ones.[8] The latter suggests that smaller firms may be underinvesting in R&D but that the largest firms have not been underinvesting. D.C. Baeder, Vice President of Exxon Research and Engineering, notes that the proliferation of research labs in recent years has made it increasingly difficult for his firm to

Table 4-5.   Cost of and Returns from Cracking Process
Innovations, 1913-1957

| Process | Cost of Innovation | | Returns from Innovation | | Approximate Ratio of Returns to Cost ($ per $) |
	Period over Which Expenses Incurred	Estimated Amount	Period over Which Returns Calculated	Estimated Amount	
Burton	1909-1917	$ 236,000	1913-1924	$150,000,000+	600+
Dubbs	1909-1931	7,000,000+	1922-1942	135,000,000+	20
Tube & Tank	1913-1931	3,487,000	1921-1942	284,000,000+	80+
Houdry	1923-1942	11,000,000+	1936-1944	39,000,000	3.5
Fluid	1928-1952	30,000,000+	1942-1957	265,000,000	9
TCC	1935-1950	5,000,000+	1943-1957	71,000,000+	16
Houdriflow	1935-1950		1950-1957	12,000,000	

Source: J. Enos, *op. cit.*

come up with a novel invention, and as a result, R&D returns fell significantly.[9] In 1966, Exxon Research and Engineering studied 100 of their inventions and concluded that real novelty was a crucial factor in obtaining a commercial success and that small returns were generally obtained from common innovations based on well known science. Exxon apparently concluded that their internal organizational structure was not conducive to selecting high risk and high payoff R&D projects and has organized a new Corporate Research Lab with favorable initial results.[10] This represents an interesting experiment but, at present, little is known concerning the relationship between a firm's internal structure and its technical progress.

Coal. In sharp contrast to the petroleum refining industry, coal producers have not been heavy spenders in R&D. Although official data is lacking, knowledgeable individuals suggest that private R&D expenditures total less than $10 million annually. The federal government, via the Office of Coal Research (OCR) and the Bureau of Mines, finances a substantial amount of coal research. OCR's total estimated budget for 1972 is about $52 million and the Bureau of Mines allocated some $53 million in 1972 for mineral research.

Some of the largest coal companies who are engaged in coal research are owned by oil firms. For instance, Consolidation Coal, owned by Continental Oil, has a favorable image as a technological leader.[11] Consolidation has been heavily engaged in the coal gasification area. Gulf Oil, owner of Pittsburgh and Midway Coal, is building a pilot plant designed to refine coal into a sulfur free solid that can be heated or burned into an oil. The project is financed by the OCR.

In summary, most of the R&D activity in coal is supported by the federal government. Major projects include coal gasification and means of controlling stack emissions. Unlike the situation in oil, much of the coal research is basic research and private firms are not as likely to be able to appropriate the returns. Hence the heavy role of government support. The total amount of R&D activity is substantially below the level of the petroleum refining industry.

Firm Size and Innovation

Petroleum Refining. In assessing the economic impact of R&D, economists distinguish between innovation and invention. Innovation occurs when an invention is first applied commercially. The distinction is made on the grounds that an invention has little economic significance until it is applied.

Under certain assumptions, one would expect the share of the innovative activity attributable to the top four firms to approximate their share of the market, measured in this case by daily crude oil capacity.[b] This relationship is examined by identifying major innovations and the first firm to adopt the new process during the period 1919-58. The results are presented in Table 4-6. By measuring firm size in terms of daily crude oil capacity, the proportion of the innovations first introduced by the top four firms can be determined.

The results from this relatively simple test are reported in Table 4-7 for two periods, 1919-38 and 1939-58. In the case of petroleum, the top four firms were responsible for a share of total innovations which exceeded their output share. These results are limited in that they only compare the top four firms to all others and do not provide an indication of whether top firms do proportionately better than somewhat smaller firms. Further analysis, based on a regression of innovative activity on firm size indicates that the maximum innovative activity relative to firm size, is associated with a firm with about 200,000 barrels of crude capacity in 1919-38 and 300,000 barrels in 1939-58, which is equivalent to the sixth largest firm. While earlier results indicated superior performance for the top four firms relative to all others, these results indicate that the top four were less innovative than somewhat smaller firms. The results also indicate that large firms were relatively more important as innovators in the 1939-58 period than in the earlier span.[c]

---

[b]Crucial assumptions include: (1) the largest firms devote the same proportion of their resources as smaller firms to R&D activity; (2) results are equally obtainable for large and small firms; and (3) the ability to apply the results in an efficient manner is the same.

[c]In the case of chemical innovations by big oil companies, a review of the evidence does not indicate that the biggest companies did more than their share of the innovating. Unpublished data derived by Frank Husic, University of Pennsylvania, and made available to Prof. Mansfield.

Table 4-6.   Innovations and Capacity (or Output) of Largest
Four Firms, Petroleum and Bituminous Coal Industries, 1919-1958

Item	*Petroleum*[a]		*Coal*[b]	
	Weighted[c]	*Unweighted*	*Weighted*[c]	*Unweighted*
		(percent of industry total)		
1919-38				
Process innovations	34	36	27	18
Product innovations	60	71	–	–
All innovations[d]	47	54	27	18
Capacity (or output)	33	33	11	11
1939-59				
Process innovations	58	57	30	27
Product innovations	40	34	–	–
All innovations[d]	49	43	30	27
Capacity (or output)	39	39	13	13

[a]Crude capacity is used to measure size of firm; it refers to 1927 in the earlier period and to 1947 in the later period.

[b]Annual production is used to measure size of firm; it refers to 1933 in the earlier period and to 1953 in the later period.

[c]In the columns headed "weighted," each innovation is weighted in proportion to its average rank by "importance" in the lists obtained. It was suggested that total savings be used to judge the relative importance of processes and that sales volume be used to judge the relative importance of products.

[d]The unweighted average of figures for process and product innovations.

Source: See Appendix G by Mansfield.

Coal. A similar analysis was conducted for coal. Major coal preparation innovations during the period 1919-1958 and the innovating firm were identified and related to firm size, measured in terms of annual production. The list of innovations is presented in Table 4-8. The results indicate that (1) the top four coal producers accounted for a disproportionately large share of total innovations, (2) the largest four producers however did less innovating than slightly smaller firms, and (3) the smallest firms were less important innovators in the period 1939-58 than in the 1919-38 span. The size of firm where the maximum number of innovations took place was 3.6 million tons of coal in the earlier period but increased to a firm size of about 7.8 million tons in the later period. These results are remarkably similar to the findings reported in the case of petroleum.

Other Industries. The relationship between firm size and innovative activity in petroleum and coal can be compared with results reported for the

Table 4-7. Innovations and Innovators, Petroleum Industry, 1919-38 and 1939-1958

Innovation	*Innovator*
1919-1938	
Burton-Clark cracking	Standard (N.J.)
Dubbs cracking	Shell
Fixed-bed catalytic cracking	Sun
Propane deasphalting of lubes	Union
Solvent dewaxing of lubes	Indian
Solvent extraction of lubes	Associated
Catalytic polymerization	Shell
Thermal polymerization	Phillips
Alkylation (H_2SO_4)	Standard (N.J.)
Desalting of crude	Ashland
Hydrogenation	Standard (N.J.)
Pipe stills and multidraw towers	Atlantic
Delayed coking	Standard (Ind.)
Clay treatment of gasoline	Barnsdall
Ammonia	Shell
Ethylene	Standard (Ind.)
Propylene	Standard (N.J.)
Butylene	Standard (N.J.)
Methanol	Cities Service
Isopropanol	Standard (N.J.)
Butanol	Standard (N.J.)
Aldehydes	Cities Service
Napthenic acids	Standard (Calif.)
Cresylic acids	Standard (Calif.)
Ketones	Shell
Detergents	Atlantic
Odorants	Standard (Calif.)
Ethyl chloride	Standard (N.J.)
Tetraethyl lead as antiknock agent[a]	Refiners
Octane numbers scale[a]	Ethyl
1939-1958	
Moving-bed catalytic cracking	Socony
Fluid-bed catalytic cracking	Standard (N.J.)
Catalytic reforming	Standard (Ind.)
Platforming	Old Dutch
Hydrogen-treating	Standard (N.J.)
Unifining	Union; Sohio

Table 4-7 (cont.)

Innovation	*Innovator*
Solvent extraction of aromatics	Standard (N.J.)
Udex process	Eastern State
Propane decarbonizing	Cities Service
Alkylation (H F$_1$)	Phillips
Butane isomerization	Shell
Pentane and hexane isomerization	Standard (Ind.)
Molecular sieve separation	Texaco
Fluid coking	Standard (N.J.)
Sulfur	Standard (Ind.)
Cyclohexane	Phillips
Heptene	Standard (N.J.)
Tetramer	Atlantic
Trimer	Atlantic
Aromatics	Standard (N.J.)
Paraxylene	Standard (Calif.)
Ethanol	Standard (N.J.)
Butadiene	Standard (N.J.); Shell
Styrene	Shell
Cumene	Standard (Calif.)
Oxo alcohols	Standard (N.J.)
Dibasic acids	Standard (Calif.)
Carbon black (oil furnace)	Phillips
Glycerine	Shell
Synthetic rubber	Standard (N.J.)
Ethylene dichloride	Standard (N.J.)
Diallyl phthalate polymers	Shell
Epoxy resins	Shell
Polystyrene	Cosden
Resinous high-styrene copolymers	Shell
Polyethylene	Phillips

[a]Innovations excluded from table because innovator had no crude capacity or because it was engaged primarily in another business.

Source: E. Mansfield, *op. cit.*

iron and steel, chemical, pharmaceuticals, and railroad industries. In the iron and steel industry, small firms were the major innovators during the 1919-38 and 1939-58 periods. Thus, the biggest oil and coal firms were more innovative, relative to size, than were the top four firms in iron and steel.

A similar conclusion is warranted in the chemical industry when one

Table 4-8.   Innovations and Innovators, Bituminous Coal Preparation, 1919-1938 and 1939-1958

Innovation	*Innovator*
1919-1938	
Simon-Carves washer	Jones & Laughlin; Central Indiana
Stump air-flow cleaner	Barnes
Chance cleaner	Rock Hill
"Roto Louvre" dryer	Hanna
Vissac (McNally) dryer	Northwestern Improvement
Ruggles-Cole kiln dryer	Cottonwood
Theolaveur	American Smelting
Menzies cone separator	Franklin County
Deister table	U.S. Steel
Carpenter dryer	Colorado Fuel & Iron
Froth flotation	Pittsburgh
1939-1958	
Raymond flash dryer	Enos
CMI drying unit	Hanna
Link-Belt separator	Pittsburgh
Bird centrifugal filter	Consolidation
Baughman "Verti-Vane" dryer	Central Indiana
Vissac Pulso updraft dryer	Northwestern Improvement
Link-Belt multi-louvre dryer	Diamond; Elkhorn; Bethlehem; Eastern Gas & Fuel
Eimco filter	United Electric
Dorrco Fluosolids machine	Lynnville
Parry entrainment dryer	Freeman
Heyl & Patterson flud bed dryer	Jewell Ridge
Feldspar type jig	Northwestern Improvement
Bird-Humboldt centrifugal dryer	Cinchfield
Wemco Fagergren flotation unit	Hanna; Sevatora; Diamond
Continuous horizontal filter	Island Creek
Cyclones as thickeners[a]	Dutch State Mines

[a]Omitted from table because innovator was not a domestic firm.
Source: E. Mansfield, *op cit.*

considers process innovations. But in the case of product innovations, the top four firms in chemicals were more active than smaller firms. As a result, the largest oil and coal firms did not exceed the performance of top chemical firms in the case of product innovations. In pharmaceuticals, the top four firms did not account for a disproportionately large fraction of innovations in the 1935-49 or 1950-63 period. In contrast to pharmaceuticals, the largest oil and coal firms are more innovative. Finally, on the basis of a small sample of railroad innovations since 1920, the biggest oil and coal companies do not appear to be more innovative than do the largest firms in the railroad industry.

Thus, the innovative record of the largest oil and coal firms appears to be somewhat better than the same relationship in the iron and steel, drug, and chemical industry, when process innovations are considered. In addition, the top four oil firms had innovative records about the same as leading firms in the railroad industry.

Diffusion of New Techniques

The final step of the process of technological change is the diffusion of the invention. Diffusion describes the process by which an innovation, once adopted, spreads throughout the industry.

Petroleum. Diffusion rates in petroleum are examined by reference to the experience with the adoption of catalytic cracking capacity rather than thermal cracking. The growth of catalytic capacity is indicated in Table 4-9. From the date of initial commercial adoption to the time when more than one half of U.S. cracking capacity was catalytic, some 16 years had elapsed. Clearly, long periods of time are required for substantial amounts of innovation to occur.[12]

Coal. Evidence of diffusion in coal is based upon the experience with (1) the shuttle car, (2) the trackless mobile loader, and (3) the continuous mining machine. In Figure 4-1, the percentage of major coal firms utilizing these innovations at various points in time is indicated.

In the case of shuttle cars, five years elapsed before half of the major firms innovated; seven years in the case of trackless mobile loaders; and four years in the case of continuous mining machines. Some of the differences can be explained in terms of differences in profitability of innovating. The data does suggest that in coal, as is true in most industries, larger firms tend to begin using new techniques more quickly than smaller firms. It should be pointed out that this finding does not necessarily mean that larger firms are more progressive. If very large firms were progressive only in proportion to their market share, they would be first in more cases, thus making them quicker in the diffusion process than smaller firms.[d]

Economists have been divided in arguing out the theoretical relationship between market structure and rates of diffusion. Some argue that competitive structures lead to faster diffusion; others the opposite.

Empirical studies of this question have been conducted and are based on experience with 12 innovations in four industries: coal, iron and steel, brewing, and railroads. The findings, reported by Mansfield, indicate that, when

[d]See Mansfield Appendix G.

Table 4-9. Growth of Catalytic Cracking Capacity As A Percentage of All U.S. Cracking Capacity

Year	Catalytic Cracking as a Percentage of All U.S. Cracking Capacity
1937	0.1
1938	0.6
1939	1.1
1940	5.6
1941	6.6
1942	7.0
1943	9.9
1944	21.4
1945	28.5
1946	30.0
1947	32.2
1948	34.0
1949	37.4
1950	41.7
1951	43.8
1952	46.7
1953	61.2
1954	67.5
1955	74.7
1956	79.3
1957	81.9

Source: *Oil and Gas Journal*, 1937-57.

profitability and size of investment required were held constant, diffusion rates tended to be higher in more competitive industries.[13] It must be pointed out however that the empirical results were not statistically significant. Similar findings were reported in a study of diffusion of numerically controlled machine tools in the tool and die industry.[14] Further study of the diffusion of numerically controlled tools in ten industries by Romeo indicated a positive and statistically significant relationship between the diffusion rate and the number of firms in an industry.[15] While these findings are clearly tentative, the evidence suggests that diffusion rates are greater in industries with competitive-like market structures.

Summary

R&D activity, whether measured in terms of inputs or outputs, does

Figure 4-1. Growth in the Percentage of Major Firms That Introduced Three Innovations, Bituminous Coal Industry

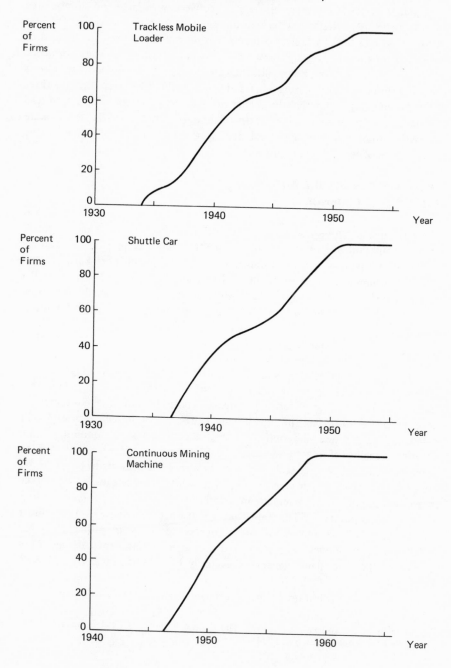

not increase in proportion to firm size in the case of petroleum refining. There does appear to be a tendency for very large firms to engage in more basic, long run, and risky R&D. However, differences between very large and simply large petroleum firms in R&D activities are not great. During the period 1969-1971, petroleum R&D outlays failed to increase from previous levels. In the case of coal, most of the R&D outlays have represented government sponsored projects. In the case of innovation, test results indicate that while the top petroleum firms do proportionately more innovating relative to all other firms, the top four firms were less innovative than slightly smaller firms. A similar relationship is observed in the case of coal innovations. Top oil and coal firms appear to be more innovative than top iron and steel and drug firms and about equal to the performance of top railroad companies.

ECONOMIC STRUCTURE AND
POLITICAL INFLUENCE

The relationship between economic and political power has been the object of much speculation but there have been few attempts to systematically examine the relationship. In a paper commission by the EPP, Professors Salamon and Siegfried attempt to test certain hypotheses and to apply the results to the energy industry.[e] Their results, while hardly conclusive, are a valuable contribution to our understanding of the interaction between the economic and political systems.

Salamon and Siegfried argue that the distribution of both incentives and relevant resources favor large, corporate interests over individual consumer-taxpayer interests in gaining access to and influencing policy-making. Political involvement can be viewed as an investment decision to be undertaken only when a satisfactory rate of return can be expected. The individual consumer has little incentive to make the investment necessary to influence government policy-making because the individual captures only a small amount of any benefits. Because of the "free rider" problem, most of the benefits flow to others. In the case of large scale corporate interests, the benefits to be gained by the corporation are much greater and, as a result, the incentives to undertake investment in political involvement are greater.

Corporate interests also have a distinct advantage over consumer interests in the distribution of the resources necessary to influence policy making; namely, money, expertise, and access to government officials. Consumers, on the other hand, generally have only the vote as a means to influence decisions.

The distribution of incentives and resources in favor of large scale

---

[e]See Appendix H: L. Salamon and J. Siegfried, "The Relationship Between Economic Structure and Political Power: The Energy Industry," (Ford Foundation: The Energy Policy Project Manuscript, 1974).

corporate interests becomes more significant in light of the highly fragmented nature of policy-making in government. Policy-making, rather than being an integrated and centralized process, is actually a series of "policy subsystems." The important factor in influencing decisions is access to the subsystems rather than access to the general political system. Individual consumers are generally limited to the latter but corporate interests are much more able to gain entry at the subsystem level.

The Hypothetical Relationship

The economic structure of an industry, it is reasoned, is a partial determinant of the extent to which the industry is able to exercise political influence. Salamon and Siegfried have identified four aspects of structure as factors affecting an industry's political influence.

Large absolute firm size is predicted to be an important source of political power. Large firms are likely to have greater economic resources than smaller firms. In addition, political activity involves high fixed costs, and large firms are better able to absorb these costs. Finally large firms are more prestigious and have a larger pool of expertise than smaller firms and, as a result, find it easier to gain access to the various subsystems of policy making. Thus, large absolute firm size is predicted to increase political power.

In a similar manner, large industry size enhances political influence. Large industry size, independent of average firm size, results in larger pools of resources, contacts, and overall resources than smaller industries. A direct relationship is predicted.

The degree of geographic dispersion in an industry can affect the extent of political influence but ambiguity exists as to the nature of the relationship. It could be argued a priori that an industry with little geographic dispersion lacks broad based support and, consequently, will have less political influence. Thus, a direct relationship would be predicted, i.e., the less (more) dispersed the industry, the less (more) the industry's political influence. It could however be reasoned that unless an industry is sufficiently important in an area, it will be unable to attract the attention of a congressman or senator. This line of reasoning predicts an inverse relationship between geographical dispersion and political influence.

Market concentration levels are, on net, predicted to be directly related to an industry's political influence. High concentration tends to yield high profits, an important political resource. In addition, a highly concentrated industry, by having fewer firms, will be able to avoid many of the problems associated with large numbers of trying to reach an industry position in political matters. In the economist's term, the transaction costs of reaching an agreement are reduced.

The impact of profit levels on the extent of political influence is not clear. On the one hand, high rates of return make an industry very visible in the public's eye and susceptible to government control if its political influence were used to obtain tax advantages. In this case, high profits would reduce an industry's willingness to exercise its political influence.

On the other hand, high profit firms have more to gain from tax reductions and thus, it could be argued, are prone to exercise their political power.[f] If correct, profits and political influence would be directly related.

Empirical Results

The model[g] tested by Salamon and Siegfried is:

$$(0.52 - t) = f(F, I, C, D, R)$$

where

t = the effective average corporate income tax rate

F = firm size (millions of dollars)

I = industry size (number of employees)

C = market seller concentration

D = index of geographical dispersion

R = the "true" accounting profit plus interest on total assets.

The dependent variable $(0.52 - t)$ represents the deviation of an industry's effective tax rate from 0.52, the nominal corporate income tax rate on earnings in excess of $25,000 in 1963. The deviation is a proxy for an industry's political influence; a greater deviation indicates more political influence. The major hypothesis to be tested is that the extent of the deviation is affected by political factors arising from the economic variables included in the model.

Firm size and seller concentration are expected to increase political influence and thus their coefficients are predicted to be positive. Ambiguity exists in the case of the remaining variables, consequently a two-tail test of significance is utilized.

Tests indicate that a multiplicative regression model is more appropriate than an additive model. The empirical tests are based on regressions of the industry and firm characteristics of 110 IRS "minor industries" for 1963 on the deviation of the effective tax rate from 0.52.

Expressing the model in terms of natural logarithms, multiple linear regression techniques can be used to estimate the equation. The results, based on a least square regression are:

---

[f]This assumes a profit maximization goal.

[g]A full explanation of the model is presented in Appendix H.

$$ln(0.52 - t) = 3.541 + 0.084 \ ln \ F - 0.095 \ ln \ I - 0.512 \ ln \ C - 0.017$$
$$(2.42)^{**} \qquad (1.72) \qquad (2.94)^{**} \qquad (0.15)$$

$$ln \ D - 0.301 \ ln \ R$$
$$(2.46)^{*}$$

$$R^2 = 0.158 \qquad N = 110 \qquad F = 3.89 \qquad (t \ \text{Values})$$

*Significant at the 0.05 level
**Significant at the 0.01 level.

The firm size variable is positive and statistically significant and indicates that increases in firm size are associated with larger deviations in the effective tax rate; a one percent increase in absolute size leads to a 0.08 percent increase in the deviation. This result lends some support to the popular idea that larger firms are more able to influence policy directions for their own advantage. The market concentration variable, hypothesized to have a positive impact, is significant but carries a negative sign. The profit rate variable is negative and statistically significant and supports the argument that high profit industries have more to lose from reductions in their tax rate because of the possibility of attracting public attention and thus are less likely to engage in efforts to lower the rate than are lower profit industries.

To summarize: the model linking structural features and interindustry variation in the effect of political influence on public policy receives limited, but important empirical support. The most important finding is the positive relationship between absolute firm size and the deviation of the effective average corporate income tax rate from the norm.[h]

Application to Energy Industry

As discussed earlier, a substantial number of mergers have occurred in the energy industry. In addition to the competitive effect of such mergers, society is concerned with the impact of the mergers on the distribution of political influence. An evaluation of the political impact of the mergers with the Salamon-Siegfried model is complicated by the fact that any given merger simultaneously affects several of the independent variables in the model. Predictions as to the impact of energy industry mergers on political influence, via their predicted impact on effective tax rates, would contain an unavoidable element of speculation.

There is evidence to indicate that the petroleum industry has already

---

[h]This finding is more impressive in light of several biases in the test which run against observing a positive relationship. See Appendix H.

been the beneficiary of favorable tax policy. The tax rate predicted by the Salamon-Siegfried model and the actual tax rate for four energy related industries is presented in Table 4-10. Observed tax rates are substantially below the predicted values. In fact, of the 110 industries used in the analysis, only three other industries were able to escape taxation to a greater extent than the energy related industries. This suggests that substantial political influence exists with the current size distribution of firms in the petroleum industry.

Motor Fuel Taxes
The finding that firm size is positively related to political influence is examined further by analyzing the relationship between the petroleum industry's economic strength in a given state and the level of state excise taxes on motor vehicle fuels in the state. It is hypothesized that tax levels vary inversely with the relative size of the petroleum industry in a state and with the extent to which large firms dominate the petroleum industry in the state.

Multiple regression results offer some support for the hypothesis. The results indicate that excise taxes are lower in states where petroleum refining is relatively more important and in those states where the dominance of large refining establishments is greater.[i] These findings are consistent with the earlier results which indicated a positive relationship between absolute firm size and political influence.

The Salamon and Siegfried results suggest a relationship between absolute firm size and political influence. This finding is based on regression results for 110 IRS minor industries in the mining and manufacturing sectors and can only be considered as tentative and the impact of firm and industry size on state gasoline excise tax levels. As a result, mergers, which by definition increase firm size, would be predicted to increase an industry's political influence. Such predictions are difficult to make because of two factors:

Table 4-10. Predicted and Observed Effective Average U.S. Corporation Income Tax Rates for Four Selected Energy Related Industries for 1963

IRS Code	Industry Description	Predicted Tax Rate	Observed Tax Rate
1100	Coal mining	32.3	18.2
1310	Crude petroleum, natural gas & liquids	41.9	24.1
2911	Petroleum refining without extraction	34.5	25.0
2912	Petroleum refining with extraction (the "majors")	37.4	19.8

Source: Salamon and Siegfried, *op. cit.*

---

[i]See Appendixes.

(1) large-scale mergers would fall beyond the range of the data used to estimate the Salamon-Siegfried model and (2) the petroleum industry already appears to possess significant political influence and it is not clear if any marginal addition to that influence would have any significant effect. However, mergers of petroleum firms with coal and/or uranium firms may indeed have real political consequences. In such cases, the danger is that the political influence of petroleum firms will be transferred to these other areas. This possibility, while speculative, suggests that such mergers contain political consequences independent of any economic effects.

NOTES

1. For a policy makers discussion of the problem in regards to the energy sector see: Statement of Dr. H. Michael Mann, Director, Bureau of Economics Federal Trade Commission before the Subcommittee on Anti-trust and Monopoly, June 27, 1973.
2. Various estimates of the cost of these policies are available. See: M.A. Adelman, "Efficiency of Resource Use in Petroleum," *Southern Economic Journal*, October 1965, p. 105; W.F. Lovejoy and P.T. Homan, *Economic Aspects of Oil Conservation Regulation* (Baltimore: The Johns Hopkins Press, 1967), and Cabinet Task Force on Oil Import Control, *The Oil Import Question* (Washington: Government Printing Office, February 1970); Charles Chicchetti and Wm. Gillen, "The Mandatory Oil Import Quota Program: A Consideration of Economic Efficiency and Equity," *Natural Resources Journal*, 1973.
3. E. Mansfield, *Industrial Research and Technological Innovation* (New York: W.W. Norton, 1968).
4. John Enos, *Petroleum, Progress, and Profits* (MIT Press, 1962), p. 234.
5. Rankings were obtained from five R&D executives in the petroleum industry and seven professors of chemical engineering at major universities. Close agreement as to rankings occurred.
6. Natural Science Foundation, *Research and Development in Industry* for 1960, 1963, 1966, and 1969, Washington.
7. John Enos, "Invention and Innovation in the Refining Industry," *The Rate and Direction of Inventive Activity*, National Bureau of Economic Research, 1962.
8. Edwin Mansfield, *Industrial Research and Technological Innovation* (New York: W.W. Norton, 1968).
9. D.L. Baeder, "Research and the Emperor's New Clothes," paper given at the Sixth Industrial Affiliates Symposium, Stanford University, May 15, 1973.
10. *Ibid.*
11. See: *Coal Age*, October 1972, p. 139.
12. Enos estimates an average lag of 11 years between invention and dillusion for major cracking processes but that the average lag in petroleum

may be somewhat shorter than the average lag for 35 innovations in other industries. Enos, *op. cit.*

13. See: Edwin Mansfield, *The Economics of Technological Change* (New York: W.W. Norton, 1968), pp. 114-125 and Edwin Mansfield, "Technical Change and the Rate of Innovation," *Econometrica,* 1961.

14. E. Mansfield, J. Rapaport, J. Schnee, S. Wagner, and M. Hamberger, *Research and Innovation in the Modern Corporation* (New York: W.W. Norton, 1971).

15. Anthony Romeo, "Interindustry Differences in the Rate of Diffusion of an Innovation." Unpublished Ph.D. dissertation, University of Pennsylvania, 1973.

Chapter Five

Public Policy

The present political climate is not only unsuited for, but also threatening, to formulating a long run, rational energy policy in the U.S. In a democracy, economic policy is formulated in the political process. This is as it should be but situations, where political rewards are such that short run policies with vast amounts of popular appeal are favored over alternative policies, are indeed dangerous. The strong belief that certain products, including energy should be available at low prices is held by many and leads to much rhetoric and confusion of basic economic issues. Increased energy supplies in the long run are available only at higher costs and the feeling that energy prices should be rolled back to some lower level does not negate this fact. Situations in which political leaders offer blatant appeals to the public's ignorance of basic economic factors is not only irresponsible but creates a real possibility that legislative actions will lead to even greater problems in the future. This is a classic example of the conflict between correct economic policies and rewarding political strategies. It is difficult to underestimate the risks and dangers if policy makers were to succumb to short run political objectives in formulating energy policy.

The present atmosphere is also diverting attention from the need to recognize that energy policy means more than oil policy. Domestic and international events have focused attention on oil and made energy problems appear to be limited to oil. This neglects problems such as natural gas pricing methods, the development of synthetic fuels, and the entry of oil firms into coal and uranium. Events have combined to focus attention on oil but clearly the proper response is an energy policy rather than simply an oil policy.

FINDINGS

The structure of the market in each fuel is, with one exception, only moderately concentrated but there is evidence of a trend towards higher concentration levels. In spite of the trend, concentration levels have not reached a level where one would feel certain that market forces, if allowed to operate without government interference, are incapable of allocating resources efficiently. Present market structures are not monopolistic. The exception to this occurs in the case of uranium where concentration levels are high due to the current small size of the market for uranium. In the case of oil, it is clear that a traditional structural analysis fails to reveal the complete nature of competition. The extent of seller interdependency, as represented indirectly by concentration levels, is significantly understated due to the large number of joint ventures among the top firms and the impact of OPEC's actions on domestic price levels is a crucial factor. Thus, the conventional interpretation of concentration ratios is not warranted in the case of oil.

Coal

Coal production has historically been highly fragmented with small producers being an important source of coal. The situation has changed dramatically in recent years. Electric utilities are currently the sole major user of coal. Small coal producers are at a distinct disadvantage in meeting the demands of utilities because of the need to offer long term supply contracts to utility buyers. Large producers can do this and, in addition, are able to attain economies in transportation with the use of unit trains. As a result, small producers are restricted to dwindling spot market sales of coal. Consequently, changes in the end use pattern of coal has placed small producers at a disadvantage and has allowed large producers to capture increasingly large amounts of the market.

Concentration levels in coal production are best described as moderate. The most important finding is the existence of an upward trend in concentration. Concentration levels in 1970, however, have not reached a level where they would be considered likely to lead to significant reductions in the forces of competition. The upward trend however cannot be simply dismissed as of little concern.

Entry barriers in coal have also been affected by changing consumption patterns. In order to compete for long-term contracts for utility sales, very large coal reserves must be accumulated. Capital requirements are also increased by the existence of scale economies in mining as reported by various engineering estimates. Even in light of these factors, however, it is likely that entry barriers are only moderate. It is no doubt true that many coal firms find the capital costs of competing for long term, utility contracts far greater than they are accustomed to. This may create a serious obstacle to existing firms but does not necessarily mean that entry barriers, as judged by noncoal firms, are high.

There is little reason to believe that coal consumption patterns will change in a manner that again favors small producers. If anything, innovations such as goal gasification and liquefaction plants will confer even greater advantages on large mines. This does not mean, however, that efficiency in coal production necessitates very high concentration levels in the future. Large absolute size is required to meet the coal needs of a gasification plant, but evidence indicates that such a mine would not be large in a relative sense. Some upward trend in concentration will persist in the long run but the trend may be offset, to some extent, by increased output from Western coal fields where the major owners of reserves are different from the dominant operators in Eastern coal fields.

The major findings are: (1) concentration levels are moderate but increasing at a fast rate; (2) much of the increase is due to shifts in the end use pattern of coal; (3) oil companies have significant positions in coal but oil companies have not monopolized coal production; (4) a shift towards greater production from Western coal fields seems likely; and (5) the charges of a price and output conspiracy following oil company entry into coal are unfounded.

In conclusion, workable competition in domestic coal production appears to be the case. The major danger to competition lies in the continuation of the observed trend towards higher concentration. It does appear that the changing economics of coal production makes it unlikely that the traditional small coal producers will be able to remain competitive. Consequently, it is likely that noncoal interests will be attracted to the industry and account for much of the expansion in coal production.

The prospects for maintaining workable competition in coal are good. Shifting production from eastern to western coal fields in response to the demand for low sulfur coal, will cause some firms, currently not among the top producers, to become more important producers. The U.S. government, as a landowner, controls access to significant amounts of the western coal deposits. The manner by which these reserves are opened to commercial operations can have important competitive implications; the use of a royalty charge favors smaller operations who are not as able to make the large capital payments associated with a bonus bid system. The royalty payment system has clear advantages in terms of promoting competition, but may lead to some resource waste by making it unprofitable to recover low-grade coal seams.

Natural Gas

The price of interstate sales of natural gas is determined administratively by the Federal Power Commission. Consequently, demand and supply forces are limited to an indirect influence on price levels. The major policy question in natural gas concerns the issue of whether to deregulate prices. This topic is the concern of a separate study commissioned by the Energy Policy Project.

The evidence presented in this book indicates that levels of seller concentration in natural gas production are low with no clear trend towards higher concentration. There is limited evidence to indicate that concentration levels in intrastate markets, where gas prices are not regulated, are not significantly above levels in the interstate and regulated markets. This implies that deregulation, per se, is not likely to cause significant upward movement in concentration. In addition, the size distribution of natural gas production indicates the existence of numerous, relatively small producers and suggests that entry conditions are easy.

In conclusion, the structure of the natural gas market, presently a regulated market, is characterized by (1) low, and relatively stable levels of seller concentration and (2) the presence of numerous small firms. These structural features hardly indicate monopolization.

Uranium

Concentration levels are high in the case of uranium mining and milling and suggest that competition is not likely to be effective. A significant amount of excess capacity exists due to the slower growth of nuclear power and has resulted in a correspondingly slow growth of demand for uranium. Only recently has the AEC stopped its stockpiling of uranium fuels and uranium producers now must rely solely upon private sales. Whether this will force a rationalization of capacity remains to be seen. In any case, present concentration levels raise serious doubts as to the effectiveness of competition. Present high levels of concentration can be expected to decline somewhat as the size of the market for nuclear fuels grows but, given the uncertainty with the rate at which nuclear power will develop, it is difficult to predict the timing and extent of concentration reductions and the type of market structure most likely to result.

Oil companies have penetrated heavily into all phases of uranium production. Companies with oil operations account for large proportions of total drilling activity, ownership of uranium reserves, and mining and milling. Such involvement virtually assures that oil interests will continue to be the dominant force in uranium for a long time to come.

The major findings are (1) the uranium market currently has excess capacity, (2) over longer periods, the market is expected to expand and absorb the excess capacity, (3) at the present, concentration levels are high and suggest that competition is weak and (4) oil interests have dominant positions in uranium production.

Oil

Given its relative importance as an energy source, it is not surprising that a major part of this book has focused on oil. The Middle East embargo, rising prices, and shortage of crude oil and petroleum products have combined to focus nearly all policy discussion on oil.

This book focuses on an evaluation of competition in domestic energy production. It is clear that recent events in world oil markets have had and will continue to have important ramifications on the domestic market.

Consequently, policy for the domestic industry must be constructed in a manner that recognizes recent events such as the Arab embargo. Policy choices must be determined in terms of both economic and political models.

The pattern of resource allocation in domestic crude oil production has been influenced by market forces and a multitude of government policies. Many of the government policies, such as demand prorationing and import quotas, had the effect of reducing the influence of free market forces on crude oil prices. The pattern of government intervention has generally resulted in lesser amounts of competition in the market.

The findings with respect to market structure indicate (1) moderate levels of seller concentration in 1970, (2) a rapid increase in concentration over the 1953-1970 period, (3) intensive use of the joint venture form of organization both in bidding for leases on public lands and in production, and (4) some evidence to indicate that the use of joint ventures lowers entry barriers to crude oil and natural gas production. While the conventional interpretation of concentration levels does not apply to oil because of the seller interdependency created by joint ventures, it does appear that, on net, the structure of domestic production does not indicate monopolization.

A review of conduct in the oil industry suggests a different conclusion. The essence of monopoly power is control over supply as a means of making goods more scarce and gaining excess profits. The analysis of structure suggests that such power is not within the domain of private firms. However, monopoly power or control over supply has been exercised by the federal government on behalf of the industry in the form of (1) market demand prorationing, which limits domestic output and (2) oil import quotas which limit the amount of foreign oil entering the domestic market.[1]

Demand prorationing has been responsible for allowing many marginal and noneconomic resources to survive in the oil industry. Today, rising demand has eliminated the excess capacity that resulted from and was protected by demand prorationing. However it does not follow that demand prorationing is a nonissue. The issue is the premise that underlies its use, namely that the competitive mechanism cannot be relied upon to allocate resources in crude oil production. If periods of excess production capacity were to occur in the future, market demand prorationing could easily be used to support crude oil prices.

The protection from foreign competition due to import quotas has led to inefficiency in domestic oil production. The present fee system, given its current structure, offers only limited amounts of protection to the domestic market. Clearly, the elimination of quotas has been procompetitive. Demand prorationing and import quotas had the effect of supporting and creating monopoly power at the crude oil stage. Such power is usually associated with a tight oligopolistic market structure characterized by high concentration levels and entry barriers. In oil, public policy has been a major source of such power. Thus the government has been the instrument of monopoly power on behalf of the industry, in the absence of a private ability to successfully control supply. While demand prorationing and import quotas are, for all practical purposes, currently nonoperative they are still significant for two reasons. One, they indicate the ability of the industry to influence government policy makers and two, the policies could easily be restored if the need arises to support domestic prices.

Currently, oil companies are benefiting from the actions of another government body exercising monopoly-like control over output and crude oil prices, namely, OPEC. As a result of OPEC's cartel actions, world prices have risen rapidly and, in order to attract investment, domestic prices on new crude oil have had to rise to meet the higher world prices. Consequently, oil company profits have increased at an equally rapid rate. But it should be noted that this is a response to OPEC's actions and not to private monopoly power.

In spite of the recent explosive rise of oil company profits, long-run profit performance of the oil industry is about the same as that of other manufacturing industries. The rate of return record provides no clear evidence of monopoly profits, but this is not conclusive evidence that monopoly control by government actions over supply was insignificant. Government policies which restricted supply in order to support domestic prices increase profitability only at the time at which they are instituted. Subsequent to their initial application, the lack of entry barriers causes marginal, higher cost, resources to enter and costs rise to meet the higher price level thus the profit gains are eliminated. But, the result is an inefficient use of resources.

To summarize: there is evidence to indicate that government instituted, and industry supported, policies have led to monopoly-like control over domestic crude oil output and prices. The actions of another government body, OPEC, have recently affected events in the domestic market.

Conclusion
On the basis of the findings, it can be concluded that the structure of the energy industry and the individual fuels, with the exception of uranium, is not monopolistic. The structure is not perfectly competitive but it appears to be sufficiently competitive to yield competitive performance. The long-run preservation of competition requires that entry barriers be low. In this way, any monopoly power exercised by established firms will attract new firms and thus lead to a restoration of competition. The use of joint arrangements in bidding for federal oil and leases acts to lower entry barriers and, on net, is procompetitive. The effect of joint activities in other areas is not easily determined and no certain conclusion can be reached.

Evidence of private monopolization of energy resources by oil companies is lacking. Oil interests have established substantial positions in coal and uranium but concentration levels remain low. In addition, oil interests have been aggressive entrants while other potential entrants, such as chemical companies, have not shown much interest. There is evidence to indicate that government decision-makers have historically exercised monopoly power for the industry over domestic output levels. Thus, they have reduced the influence of free market forces on the pattern of resource allocation in the production of crude oil.

COMPETITION POLICY

This book is addressed to a limited aspect of energy policy, namely the state of competition in domestic production of fossil and nuclear fuels. The findings are applicable to the question of the effectiveness of competition in these markets and the role of public policy in ensuring an effective working of market processes.

It should be emphasized at the outset that a lack of meaningful data has been a major constraint. Deficiencies such as limited information concerning the ownership pattern of fuel reserves, and the nearly total lack of insight into the internal operations of major energy companies, are serious obstacles to formulating long-run public policies. This lack of information to policy-makers makes it only too likely that past errors in policy selection will be repeated. This is a risk and cost that society currently bears but, with increased flows of information to policy makers, such risks could be substantially reduced.

The general conclusion of this book is that government imposed restrictions on free markets are responsible for many of the energy problems facing the U.S.[2] Government intervention has ranged from the adoption of administrative decision-making as a substitute for market processes in determining price, as in the case of natural gas, to attempts to influence the pattern of resource allocation by various subsidy programs. Examples of the latter are market demand prorationing and the use of import quotas in oil.

A major effect of government intervention has been to change the incentives or profit opportunities available to firms. In other words, firms have faced a different set of incentives that would have existed without government intervention. As an example, the use of foreign tax credits had the effect of causing investment in new oil capacity to be more attractive, i.e. profitable, in foreign markets relative to the U.S. Oil company investments responded accordingly, leaving the U.S. in the position of having a shortage of refining capacity. This movement was advanced further by the fact that after years of stimulating U.S. oil production through tax subsidies, the remaining prospects in the U.S. were poor relative to foreign oil prospects. Resource allocation problems exist in most of the energy sources not because market mechanisms failed to operate but because of government policies, both federal and state, imposed on the market.

The higher prices and shortages of energy in the U.S. are easily observed. In such situations, a strong tendency exists among the public, to respond by imposing even more comprehensive controls and restrictions on the market mechanism. This presumes that observed problems were due to the inability of market forces to carry the allocation function. Much of the problem, however, is traceable to government policies rather than to market failure. In the case of oil, market forces influenced resource allocation, but government-induced change in the set of incentives in the market had the effect of causing investment to flow away from the U.S. market. The problem lies more with the structure of incentives resulting from government intervention than with an inability of markets to allocate resources.

The unacceptable past performance of policy-makers seems to be scant support for the position that greater regulation and less reliance on the market mechanism is now required. The inability of the FPC to set natural gas prices at a level that approximates a market clearing level is a prime example of government's contribution to present energy problems. In addition, the Congressional program, administered by the FEO, to reallocate crude oil supplies from majors to independents contained strong disincentives for increased imports of oil. Such examples illustrate the difficulty, and perhaps impossibility, of handling such complex production and distribution problems by administrative decision-making. However, the lessons of history appear to receive little attention as the public continues to support increased regulation as the solution to present problems.

A final point is appropriate. There is much current talk by the oil companies of a desire to scrap price controls and return to free markets as a solution to energy problems. It should be pointed out that the oil companies' definition of a free market differs substantially from and has little resemblance to the concept of free markets or competition in economics. Oil markets in the U.S., characterized by extensive government intervention to suppress or alter the price mechanism, are prime examples of what free markets are not. In an economic concept, prices set by freely fluctuating demand and supply forces are the essence of free markets. Oil markets bear scant resemblance to this concept.

Economic models are only partial guides to policy formulation for several reasons. In the area of energy two of the more important are: (1) a lack of public data to examine the current situation and (2) the important role of noneconomic forces on markets.

Intrafuel Competition

Coal. Coal production was found to be effectively competitive and the upward trend in concentration mainly due to the industry's adjustment to changing demand patterns. Thus,

1. no divestiture of existing producers is necessary;
2. antitrust action should not be taken to halt the upward rise in concentration unless it can be demonstrated that efficiency gains are not present;

3. small coal producers may continue to exit the industry—this is consistent with workable competition;
4. access to coal reserves on public lands should be designed to maintain and promote competition.

Uranium. The current situation in the uranium market is similar to an "infant industry" situation. The current market size is small and, consequently, concentration levels are high. The major policy question in this area concerns the creation of a competitive market for uranium enrichment services. The realization of this goal depends upon the type of technology selected for the new plants.

In order to promote competition, it is recommended that

1. the centrifuge technology be adopted for new enrichment plants;
2. the AEC ban on imports of U_3O_8 be lifted.

Natural Gas. The major policy problem in this area concerns the deregulation of prices. This question is outside of the scope of this book but it should be noted that this study, like most others before it, finds that the current structure of natural gas production is consistent with effective competition.

Petroleum. It was reported that, while the structure of domestic crude oil production fails to indicate private monopoly, an analysis of conduct indicates that government policies have led to monopoly-like control over domestic output. The following policies are recommended:

1. the phasing out of incentives for increased production;
2. the elimination of authority for government controls over domestic production;
3. a ban on the use of joint ventures involving dominant firms in bidding for oil and gas leases. An outright ban on all joint ventures is not warranted.

Independent Operators.[a] The independent or nonintegrated operator has been a major source of competition in petroleum especially at the refining and marketing stages.[3] To an important extent, independents have been able to survive by purchasing crude oil and/or gasoline due to capacity imbalances between stages of integrated firms because tax advantages caused majors to emphasize crude production as the main profit center. In addition, independents received preferential treatment under the Mandatory Oil Import Control Program due to the sliding scale of allocating import tickets. On the other hand, prorationing, by keeping domestic oil prices higher than the market level up to 1971, had a greater adverse effect on independent refiners because

[a]For a discussion of the future prospects for the independent sector see: Appendix F by John Lichtblau.

integrated refiners benefited from the enhanced value of the depletion allowance. One question concerns the impact of the fee system for influencing the imports on the independent sector.

The immediate effect is to remove a source of advantage independent refiners enjoyed over major firms. The fee schedule discriminates against petroleum products to a greater extent than it does against crude oil imports. The purpose is to stimulate domestic refinery construction. Import competition in products can be expected to be greater under the fee system because of the relatively low fee structure and the elimination of the severe restrictions on gasoline imports present under quotas. Thus, companies with foreign refining capacity may have an advantage over companies with only domestic refining capacity.

Developments in international markets indicate that major companies have lost and will continue to lose much of their control over world oil production and price levels to host governments. In addition, if Congress reduces the percentage depletion allowance below the current 22 percent level, relative posttax profitability may shift in favor of refining and marketing. Any tendency of the majors to take profits at the crude stage and to subsidize downstream operations would be weakened.

Independents should benefit from these changes if they can obtain crude oil supplies and capital necessary for expansion. These are significant restraints, given events in the Middle East. It appears that the independent sector has greater opportunities, but serious doubts exist as to whether crude oil availability will allow them to respond. Much depends on decisions by Middle East producers. In addition, Congress may attempt to legislate their existence by guaranteeing independents access to crude oil stocks of major companies or independent crude producers.[b] Thus, crude oil availability is likely to be determined by political as well as economic factors.

Vertical Integration. It is commonly believed that antitrust action should be taken to break the integrated structure of major petroleum firms. While such action has much popular appeal, for a variety of reasons, it is not clear that it is warranted or, if taken would be effective. Vertical integration, per se, does not create market power; the existence of monopoly power depends on the structural characteristics at each point in the integrated chain. If competitive conditions prevail at a given stage, economic analysis reasons that the simple fact that some or all of the firms operating at a given stage also operate at other stages doesn't affect the extent of competition. However, if vertical integration

---

[b]Congress has passed legislation designed specifically to ensure that independent refiners gain access to price controlled and low cost domestic crude oil supplies. The Emergency Petroleum Allocation Act of 1973 calls for the "preservation (of) the competitive viability of independent refiners, small refiners, nonbranded independent marketers', and branded independent marketers." See: Appendix F by Lichtblau.

encompasses a stage where market power exists, integration can be used as a vehicle to transmit that power to other stages, as in the case of market power at the crude oil stage due to tax provisions. The proper policy response is to eliminate the source of the market power. It should be recalled that government policy has been the mechanism by which monopoly power has been created.

Efficiency arguments in favor of extensive vertical integration in oil do not appear overwhelming. Independent operators have existed alongside integrated firms at each stage. In fact most of the pipelines represent joint ventures rather than wholly owned subsidiaries. It seems that economic reasons for the extent of vertical integration are lacking. A program of divestiture aimed at separating the various functions would receive popular support, but whether the gains in competition would be sufficient to justify such action is debatable. It could be argued that given the elimination of government controls on the market mechanism, present market structures are consistent with effective competition. The elimination of vertical integration without other changes in the oil industry may produce only marginal gains in competition. For instance, divesting crude oil production from downstream operations would not have reduced the market power that existed at the crude stage. In any case, courts are not likely to find such massive restructuring as acceptable when the gains are questionable. Certainly, other drastic policy choices should be considered first.

Interfuel Competition

The general policy conclusion with respect to interfuel competition is that oil entry into coal and uranium has not resulted in monopolization of the nations energy supplies, and antitrust action to halt further entry and/or require divestiture, is not economically justified on the basis of the current situation.

As reported in this study, increased interfuel competition has led to oil company entry, via merger and *de novo* entry, into coal and uranium. Unfortunately, the exact extent of oil's position in these areas cannot be ascertained because public information about the pattern of reserve ownership is not available. The simple entry of oil firms into coal and uranium, when viewed from a structural setting, doesn't appear to present substantial threats to competition. Concentration levels in a combined fuels market, an energy market, remains low.

However, when viewed from the perspective that oil firms are entering from an industry where policies supported by the industry and implemented by government decision-makers have been successful in off-setting the forces of free competition and fostering great inefficiencies in resource use, one has to be concerned that the oil companies' concept of free markets and competition will be transferred to coal and uranium. The demonstrated ability of government decision-makers to adopt policies favoring the private interests of the companies is cause to be concerned with oil's entry into other fuels.

An outright ban on the entry of oil firms into coal and uranium probably has little economic justification. Such entry into coal, while posing a danger to interfuel competition, has also had procompetitive effects in several instances. In any case, oil companies do not dominate coal production. The greatest danger to interfuel competition occurs in uranium. Whether oil's penetration will be reduced when the size of the uranium market increases is not clear but oil domination of uranium in the future seems to be more likely than in the case of coal.

In developing a policy to promote interfuel competition, the following steps are recommended:

1. the collection and public dissemination of complete information on the extent of involvement by oil companies in other fuels;
2. *de novo* entry should be preferred over entry by merger;
3. nonoil entry should be a clear policy goal;
4. leading firm mergers should be prohibited;
5. significantly more restrictive policy is applicable in the case of oil entry into uranium, unless it can be shown that nonoil entry is not available;
6. if oil's penetration into uranium increases, antitrust action to achieve divestiture would be required.

This book reports that the energy company development has not substantially altered the structure of the energy market and that the current structure is compatible with effective competition. This finding, by itself, does not necessarily mean that effective competition exists either in the energy market or in the market for individual fuels. As was discussed earlier, the record of various government interventions into the crude oil market indicates that institutional features are major determinants of competition in energy production. At this time, the greatest threat to effective interfuel competition is more likely to be in the form of government imposed, and industry supported, policies rather than monopolization by private energy companies.

The U.S. faces a crucial decision as to whether resource allocation in the production of energy is to be determined by free market forces or by a combination of company-government decision-making. There is a danger that public debate on the choice will never materialize and that a choice will be made implicitly by incremental policy choices.

The competitiveness of domestic energy production is affected by events in the world oil markets. The role of the international oil companies is changing at a fast rate and they can be expected to yield greater control to host governments in the foreign lands where they operate. The loss of control and profits from crude production is certain to place greater emphasis on refining and marketing activities. Because events in the world market are politically determined to a large degree, predictions are necessarily risky. What is clear however is that the U.S. is richly endowed with energy sources. The task is to decide the manner in which resource allocation decisions in developing the resources will be made. History indicates that government has consistently intervened to weaken market forces in domestic oil and natural gas production and that such intervention has contributed heavily to present and future problems. U.S. energy supplies are not dominated by a few firms and, with appropriate public policy, effective competition in energy markets can be realized.

Long Run Role of Competition

The manner of decision-making, market or nonmarket, in regards to resource allocation in energy becomes very important when one considers the long run. The very-long-run problem is to ensure that the U.S. makes the transition to alternate energy sources in an optimal manner. This means changes in resource allocation in the development of new energy sources in a manner and sequence consistent with efficient utilization of present and future energy supplies. Development should follow the least cost path. Development involves a waste of resources if the least cost path is not followed. For example, the timing of nuclear power, if based on its economic viability relative to fossil fuels, might have been quite different. Nationalism and regulatory policies favoring capital intensive production have combined to cause a premature development of nuclear power.[4] Professor Nordhaus has attempted to simulate the least cost path of new energy development in the U.S. and concludes that, on economic grounds, nuclear power is not appropriate until early in the 21st century.[5]

Market prices, which reflect relative scarcity values, can provide the information necessary for optimal decision-making in developing energy resources. Professors de Chazeau and Kahn note:

> It is the crucial strength of a private enterprise system that it permits flexible adaptation to unknown future problems as they evolve by permitting the individual and firms to undertake whatever market function opportunity recommends without arbitrary disbarment, excepting only a clear transgression of the public interest.[6]

To carry out this function, effective competition must exist; price distortions due to market power or government intervention lead to inefficient utilization of scarce resources.

The Nordhaus study was designed to "find the optimal path for consuming scarce resources and the prices associated with this path."[7] The study compares the set of relative prices of energy sources consistent with optimal resource development to actual market prices. Using 1970 as a base, final demand prices consistent with optimal resource development were reasonably close to market prices with the major exception of petroleum prices which, for the U.S. in 1970, are reported as some 269 percent of the price predicted in the model.[8] Another exception is natural gas where the actual 1970 price, relative to the efficiency price, was underpriced by at least 20 percent.[9] In both of these cases, government intervention, either to prevent price from allocating resources or to influence the pattern of resource use, has suppressed market forces. It is interesting to note that the discrepancy, between the market and optimum price of crude oil in Western Europe was significantly less than for the U.S. The reason, according to Nordhaus, was the lack of import quotas and market impediments in European markets.[10] The finding of Nordhaus that in 1970 market prices of energy sources, except petroleum and natural gas, were fairly consistent with the prices associated with optimal utilization of energy sources indicates the ability of competitive markets to provide prices which reflect true scarcity values. The deviations of actual and optimum prices indicate the distortions, and costs, resulting from government suppression of forces.

From his analysis, Nordhaus makes the following point:

> *As a long run policy* it would be unwise to jack up the prices of energy products in the interest of artificially preserving energy resources. Nor does a more drastic policy of permanent rationing of energy resources make sense.[11] (emphasis in original)

The results are heavily dependent on two assumptions: (1) the existence of free trade in energy resources and (2) the pattern of optimal resource use depends partly on unproven technologies. The former is especially significant because of recent Arab actions. Combining such political actions with economic factors would undoubtedly lead to differences in the optimal pattern of energy prices and development in the U.S.

Alternative Explanations[12]

The principal characteristic of the energy crisis, in the short run, is the widespread shortage of energy. At present prices, shortages exist in natural gas, crude oil, gasoline, and occasionally other petroleum products. In addition to the shortages, rapidly rising prices of all energy supplies are a major ingredient of the energy crisis.

A possible explanation of the shortages and higher prices is a conspiracy and monopolization of the nation's energy supplies by American oil companies. The evidence presented in this book indicates that the structure of the energy market and its subsectors is not monopolistic. Yet shortages and higher energy prices persist and there is a need for alternative explanations. To a significant extent, current conditions can be explained in terms of past and present policies in combination with events over which the U.S. has little or no control.

Natural Gas Shortages. Present buyers of natural gas can generally obtain the quantities they wish at current prices but potential buyers of gas, not currently buying gas, are often unable to get commitments for new gas supplies. Shortages are generally estimated to be about 25 percent of domestic demand.[13]

The explanation of the shortage is generally believed to be with the pricing decisions of the Federal Power Commission. Prices set by the FPC, it is argued, have been held below what economists call the equilibrium level, i.e., the level which clears the market or equates the demand and supply. At prices below the equilibrium price the quantity demanded exceeds quantity supplied—the result is a shortage. Thus, there is no need for a conspiracy theory to explain the shortages of natural gas.

The solution to the energy crisis in natural gas is straightforward: remove price controls and permit the market to allocate scarce resources. Prices will be higher, but after an adjustment period, there will be no shortage. Higher prices provide an incentive both for consumers to economize on their use of gas, and for producers to expand output. MacAvoy and Pindych suggest a phasing out of FPC control of natural gas by 1979, when their econometric analysis indicates that supply would equal demand at a wellhead price of 62¢ per mcf.[14]

Crude Oil. Crude oil shortages are deeply rooted in past public policies. The federal government has been providing tax subsidies for oil and gas production for a long time. Prime examples are the depletion allowances and provisions for expensing of intangible drilling costs.

The economics of such subsidies is to initially increase profits in the industry. The supply curve shifts to the right causing prices to fall. As a result of a government "low price" policy for oil, consumers increase consumption because petroleum and petroleum products are now bargains relative to unsubsidized prices and in addition, the development of alternative energy sources that are either not subsidized or receive a lower subsidy, are delayed.[c] The effect is to speed up the rate at which the U.S. has depleted its crude oil supplies.

A faster rate of depletion of domestic crude reserves was one effect of the oil import quota program. By restricting imports, additional domestic production was substituted for what might have been obtained from foreign sources. The effect of both policies, tax subsidies and the import quotas, has been to stimulate domestic output and to accelerate the energy crisis of the 1970s. Domestic crude oil production during the 1950-1970 period increased at a compound annual rate of about 3 percent, but domestic production peaked in November 1970 at a rate of 10.0 million barrels per day. With the exception of the period of import quotas, U.S. policies pursued a low price policy and even the quotas, while having an upward effect on domestic prices, acted to increase domestic production. These policies contributed to the current crude oil shortages.

Further shortages were caused by the Arab petroleum exporting countries when, on October 17, 1973, they instituted a five-month embargo on

[c]Oil shale and coal receive relatively low subsidies.

shipments to the U.S. and other countries. Some Arab countries cut production by as much as 25 percent to support the embargo. The effect was to create further shortages and to cause world oil prices to skyrocket. Prices of crude oil from non-Arab countries such as Canada and Venezuela rose to the higher level and the effect was transmitted to the U.S. market.

Prior to the Arab embargo and the higher prices in the world oil market, the U.S. price control system, which was abandoned for most products on April 30, 1974 but continued for petroleum products until February 1975, appears to have been responsible for increased shortages. The price control system provided for a "pass through" of higher cost foreign crude but over time, various exceptions to this provision made it difficult to actually pass the higher costs to consumers. Thus, the incentive to import was reduced.

Further, the Federal Energy Office required companies with more than average crude supplies to sell part of their "excess" to crude deficit companies. The required selling price was set at about $8 per barrel; a price which was between $1 and $5 less than the cost of marginal supplies of foreign crude delivered to the U.S. Crude surplus firms were thus forced to subsidize competitors who were crude short. One way to avoid this situation was to sell crude abroad rather than to deliver it to the U.S. where the firm would be forced to sell crude to competitors. As a result, the price control system reduced the supply of crude oil available to the domestic market.

A final factor contributing to shortages, and thus to higher prices, has been the delays in leasing lands for exploration and delays in producing from known fields in the Santa Barbara Channel and the Prudhoe Bay area. The environmental concerns and the delays may be wise public policy but it cannot be denied that they contributed to the shortage of crude oil.

In summary, the observed crude oil shortage can be explained in terms of faulty and conflicting public policies pursued by the U.S. over the last half century, plus the fact of the five month Arab embargo. There is no need for a conspiracy theory to explain the shortage.[d] The most important step that can be taken to eliminate the crude oil shortage is to eliminate price controls. At higher prices, supplies would increase and less would be demanded. Prices would increase. The long-run price effect is uncertain. But without price control, there would be no shortage of crude oil.

Refining Capacity. The U.S. currently faces a shortage of refining capacity. In fact, a shortage has existed since May 1973. In 1971, U.S. oil refineries operated at an average 86 percent of capacity. In 1972, the last year of the oil import quota system, refineries operated at 88 percent of capacity. With the end of the quota system effective May 1, 1973, refinery operations quickly

---

[d]One may charge that the Arab embargo is a conspiracy. However, our concern in this study is with a possible conspiracy among American oil companies and not among foreign governments.

expanded to 94 percent of capacity.[15] The American Petroleum Institute reported U.S. refineries operating at 95.4 percent of capacity in the last week of October 1973, according to the API old definition, and 99 percent of capacity according to the API new definition.[16]

During 14 years of import quotas restricting the availability of crude oil in the U.S., some new refinery capacity was shifted abroad from the U.S. The oil industry has been well aware of impending, then actual declining domestic supplies of crude oil. Government policy in the U.S. until 1973 has been to reaffirm the barriers against import of foreign crudes. While U.S. refinery capacity has increased in the U.S. year by year, the expansion has been less than what would have been true in the absence of the quota system.

To compound the problem of inadequate refinery capacity, American refineries in general have been constructed to process the so-called "sweet" crudes (low sulfur content crudes). In general, U.S. crude oil production has been low in sulphur content compared to most foreign crudes. With the end of the import quota system, free importation became the rule. However, Persian Gulf crudes are characteristically "sour" crudes (high in sulfur content) and it was the Persian Gulf area that was undergoing a rapid expansion of crude oil production. However, the "sour" crudes tend to corrode equipment in refineries constructed to handle "sweet" crudes only. Therefore, American refineries were somewhat constrained in their ability to import and refine foreign crudes even after the import quota system was eliminated.

The environmental concern contributed to the refinery capacity problem in the U.S. New refinery construction, as well as major plant expansions in existing refineries, required environmental clearance and a consequent delay. Without questioning the legitimacy of the environmental concern, it seems obvious that new environmental requirements have contributed to the refinery capacity problem.

The refinery bottleneck is explainable in terms of the factors listed above. Again, we need not resort to a conspiracy theory to explain the problem.

In conclusion, the shortages and rapid rise of energy prices are not attributable to a successful monopolization and conspiracy by U.S. oil companies. Much of the shortages, the rapid rise in prices, and the very high profits in 1973 and 1974 find their roots in past and present policies which promoted low energy prices and accelerated the rate at which domestic energy supplies are depleted. In addition, the actions of the Arab countries, while outside of the direct control of U.S. policy, had a dramatic effect on the domestic market.

NOTES

1. Import quotas were strongly supported by the coal industry and by the oil industry with the exception of the international majors. Texaco, Gulf, and Standard Oil of California all strongly supported free trade in oil and opposed the import quota system in testimony in 1956

hearings on possible import quotas. Hearings held by the Office of Defense Mobilization, 1956, pp. 550, 589-90, 594, 603-04.

2. A similar conclusion has been expressed by Prof. Houthakker. See: Hendrik Houthakker, "Comments and Discussion," *Brookings Papers on Economic Activity* (Washington: The Brookings Institution, 1973), pp. 571-72.

3. De Chazeau and Kahn, *op. cit.*

4. See: Hendrik Houthakker, "Comments and Discussion," *Brookings Papers on Economic Activity 3* (Washington: The Brookings Institution, 1973), p. 572. Given the recent Arab embargo, it is clear that efficiency can not be the sole factor determining U.S. energy policy. *Ibid.*, p. 573.

5. William Nordhaus, "The Allocation of Energy Resources," *Brookings Papers on Economic Activity 3* (Washington: The Brookings Institution, 1973), pp. 529-70.

6. De Chazeau and Kahn, *op. cit.*, p. 567.

7. Nordhaus, *op. cit.*, p. 566.

8. *Ibid.*, p. 557.

9. *Ibid.*

10. *Ibid.*

11. *Ibid.*

12. The following section draws from: Mead, *Competition . . .* , *op. cit.*

13. Mead, *Competition. . . .* , *op. cit.*

14. P.W. MacAvoy and R.S. Pindych, "Alternative Regulatory Policies for Dealing with the Natural Gas Shortage," *The Bell Journal*, Autumn 1973, pp. 454-98.

15. U.S. Department of Commerce, *Survey of Current Business*, April 1974.

16. American Petroleum Institute, Weekly Statistical Bulletin, week ended December 7, 1973.

Competition in the U.S. Energy Industry—Supplementary Studies

Thomas D. Duchesneau, *Editor*

Index of Appendixes

Appendix A

The Geographic Scope of Energy Markets: Oil, Gas and Coal

Thomas F. Hogarty*

The concept of a "market" is one of the most frequently used items in the economists' analytical tool kit; however, applying this concept to real world markets can prove difficult. Thus, the market for U.S. corporate bonds encompasses not only Wall Street, existing bondholders, and firms with bonds outstanding, but also potential bondholders and corporations located in other cities, states, and nations. To be sure, the prevailing price of bonds is determined at the margin by those buyers and sellers active in the market on any given day; however, failure to consider potential sources of supply and demand in any attempt to explain bond prices would clearly be erroneous. In sum, a market exists whenever and wherever buyers and sellers can interact freely such that the prices of identical commodities tend to equality easily and quickly.[1]

This chapter analyzes and describes the geographic scope of the markets for energy, namely, the markets for uranium, bituminous coal and lignite, natural gas, and crude oil. Estimates of the geographic scope of each of these markets are provided, both in terms of their current scope and likely configuration in the foreseeable future.[2]

COMMODITIES AND MARKETS

In the strictest sense of the term, a commodity is characterized by its physical properties and the time and location at which it is available.[3] The price of a commodity is the amount which must be paid now to secure ownership (present

*The author is a faculty member of the Department of Economics, V.P.I. and S.U. This chapter represents an extension of work completed earlier by Kenneth Elzinga and the present author [7].** In fact, the techniques and terminology used here are identical to those developed in the earlier study.

**See Reference No. 7. Numbers in brackets will refer to references, which follow the list of notes.

and/or future) of 1 unit of that commodity. Thus, a specific commodity might be No. 2 Red Winter Wheat available in Chicago no later than September 1, 1975. Moreover, the price of No. 2 Red Winter Wheat will vary as the location and time availability also vary. Thus, in the strictest sense, commodity and price are precise concepts, not readily amenable to empirical examination unless—as in the case of wheat—an organized, central market place exists.

A related concept is that commodities are identified in terms of the utility characteristics possessed by each. Thus, restaurants and kitchens provide similar utility characteristics while differing greatly in physical attributes.[4] Consistent with this view is equality of prices for physically dissimilar and otherwise distinct commodities. For example, two used Chrysler sedans, identical except that one has both air conditioning and greater mileage recorded on the odometer, might sell for approximately the same price. Moreover, these prices might also be the same even if one was purchased in Chicago in May and the other purchased in New York in June. In sum, the strict definition of a commodity is theoretically unnecessary and empirically cumbersome. So long as the commodities are highly similar in the utility characteristics they provide, their prices will tend toward equality and they can be considered as part of one market.[5]

This discussion also shows, however, that any attempt to delineate geographic markets must presume agreement on the definition of the product. Furthermore, it is generally preferable to deal with a relatively short time span, both for the reason given earlier and other reasons discussed later.

ECONOMIC THEORY OF MARKET AREAS

If raw materials are geographically concentrated and firms tend to locate close to materials sources, then the space dimension of markets ceases to concern us. Specifically, with all producers located at a single point, the industry or markets in question will be national in scope. Interregional differences in prices will simply reflect transport and similar costs. Of course, an industry organized as a single plant monopolist will have interregional price differences not attributable to transport costs; i.e., at least some price discrimination is likely.

Other than these two limiting cases, the simplest[6] situation is one in which there are two sellers surrounded by many buyers. This was the case first examined some 50 years ago[7] and, since that time, described as the law of market areas. This law, as reformulated, states that—for homogeneous goods produced and shipped over a featureless plain—the market area of an individual firm will be discrete, the exact shape depending on: (1) the firm's costs and prices, with lower costs and prices meaning a larger territory for the firm; and (2) transportation rates, with (discriminatory) low rates and rates not proportional to distance shipped enlarging the firm's market area. As the number of firms increased, the shapes of the individual firm market areas would be altered;

however, the aggregate effect might not be space filling.[8] Thus, with free entry, each firm might be a spatial "monopolist," with specific but small areas of customers not served by any firm. These "remote" customers would not purchase enough to make serving them worthwhile.

This basic model would suggest approximating the firm's market area by a circle whose radius would comprise the maximum distance the firm's products were shipped. Thus, the market area of an individual cattle rancher would probably comprise a circle with a radius of (at most) several hundred miles. That is, 90 or more percent of that rancher's cattle shipments would occur within such a circle.

On the other hand, these remarks offer little guidance on methods for aggregating the market areas for individual firms. For example, while an individual rancher might ship his cattle less than a thousand miles, we cannot thereby infer that the market is regional. The reason is that even small changes in output or price at one point affect prices and outputs at adjacent points, which further influence prices at subsequent points, etc. In sum, a chain reaction effect is likely so long as the markets are reasonably competitive and efficient.[9] Therefore, a procedure which merely calculates market areas for individual firms is inappropriate.

Aggregation problems aside, a common difficulty in geographic market analysis is the tendency for firms to compete in terms of quality of product and promotional activity in addition to price. This tendency, together with the existence of elaborate transportation networks, effectively eliminates the possibility of discrete and regular (e.g., circular) market areas even for individual firms. Thus, the market areas—both for individual firms and groups of firms—will be shaped primarily by the nature of the (surface) transportation network. As product variations and promotional activity increase, consumers will find it more costly to evaluate the relative merits of individual products. This will produce large overlaps among the areas of individual firms. Besides varying directly as information costs, the amount of overlap will be greater as the mobility of customers increases and smaller as the density of firms increase. Of course, large amounts of overlap likely indicate not merely "noise" or lack of precise measurement, but rather elements of the chain reaction process discussed above.

The above distinction between the market areas of individual firms and geographic market areas can be summarized as meaning that both supply and demand elements must be simultaneously considered in delineating geographic market areas. The impropriety of ignoring either supply or demand elements is readily illustrated.

Consider the market for beer. In the past—and to some extent today—this market was segmented by the qualitative categories "premium," "popular," and "local." As a first approximation, these categories represented not merely price differentials,[10] but also rough geographic market areas for individual brands or firms. Since the shares of total beer consumption attributable to these three categories varied over time and among areas, could one speak at all of a geographic market area for beer? Consider, for example, the state of Wisconsin. The vast bulk of beer consumed in Wisconsin was produced there; hence, Wisconsin might be termed a distinct market. This reasoning, usefully termed "Little In From Outside" (LIFO)[11], fails to recognize the importance of exports. In the case of Wisconsin, some 70 percent of beer produced there has traditionally been exported to other states. An equally significant, but contrary, type of error is the frequently encountered "Little Out From Inside" (LOFI)[12] argument. In this instance the error consists in an unduly literal application of the firm market area analysis to entire markets and industries. Thus, in a prominent antitrust case involving bank mergers,[13] the argument for a distinct geographic area was that little business done by the merging banks was nonlocal. The argument was erroneous because much of the bank business conducted within the hypothetical market area involved banks from other areas. In sum, the LIFO argument overlooks exports; the LOFI argument overlooks imports. Only by simultaneously satisfying both the LIFO and LOFI criteria do we ascertain with a reasonable degree of certainty that a given hypothetical market area is, in fact, geographically distinct.

METHODS OF ESTIMATION OF
GEOGRAPHIC MARKET AREAS

While theoretical work on geographic market areas is at least 50 years old, empirical work on this topic is relatively recent.[14] Paradoxically, almost none of the empirical work has been done by scholars specializing in regional/urban economics. These specialists have tended to concentrate efforts on phenomena such as optimal plant location, while specialists in transportation economics tend to abstract from decisions involving plant location. There are at least two explanations for this state of affairs.

One explanation is the inherent difficulty of estimating market areas. While few experiments in economics are controlled, the role of judgment (extent of uncertainty) in interpreting results in this particular area is well above average. Put another way, estimating geographic market areas is comparatively less scientific than most empirical work in economics. Another explanation is that, since the scope of geographic market areas is strongly affected by existing plant locations and transportation networks,[15] study of the latter phenomena assumes greater importance. Changes in plant location and transportation networks tend to be *very long-run* in nature, while the study of geographic market areas is primarily concerned with short and intermediate length periods.

Therefore, while it is true that changes in plant location, sizes of plants, transportation networks, etc. may change the geographic scope of markets, abstracting from these factors in the present context is a minor problem inasmuch as the location of mineral deposits is relatively fixed while even oil refineries and electric utilities change size and location only over long periods of time. Similarly, technological changes in transportation occur infrequently; hence, major changes in the transportation network can be ignored as well.[16] However, we shall later indicate the effects of major changes in transportation facilities and their interaction with other long run changes.

Various means of estimating market areas are available. First, interregional price differences may effectively delineate geographic market areas. For example, the Federal Trade Commission Report on the Cement Industry[17] notes (pp. 26-7) that, in the cement industry during the mid 1960s, metropolitan markets tended to be distinct markets. Each metropolitan area had distinguishing supply and demand characteristics and the measure of these was differences in prices. Thus, in the New York metropolitan area, transactions prices for cement were substantially below the FOB mill prices of producers located in the Lehigh and Hudson Valleys.[18] Prices in markets adjacent to the Lehigh and Hudson Valley mills (e.g., Philadelphia, northern New Jersey, and Connecticut) were affected by developments in New York City, "but not in direct proportion" (p. 27).

This brief example illustrates the problems involved in the use of price data. Thus, the comparison of FOB mill and transactions prices involves measurement problems; in addition, the differential prices might merely indicate price discrimination by a monopolist among classes of customers operating in the same market. These problems aside, the principal reason for eschewing the use of price data are their general unavailability.

In some circumstances the pattern of seller or buyer locations may be indicative of market size. As noted earlier, if all domestic sellers are located at a specific point, the market is necessarily national in scope. For example, Philip Morris presently is completing a cigarette plant in Richmond, Virginia; this plant will ultimately be able to produce 25 percent of total domestic cigarette consumption.[19] Presuming all cigarettes to be produced in the Virginia-North Carolina area, consumption patterns in themselves would be sufficient evidence of a national market for cigarettes.

Assuming estimates of geographic market areas must be made for, say, hundreds of commodities (industries), a third alternative would comprise estimates of the maximum distances shipped.[20] For example, consider a representative plant whose market area is circular. We observe that a radius of 1400 miles delineates a circle within which are absorbed 90 percent of plant shipments. Assuming a central plant location (e.g., Chicago, St. Louis), we then infer that the market is national. As noted earlier, this procedure presents aggregation problems whose severity increase as any one producer (location) ceases to be representative. More specifically, there are two major problems with this procedure which tend to vitiate its usefulness.

First, presumption of a circular market area is appropriate only if air transport is the only means of shipping (in fact, air transport is very uncommon for freight). Thus, a plant located in New York City might make half its shipments to Los Angeles and the remainder to Eastern Pennsylvania and Southern New England. Clearly in this case, the distance required to account for 90 percent of shipments[21] tells us little about the geographic scope of the market. Second, failure to consider shipments from plants in other locations will mean a tendency to underestimate the geographic scope of the market. Recall our earlier discussion of the necessity to consider both supply and demand elements simultaneously. That is, both the LIFO[22] and LOFI[23] tests had to be passed. Anticipating our later discussion, we note that during 1971 refineries in California absorbed virtually all the crude oil produced in that state. On the other hand, approximately one-third of the shipments to refineries in California were produced elsewhere.

In sum, maximum distance shipped overestimates the scope of geographic market areas in the first situation and underestimates it in the second. Since neither of these situations is uncommon, the maximum distance procedure may be simultaneously afflicted with sources of upward and downward bias.

A procedure for delineating geographic market areas which avoids the conceptual and data limitations of the aforementioned methods was developed by K. Elzinga and the present author [7]. The steps of the procedure are as follows.

(1) Identify the product's major producing centers (areas). To repeat our earlier statement, if only one major producing center exists but the product is sold nationally, then the market is obviously national in scope.

(2) Next, organize the product's shipments data in terms of both production origin and consumption destination. Typically, the latter will be available in political units such as states and for most purposes this is adequate.

(3) Taking each producing area one at a time, calculate the minimum area (e.g., minimum number of states)[24] required to account for at least 75 (90) percent of shipments from the producing area. This will satisfy the weak (strong) form of the LOFI criterion. Designate this area as a hypothetical market area (HMA).

(4) Of total shipments to destinations within the HMA, do 75 (90) percent or more of the shipments originate from the designated producing area? If so, the weak (strong) form of the LIFO test is met. So long as only one producing center exists within the HMA, the LOFI test is met through step 3. Given more than one producing center, however, the LOFI test must be repeated. If the LIFO test is not met, then re-draw the HMA, determining the minimum area necessary to absorb 75 (90) percent of the

shipments from producing centers located within the new HMA. If there is no sub-national area satisfying this criterion, the market is (at least) national in scope.

(5) If both the LOFI (step 3) and LIFO (step 4) criteria are satisfied, the HMA comprises a distinct geographic market area. Assuming this area is less than national in size, the final step consists in calculating total consumption within this area to get market size in terms of volume.

The above procedure avoids the delineation errors often encountered in antitrust cases and elsewhere. For example, even if all the beer consumed in Wisconsin had been produced in the state, the above procedure would not designate Wisconsin as a separate geographic market area. The reason is that most Wisconsin produced beer is consumed in other states (LOFI test), which themselves should be included in the market area for beer. Similarly, even though crude oil produced in California is typically refined there, crude oil produced elsewhere (LIFO test) comprises a large portion of California consumption so that this state is not a distinct market for crude oil.

Two problems with the suggested procedure are: (1) it may require data difficult to gather and (2) interpretation of its results can be difficult. The first difficulty is obvious. Copious data on sales of new cars by state exist; however, since the origins of these shipments are less readily obtained, the geographic scope of the market for cars cannot be easily designated.[25] The second problem is encountered by all measures of geographic market scope. One might argue that our 75 percent "cutoff" is too low or that a 90 percent "cutoff" is too high. Unfortunately, no precise limit would satisfy all observers and we can only hope that our demarcation points will provide reliable, useful estimates. In that spirit, we shall use both the 75 and 90 percent figures, denoting them as the "weak" and "strong" tests, respectively.[26]

The nature of the procedure we suggest can be further explored through a few (highly simplified) examples.

(1) Consider the retail market for food in Sun City, Arizona. Many inhabitants of this retirement community do not have drivers' licenses but are permitted to drive golf carts within the confines of the city. Since there is only one major supermarket in the city limits, those citizens restricted to golf carts must buy at that store. Since this one supermarket charges slightly higher prices, the vast bulk of its sales are confined to residents of Sun City. However, Sun City is not a distinct retail market for food. Why? A high proportion of Sun City residents have cars and can patronize cheaper supermarkets which are nearby but beyond city limits.[27] In addition, small changes in the relative prices of one of the stores will strongly affect patronage at that and rival stores.

(2) In automobiles, several makes are produced at essentially one location. On this basis, one might expect extreme dispersion of shipments and

hence a national market. Yet, the data in Table A-1 shows that, for four makes of cars produced at single locations, only 30 states are required to absorb 90 percent of shipments. Not surprisingly, the 20 excluded states tended to be the less populous ones.[28] Of course, lacking adequate data to perform the LIFO test, we cannot know if this market is national or regional.

(3) Fragmentary data on beer shipments has been used to calculate hypothetical market areas for beer.[29] This data indicates that Schlitz shipments originating in California were typically absorbed on the West Coast while Milwaukee shipments of beer (say, Pabst) were principally confined to the Midwest. Of course, data sufficient to carry out step (4) was unavailable, but it is unlikely that this market would be designated as national.

THE GEOGRAPHIC MARKET AREAS FOR GAS, OIL, AND COAL

As noted twice earlier, a market in which all producers are located at a single point (with customers nationwide) is at least national in scope. The producers of fossil fuel tend to be somewhat scattered, but in the case of natural gas, production has a decidedly centralized character. A majority of the states have some marketed production of natural gas, but 90 percent of the marketed production originates in only 5 states and the largest producing states (Texas and Louisiana) account for more than two-thirds of domestic, marketed production (Table A-2). Three adjacent states (Oklahoma, New Mexico, and Kansas) produce about 18 percent and hence this 5-state area roughly comprises *the* producing area of the United States. If one takes account of offshore natural gas in the Gulf Coast area, the effect is to reinforce this conclusion. To be sure, 5 states are more than a single point on a map. Nonetheless, the facts that these states are adjacent and are characterized by interlocking pipelines suggests the reasonableness of defining them as a single producing area.

Table A-1. Share of Domestic Sales[a] of Selected Makes[b] of Automobiles Absorbed by Various States, 1970

Number of States	Rambler	Mercury	Chrysler	Cadillac
10	61.08	61.61	58.87	65.27
15	73.25	72.63	71.07	74.26
20	81.55	81.10	80.41	81.30
30	92.39	92.01	91.57	89.05

[a]Actually, registrations of new cars by make. Data for Oklahoma unavailable.

[b]Rambler produced in Kenosha, Wisconsin; Mercury in St. Louis, Missouri; Chrysler and Cadillac both produced in Detroit, Michigan.

Source: Slocum Publishing Co., *Automotive News*, 1970 Almanac Issue, pp. 36-7, 68.

Table A-2.   Marketed Production of Natural Gas, Five Largest Producing States, 1967-71 (million cubic feet)

Year	Texas	Louisiana	Oklahoma	New Mexico	Kansas	Total (5 states)	Total (all states)
1967	7,188,900 (39.56)	5,716,857 (31.46)	1,412,952 (7.77)	1,067,510 (5.87)	871,971 (4.79)	16,258,190 (89.47)	18,171,325
1968	7,495,414 (38.79)	6,416,015 (33.20)	1,390,884 (7.19)	1,164,182 (6.02)	835,555 (4.32)	17,302,050 (89.54)	19,322,400
1969	7,853,199 (37.94)	7,227,826 (34.92)	1,523,715 (7.36)	1,138,133 (5.49)	883,156 (4.26)	18,626,029 (89.98)	20,698,240
1970	8,357,716 (38.12)	7,788,276 (35.52)	1,594,943 (7.27)	1,138,980 (5.19)	899,955 (4.11)	19,779,870 (90.23)	21,920,642
1971	8,550,705 (38.01)	8,081,907 (35.93)	1,684,260 (7.48)	1,167,577 (5.19)	885,144 (3.93)	20,369,593 (90.55)	22,493,012

Numbers in parentheses are percentage figures. Detail may not add to total due to rounding.
Source: U.S. Department of the Interior, Bureau of Mines. *Mineral Industry Surveys: Natural Gas Production and Consumption* (various years).

One might argue that California, the sixth leading producer state, forms the nucleus for a separate market area. However, taking 1971 as an example, the ratio of production to consumption of natural gas in California was 0.28. Similar results obtain for earlier years. Finally, if we array the states in order of natural gas consumed, the results confirm our inference about the market for gas. Of the 15 largest consuming states in 1971 (1966), only 4 (5) states[30] produce more than they consume.

A major difference between gas and oil is that, at present, only oil is shipped in significant amounts by tanker.[31] Thus, in 1971, approximately 15 percent of total refinery receipts of crude oil were from foreign sources. Not surprisingly, most of this foreign crude is received by refineries in California and on the East Coast; nonetheless, significant amounts of foreign crude are also shipped to the (northern) midwest.

Refinery receipts of crude oil are classified by Petroleum Administration for Defense (PAD) Districts. Similarly, the Bureau of Mines offers a system of Refining Districts. Lacking justification for these classifications as distinct market areas, we began instead with refinery receipts of crude oil, classified by state(s) of origin and destination.[32] An immediate finding was the fact that in 1971, 76 percent of refinery receipts of crude oil in Texas were intrastate; the corresponding figure for Louisiana was 84 percent.[33] On the other hand, only 64 percent of Texas crude shipments were made intrastate while intrastate shipments of crude for Louisiana were only 38 percent of the total. Hence, taken alone, neither Texas nor Louisiana comprised distinct market areas. Successive experiments with various definitions of producing centers eventually led to specification of two distinct producing areas.[34] Producing Area I comprised the states Texas, New Mexico, Louisiana, Oklahoma, and Kansas. The other producing area (Area II) comprised a residual, namely, all other producing states plus imports. Attempts to define smaller subareas within Area II were unsuccessful; that is, our tests for distinct geographic market areas were not satisfied.[35]

Table A-3 contains the results of our geographic market area tests. Examining the 1971 data, we find that: (1) shipments from Area I to 5 states[36] absorbed 76.22 percent of shipments from Area I (LOFI criterion); and (2) of total shipments to these 5 states, 92.60 percent originated in Area I (LIFO criterion). On the other hand, while 9 states[37] absorbed 92.40 percent of Area I shipments, only 84.73 percent of total shipments to these 9 states originated in Area I. Thus, Area I is a distinct market area only on the weak form of our test. The second area failed to pass even the weak form. Since previous tests demonstrated that no subarea could be used to delineate a distinct market area, the market for crude oil is (at least) national in scope.

Of total domestic refinery receipts of crude oil, only about 64 percent originates in Area I. Thus crude oil from other sources (both domestic and foreign) comprises a substantial share of refinery receipts and, unlike the

Table A-3. Geographic Scope of Market Areas for Crude Oil,
1967 and 1971

	1971	*1967*
Minimum number of states[a] required to absorb 75 (90) percent of shipments from Area I[b] (LOFI)	5 (9)	5 (9)
Percent of shipments to these states originating in Area I (LIFO)	92.60 (84.73)	96.09 (84.55)
Minimum number of states[c] required to absorb 75 (90) percent of shipments from Area II[d] (LOFI)	13	9
Percent of shipments to these states originating in Area II (LIFO)	62.21	59.39

[a]These states (in order of importance, 1971) were Texas, Louisiana, Illinois, Oklahoma, Indiana, Ohio, Pennsylvania, Kansas, and New Jersey.

[b]Producing Area I comprised the states Texas, Louisiana, Oklahoma, New Mexico, and Kansas.

[c]These states (in order of importance, 1971) were California, Pennsylvania, New Jersey, Indiana, Ohio, Illinois, Wyoming, Utah, Montana, Michigan, New York, Minnesota, and Wisconsin.

[d]Area II comprised shipments of crude oil from foreign sources plus all producing states except the 5 in Area I.

Source: U.S. Department of the Interior, Bureau of Mines. *Mineral Industry Surveys: Crude Petroleum, Petroleum Products, and Natural Gas Liquids* (1967, 1971).

situation in natural gas, the 5 states comprising Producing Area I do not (even approximately) supply the nation's demands. In addition, the numerous other sources of crude oil cannot be used as the basis for a second geographic area. Finally, the market based on Area I shipments is not distinct under the strong form of our test. Hence, we conclude that the market for crude oil is international in scope. In view of the recent oil embargo and announced plans for self sufficiency in energy, this conclusion may seem odd. However, the severe impact of the embargo merely demonstrates the existence of an international market. Of course, import restrictions or similar measures can render the U.S. market for crude oil national. On the other hand, strenuous efforts in behalf of self sufficiency in energy could produce the "unintended" effect of making the U.S. a net exporter of crude oil.

The most complex market, in a geographic sense, is bituminous coal. Following Moyer[38] we note that the nation's coal resources are divided into 6 major fields: Eastern, Interior, Gulf, Northern Great Plain, Rocky Mountain, and Pacific Coast. Traditionally, the high quality coal has been in the East, decreasing progressively in quality as one moves west.[39]

The Eastern region consists of the Appalachian chain from Pennsylvania to Alabama, running through eastern Ohio, eastern Kentucky, Virginia, West Virginia, and Tennessee. The Interior comprised western Kentucky, Illinois, Indiana, Iowa, Missouri, Kansas, Oklahoma, and Arkansas. The Gulf and Pacific areas tend to be of minor importance, while the Northern Great Plain and Rocky Mountain coals vary substantially in quality.

On the basis of these considerations and some experimentation,[40] we devised the aggregate producing areas given in Table A-4. The Eastern Center corresponds rather well to Moyer's Eastern region while the Western Center essentially comprises the Pacific Coast and Rocky Mountain regions. Other regions are lumped together as a Central Center. Alternative alignments were tried and found to be less accurate.[41]

With these three producing areas, successive applications of our test for the scope of market areas produced the results contained in Table A-5. Turning first to 1971, we note that the Eastern area forms the basis for a market area encompassing 9 states (weak form of test); however, application of the strong form of our test fails to delineate a distinct market area in all cases.[42] We also note that 1971 was unique in that neither the Central nor the Western producing areas formed the basis of a distinct market. Only by combination of these two producing areas do we obtain anything approaching a distinct geographic market. In contrast, the years 1965, 1969-70 are characterized by three distinct market areas, at least under the weak form of the test. Our earliest year[43] produces an apparently peculiar result: there are distinct market areas in the West and East, but not in between. In sum, for 3 years we have 3 distinct market areas; for 1 year (1971) we have two distinct market areas and for 1 year (1961) we have two distinct areas plus a "no-man's land."

Table A-4. Coal Producing Areas and Corresponding Bureau of Mines Producing Districts

Coal Producing Area	Bureau of Mines Districts[a]
Eastern[b]	Districts 1-8, 13
Central[c]	Districts 9-12, 14-15
Western[d]	Districts 16-23

[a]In some cases the districts comprise entire states (e.g., District 5 is Michigan); in other instances only part of a state is included (e.g., District 2 is Western Pennsylvania).

[b]Comprises Pennsylvania, Ohio, West Virginia, Michigan, Eastern Kentucky, Virginia, Tennessee, Alabama, and Georgia.

[c]Comprises Western Kentucky, Illinois, Indiana, Iowa, Arkansas, and Oklahoma.

[d]Comprises Colorado, New Mexico, Wyoming, Utah, North and South Dakota, Montana, Washington, Oregon, and Alaska.

Source: U.S. Department of the Interior, Bureau of Mines. *Mineral Industry Surveys: Bituminous Coal and Lignite Distribution* (various years).

Table A-5. Geographic Scope of Market Areas for Bituminous Coal and Lignite, Various Years

	1971	1970	1969	1965	1961
Minimum number of states required to absorb 75 (90) percent of shipments from eastern area (LOFI)	9* (15)	8* (14)	9* (15)	9* (16)	10* (15)
Percent of shipments to these states originating in eastern area (LIFO)	92.90 (81.55)	87.29 (85.31)	86.21 (86.48)	89.82 (86.01)	92.81 (85.17)
Minimum number of states required to absorb 75 (90) percent of shipments from central area (LOFI)	7	6* (10)	5* (10)	5* (10)	5 (9)
Percent of shipments to these states originating in central area (LIFO)	69.17	77.25 (69.19)	80.08 (72.89)	76.70 (66.68)	71.51 (58.49)
Minimum number of states required to absorb 75 (90) percent of shipments from western area (LOFI)	9	7* (13)	7* (12)	7* (10)	6* (11)
Percent of shipments to these states originating in western area (LIFO)	42.55	81.17 (42.36)	99.46 (77.79)	98.97 (73.09)	96.83 (71.78)
Minimum number of states required to absorb 75 (90) percent of shipments from central and western areas combined (LOFI)	10* (19)				
Percent of shipments to these states originating in central and western areas combined (LIFO)	77.75 (53.38)				

*Denotes distinct geographic market area.

Source: See note 1, Table A-4.

One interpretation of these findings is that the results for 1961 reflect the impact of the severe recession in that year.[44] Excess capacity commonly produces expansion in the geographic scope of markets. For example, in 1965 (a prosperous year) coal shipments into Indiana came primarily from the Central producing area (64 percent) and secondarily from the Eastern producing area (36 percent). In 1961 the figures were reversed: 67 percent from the Eastern area and 33 percent from the Central area. Absent major changes in transport rates, these results imply price discounts by Eastern coal producers.

The 1971 results partially reflect the impact of clean air standards. Thus, during the 1960s, the only midwestern state receiving significant shipments of western coal was Minnesota; however, in 1970-71, significant shipments of western coal to Illinois (presumably Chicago) were observed. In fact, Detroit Edison Company alone expects to use 8 million tons of western coal *per year*, starting in 1980 [3].

The coal industry's move west is only partially due to air pollution standards [2]. First of all, nearly two-thirds of West Virginia's coal production contains no more than 1 percent sulfur.[45] Secondly, the growing importance of western coal reflects lower production costs, easier acquisition, and fewer labor problems. "The result of all this is that coal can be strip-mined in the West for $3 to $5 a ton, compared with $9 to $14 in Eastern deep mines—*enough of a saving to offset transportation costs as far east as West Virginia*" [2, p. 136, italics added].

Suppose, however, that future controls of strip mining, etc. eliminate all but the low sulfur content advantage of mining western coal. Since the Clean Air Act takes effect only in 1975 and may subsequently have reduced standards, may the expansion of the coal market's scope be merely transitory? For example, suppose all air pollution standards were drastically reduced in, say, 1976. Would the geographic scope of the coal market shrink to its size in the middle 1960s?

The answer must be: "probably not." A commonly overlooked characteristic of (certain) markets is that substantial disturbances generate forces which permanently alter demand characteristics in the market. For example, a 25 percent increase in the price of coffee in 1954 precipitated a 19 percent increase in cups brewed per pound of coffee. This increased output per pound of coffee grounds was maintained despite subsequent price reductions in 1955 and 1958. The inference is that the higher price of coffee reduced its demand in two ways: (1) a customary movement along the demand curve (and (2) a shift in the demand curve—pounds of coffee per unit time—due to increased efficiency (technological change?) in household "production" of coffee from coffee grounds [10].

Similarly, the prospect of large shipments of western coal to eastern utilities has already stimulated investment in cheaper transport. Thus, new and enlarged fleets for the Great Lakes are under construction. These ships will have more than twice the capacity of existing ships in hauling coal [3, p. 40]. Once these ships are constructed, the lower transport rates they permit will be available even if air pollution standards are relaxed. In short, the impact of antipollution laws on the geographic scope of coal markets was *not* temporary.

Returning to the results in Table A-5, a striking aspect is the fact that the Eastern Area is almost a distinct geographic market under the strong form of our tests. Thus, an average of 15 states absorbed 90 percent of shipments from the Eastern area; moreover, during 1961-70, *almost* 90 percent of shipments to these states originated in Area I. For 1971, however, a different impression arises. In this year satisfaction of the strong form of the LOFI criterion meant decisive failure to satisfy the strong form of the LIFO criterion. To illustrate, note Table A-6, which lists the 9 states comprising the Eastern market area during 1971 under the weak form of our tests. The number 10 state in order of importance, not shown in Table A-6, is Indiana, which is also the number 2 state for the combined Central plus Western regions. In fact, about one-fourth of the coal shipments to Indiana originated in the Eastern Area.

Unfortunately, we have insufficient experience with the market area tests to determine whether or not the weak or the strong form of our tests is the more appropriate. A potential problem with the strong form of our test is that possibly no markets would satisfy its strict criteria. This view appears unwarranted. For one thing, it appears that the retail markets for food and dry cleaning would be declared local even under the strict form of our tests. For another, it very well may be that only a small proportion of commodities are sold in nonnational markets. As with the case of cattle cited above, many markets are such that even small changes in price in one locality beget immediate and large changes in adjacent and even distant localities.

In any event, the important point is that the results for 1971 are more consistent with the presumption of a nationwide market for coal than those for earlier years. In addition to the impact of air pollution controls and the increasing advantages of western coal cited above, this trend may reflect the delayed impact of transport innovations such as the unit train, large hopper cars, and improved barge transport of coal.[46] Finally, a symptom of growth in market scope has been the increase in exports of coal, especially for the steel industries of Europe and Japan.

CONCLUSIONS

The test for the scope of geographic markets used in this study is new and, in view of widespread data limitations, is essentially untried. Since this application to the markets for gas, oil, and coal is the first major one, interpretation of the results requires substantial judgment in at least two of the markets. As noted earlier, the geographic market for gas would be described as national by almost any criterion or set of criteria. The geographic market for crude oil appears

Table A-6.   States Comprising Geographic Market Areas for Bituminous Coal and Lignite, Various Years[a]

	1971	1970	1969	1965	1961
Eastern	Ohio	Ohio	Ohio	Pennsylvania	Pennsylvania
	Pennsylvania	Pennsylvania	Pennsylvania	Ohio	Ohio
	Michigan	Michigan	Michigan	Michigan	Michigan
	W. Virginia	W. Virginia	W. Virginia	New York	Indiana
	N. Carolina	New York	New York	W. Virginia	New York
	New York	N. Carolina	N. Carolina	Virginia	W. Virginia
	Tennessee	Virginia	Virginia	Indiana	Virginia
	Alabama, Mississippi[b]	Indiana	Indiana	N. Carolina	N. Carolina
			Tennessee	New Jersey	Alabama, Mississippi[b]
Central		Illinois	Illinois	Illinois	
		Indiana	Indiana	Indiana	
		Kentucky	Kentucky	Kentucky	
		Missouri	Missouri	Wisconsin	
		Wisconsin	Wisconsin	Wisconsin	
		Tennessee			
Western		New Mexico	Colorado	Colorado	Colorado
		Colorado	Wyoming	Utah	Utah
		Wyoming	New Mexico	New Mexico	California
		Utah	Utah	California	Wyoming
		Minnesota	California	Wyoming	Dakotas[b]
		Dakotas[b]	Dakotas[b]	Dakotas[b]	
Central plus Western	Illinois				
	Indiana				
	Kentucky				
	Missouri				
	Wisconsin				
	Tennessee				
	Minnesota				
	New Mexico				
	Iowa				
	Colorado				

[a]States listed in order of importance (i.e., shipments received). This listing pertains only to the weak form of our tests.

[b]Data for these states combined in Bureau of Mines reports on bituminous coal and lignite.

highly concentrated geographically. Since crude oil is not a final product, but merely an input to the refining process, this is hardly surprising. First, unless the costs of transporting refined oil products are prohibitive, wide dispersion of refineries is very unlikely. Second, at one time the location of refineries tended to be raw material oriented.[47] More recently, of course, refineries have been located closer to markets. Thus, in addition to the Gulf Coast, the U.S. has substantial refining capacity in New Jersey, Eastern Pennsylvania, and the

Chicago area. Our data is revealing in this respect. The top two recipients of crude oil from the Gulf Coast are Texas and Louisiana; however, the third and fifth largest recipients are Illinois and Indiana, respectively, while New Jersey and Pennsylvania are seventh and ninth.[48] The largest recipient of foreign crude oil is California while the second through fourth largest are Pennsylvania, New Jersey, and New York.[49] Thus, refineries in the midwest are heavily "dependent" on domestic crude while coastal refineries obtain roughly half their receipts of crude oil from foreign sources. These considerations, together with the fact that a market based on Gulf Coast production is geographically distinct only under the weak form of our test, impels us to conclude that the market for crude oil is international.[50]

Examination of data on coal shipments revealed a tendency for the scope of geographic markets to increase. This trend toward increased geographic scope is "artificial" to the extent it has been induced by antipollution requirements. Nonetheless, such requirements have induced or augmented changes in transportation capabilities which will remain even if air pollution standards are relaxed. Finally, the market for coal is regional only under the weak form of our tests. For these reasons we offer the *tentative* conclusion that the industries comprising the energy (input) sector—uranium, gas, oil, and coal—are (at least) national in scope.

NOTES (Numbers in brackets refer to references, p. 217.)

1. Actually, this is a definition of an efficient market. Thus, the prices paid for 3-year-old Fords may vary substantially both within and among cities, depending on information costs, etc. For our purposes, however, the more restrictive definition is appropriate.
2. An exception is the uranium market, which is simply presumed to be at least national in scope.
3. Much of this material is taken from Gerard Debreu [5, Chapter 2].
4. In other cases, e.g., new automobiles, the distinction between utility characteristics and physical attributes is less clear. See, for example, Thomas Hogarty [9].
5. A detailed discussion of this approach is contained in Kelvin Lancaster's article [12].
6. We shall ignore other, perhaps simpler, configurations since they are relatively uncommon. Among these situations are the central market case and the situation where buyers are concentrated and sellers dispersed. For a discussion of all these cases, see Harry W. Richardson [17, Chapter 2].
7. See Frank A. Fetter [8]. Subsequent developments are contained in C.D. and W.P. Hyson [11].
8. See R.D. Dean, W.H. Leahy, and D.L. McKee [4, p. 194].
9. For the distinction between competitive and efficient markets, see the discussion in note 1.

10. A Budweiser survey found, in 90 percent of the cases, a 5 cents per 12 oz. bottle (can) difference among the three categories, taken two at a time.
11. For the origins and historical significance of this line of reasoning, see K. Elzinga and T. Hogarty [7].
12. *Ibid.*
13. For a discussion of this and other cases, see K. Elzinga and T. Hogarty, *op. cit.*
14. See [7] and [20].
15. Suboptimal plant locations will often mean a smaller scope for the market; similarly, the market areas of coastal firms will typically be larger.
16. See, for example, C. Phillips [16] and P. MacAvoy and J. Sloss [13].
17. U.S. Federal Trade Commission. *Economic Report on Mergers and Vertical Integration in the Cement Industry* [1].
18. The importance of transportation networks to the determination of geographic market areas in cement is aptly illustrated by the Report [1, p. 28, note 7]:

> Plants with access to cheap water transportation reach out farther for markets than those which do not have such access. A most notable example is that of Atlantic Cement which operates distribution terminals up and down the whole east coast. These terminals are strategically located in major metropolitan areas where cement demand is concentrated. They are supplied by barges from Atlantic's huge 10 million barrel plant located on the Hudson River. Counterparts to Atlantic's operation can be found on the Great Lakes, the Mississippi and Ohio Rivers, the Gulf Coast, and the west coast. In each instance the plants with access to water will reach for markets many hundreds of miles in distance beyond that ordinarily reached by landlocked plants.

19. *Business Week*, January 27, 1973, p. 54.
20. For a discussion and application of this procedure, see L. Weiss [20].
21. Used as an approximation to the maximum distance shipped. See L. Weiss, *op. cit.*
22. Little In From Outside, signifying a market area relatively free of "imports."
23. Little Out From Inside, signifying that market area was not merely an "export" base.
24. In calculating the minimum number of states required to account for some given percent of shipments, land area should be taken into account. For example, state A may have double the shipments of state B, but triple the land area. If state A is included before B, the minimum area criterion is, strictly speaking, violated. Fortunately, such differences are important only for political units such as Alaska and the District of Columbia. In applying both techniques to the data on energy markets, the same results were obtained. Hence, in the

market area, estimates presented below, we ignore interstate differences in land area.

25. Writers on the subject generally act as if this market were necessarily national. However, the fact that there are a variety of assembly plants scattered about the nation suggests otherwise. For purposes of computing market shares, the high concentration of this industry makes geographic market delineation largely unnecessary. Yet Volkswagens cost more in the Midwest and, were the industry populated by 100 single plant firms, its geographic scope would likely be observed to be regional in nature. This further indicates the irrelevance of ownership patterns: a market which is nonnational with 100 single plants firms remains so even if it is subsequently monopolized.

26. Given our procedure, when is a market national? It would be tempting to say that a national market requires 48 (50) states to satisfy the LIFO and LOFI criteria before we can declare the market as national. This is inappropriate since 75 percent of the states may be sufficient to account for 90 percent of the population (economic activity).

27. The prices within the city limits may be marginally higher, reflecting costs of transportation and general information (shopping costs).

28. In the case of Rambler, these states were Idaho, Montana, Nevada, New Mexico, Utah, Wyoming, Arkansas, Oklahoma (not available), Nebraska, North Dakota, South Dakota, Mississippi, Delaware, District of Columbia, South Carolina, Maine, New Hampshire, Rhode Island, Vermont, Alaska, and Hawaii.

29. For details, see Elzinga and Hogarty [7].

30. In 1971, these 4 states were (in order of importance) Louisiana, Texas, Oklahoma, and Kansas. In 1966, the 5 states comprised these 4 plus New Mexico.

31. In the future liquified natural gas may eliminate this difference so that more natural gas can be imported.

32. In some cases, states were combined; however, this typically occurred because shipments to (or from) the states in question were relatively small.

33. The source of the data used is Mineral Industry Surveys, Annual Petroleum Statement, 1971, Table 13, p. 15.

34. For example, if we combine Texas and Louisiana, then we find that only 63 percent of total shipments of crude oil from these states is destined for refineries within these states.

35. If Area II is defined so as to exclude imports, then no distinct geographic market area can be delineated.

36. Texas, Louisiana, Illinois, Oklahoma, and Indiana. The leading states in terms of oil and gas production differ for a variety of reasons, among them geological, economic, and political. Thus, in West Texas gas is more valuable than oil due to the proximity of California while in East Texas are the refineries. In addition, restrictions on oil production and other regulations beget differing production relations among states.

37. In addition to the 5 cited in the previous footnote, we had Ohio, Pennsylvania, Kansas, and New Jersey.
38. R. Moyer [15, pp. 10-12].
39. In descending order of quality, coals are classified as anthracite, bituminous, subbituminous, and lignite. The greater the amount of carbon relative to volatile matter and moisture, the greater the quality of the coal. A property not related to this ranking is whether or not a coal will produce coke when placed in a coke oven. Eastern coals are largely bituminous and coking while Western coals are generally noncoking and consist of a larger proportion of lignite.

    In recent years, sulfur content has become an additional quality characteristic. In this respect, low sulfur coal is typically that containing 1 percent or less sulfur. Such coal is relatively more common in the Western states; however, West Virginia is such that nearly two-thirds of its production contains no more than 1 percent sulfur. [See T.W. Hunter, "Bituminous Coal and Lignite," *Mineral Facts and Problems*, p. 42.]
40. For example, consider the 1971 data for District 1 (Eastern Pennsylvania) and District 4 (Ohio). If we take District 1 as a distinct area, we find that 6 states absorb 93.63 percent of shipments from this district (LOFI), but that, of total shipments to these 6 states, only 32.11 percent originate in District 1 (LIFO). District 4 was such that 2 states (Ohio, Michigan) absorbed 90.34 percent of its shipments (LOFI); however, only 47.25 percent of shipments to these 2 states came from District 4 (LIFO).

    District 5 (Michigan) made no significant shipments during 1971. Data for Districts 3 (Northern West Virginia) and 6 (West Virginia Panhandle) are combined by the Bureau of Mines. If to these we add District 2 (Western Pennsylvania) we find that 6 states absorb 92.07 percent of shipments from these three districts (LOFI). Again, however, total shipments to these 6 states typically originate elsewhere as only 30.60 percent came from Districts 2-3, 6 (LIFO). In each of these cases, use of the weak form (75 percent) for the LOFI criterion still did not beget success on the LIFO test. For example, of total shipments from Ohio mines, 67.10 percent are intrastate. Yet, total shipments to Ohio are such that nearly one-half originate in other districts. Similar results obtain when the weak form for the LOFI test is used for the other districts.

    In sum, failure to delineate smaller geographic markets resulted from inability of these smaller areas to satisfy the LIFO criterion. We shall return to this point again as it sharply points out the erroneous results obtained by use of maximum distance shipped or the LOFI criterion by itself.
41. *Ibid.*
42. For all our tests (crude oil and coal) we attempted expansion of market areas when the LOFI criterion was passed but the LIFO criterion was not. In all cases, expansion of the relevant area did not produce distinct market areas.

43. The earliest year for which coal distribution data was available was 1961. For crude oil and natural gas, our data do not go back even this far.
44. The year 1961 was not merely one with high unemployment, but also a trough in coal production versus its peak in 1947.
45. See footnote 39, *supra.*
46. For example, a new river-to-ocean shipping transfer station south of New Orleans permits midwestern coals to compete in energy markets in Florida. (See T.W. Hunter, *op. cit.*, p. 47.)
47. Manners [14, p. 194].
48. These rankings are computed on the basis of 1971 data alone.
49. *Ibid.*
50. For similar arguments, see M. Adelman, "Is the Oil Shortage Real?," *Foreign Policy*, January, 1973, pp. 69-107.

REFERENCES

[1]   A. Andersen. "On Mergers and Vertical Integration in the Cement Industry," *Economic Report*, U.S. Federal Trade Commission. Washington, D.C.: Government Printing Office, 1966.

[2]   Business Week. "The Coal Industry's Controversial Move West," *Business Week*, May 11, 1974, 134-38.

[3]   _____. "The New Great Lakes Fleets," *Business Week*, May 18, 1974,

[4]   R. Dean, W. Leahy and D. McKee, *Spatial Economic Theory*. New York: The Free Press, 1970.

[5]   G. Debreu. *Theory of Value*. New York: Wiley and Sons, 1959.

[6]   T. Duchesneau. "Interfuel Substitutability in the Electric Utility Sector of the U.S. Economy," *Economic Report*, U.S. Federal Trade Commission. Washington, D.C.: Government Printing Office, 1972.

[7]   K. Elzinga and T. Hogarty. "The Problem of Geographic Market Delineation in Antimerger Suits," *The Antitrust Bulletin*, 18, 1, Spring, 1973, 45-82.

[8]   F. Fetter. "The Economic Law of Market Areas," contained in Dean, et al. *Spatial Economic Theory*, 157-64.

[9]   T. Hogarty."Hedonic Price Indexes for Automobiles: A New Approach," *Applied Economics* (forthcoming).

[10] Hogarty and R. MacKay. "Some Implications of the New Theory of Consumer Behavior," *Amer. Journ. Agr. Econ.*, May, 1975.

[11] C. Hyson and W. Hyson. "The Economic Law of Market Areas," contained in Dean, et al. *Spatial Economic Theory*, 165-70.

[12] K. Lancaster. "A New Approach to Consumer Theory," *Journal of Political Economy*, 74, 2, April, 1966, 132-57.

[13] P. MacAvoy and J. Sloss. *Regulation of Transport Innovation*. New York: Random House, 1967.

[14] G. Manners. *The Geography of Energy*. London: Hutchinson and Co., 1971.

[15] R. Moyer. *Competition in the Midwestern Coal Industry*. Cambridge: Harvard University Press, 1964.

[16] C. Phillips. *The Economics of Regulation.* Homewood, Illinois: Irwin, 1969.

[17] H. Richardson. *Regional Economics.* New York: Praeger Publishers, 1969.

[18] U.S. Bureau of Mines. *Mineral Facts and Figures.* Washington, D.C.: Government Printing Office, 1971.

[19] _____. *Mineral Industry Surveys.* Washington, D.C.: Government Printing Office, various years.

[20] L. Weiss. "The Geographic Size of Markets in Manufacturing," *Review of Economics and Statistics*, 54, 3, August, 1972, 245-57.

Appendix B

Economies of Scale of Firms Engaged in Oil and Coal Production

Thomas Gale Moore*

The purpose of this chapter is to present a summary of what is known about economics of scale in the coal industry, in the crude oil industry, in the petroleum refining industry, and in the integration of these industries. By economies of scale is meant the relationship of costs to size. The measurement of economies of scale is often an attempt to designate the most efficient size or sizes. Most economies of scale studies have dealt with size of plant rather than with size of firm. This focus on size of plant has occurred because data limitations often make it impossible to say much about multiplant firms and their possible efficiencies.

Three approaches have been used to measure economies of scale: the engineering study, estimation of the production function, and the survivor technique. The engineering approach studies the actual technology that is most efficient in the industry and determines which size or sizes of plant can fully take account of all the possible cost saving techniques. Inevitably, the engineering study must overlook many factors, such as management efficiency, shipping costs, stockpiling costs, and the effect of variations in capacity utilization on costs. As a consequence, it will almost always indicate that the largest sizes are the most efficient. It can, however, produce dollar estimates of cost saving of large size.

Fitting production functions to industry data can produce statistical evidence on the degree of homogeneity. However, because of data limitations and sample size, the conclusions drawn are usually that constant returns to scale are or are not consistent with the evidence. Little can be said about either the size of the cost advantage of large operations or about the minimum efficient size.

*Professor of Economics, Michigan State University.

The survivor technique compares the size of plants or companies in one period with the size in some subsequent period. Those sizes that grew over the period relative to the industry are clearly efficient according to the dictates of the market place. This approach has the virtue as well as the drawback of including all the factors that affect survival. In other words it takes into account not only existing technology but the legal structure and government policies. It can be particularly misleading in some cases where government policies have a significant impact on the profitability of different size plants. For example, the oil import program allocated valuable import tickets to refiners in a way that aided the smallest refineries. Thus the true cost advantage or disadvantage of such refineries has been masked by this additional source of profit.

While there are significant differences between the engineering approach and the survivor technique, in actual practice they have given very similar results for particular industries. This chapter will report on studies using all techniques. Most such studies are weakest in their identification of economies of multiplant firms. No studies have been made of possible gains from the integration of firms operating in several industries such as coal and oil.

COAL MINING

The concept of economies of scale in mining is unclear. Each coal mine is a unique resource which may need or call for different scales of output. Thus mines with thick seams and plentiful coal may call for larger operations than those mines with smaller seams. Deeper seams call for more capital per unit output, ceteris paribus, than shallower. On the other hand, if the seams are close enough to the surface, strip mining, which is more capital intensive, is called for. Even here the natural resource base may be the basic determinant of the optimum size.

Because the mine is the natural unit in coal mining, little attention has been paid to the concept of economies of scale. One study [Moyer, p. 105-106] examined the relationship of output per man-day by size of mine. On the basis of statistics for Illinois mines in 1959, Moyer found that underground mines smaller than 50,000 tons per year had a significantly lower output per man-day than larger mines. For strip mines, he found that the larger the mines the greater the productivity per mine. This data, however, does not prove that there are significant economies of scale in coal mining. Larger mines may have higher labor productivity because they use more capital, yet their costs may not be significantly lower than smaller operations. Larger mines tend to be more heavily unionized with the result that wages are higher, thus leading firms to increase the proportion of capital to labor. Moyer did find that larger mines, both strip and underground, tended to have better resource bases, that is, thicker seams, and larger stripping ratios, than did the smaller operations.

Recent statistics indicate that productivity continues to be positively correlated with size. As can be seen from Table B-1, productivity in terms of output per man-hour increases steadily with size of mine for both underground mining and strip mining and for both 1968 and 1970. Data for auger mines (not included in the table) has the same pattern. Whether this indicates some basic advantage of large scale mining over small cannot be determined from this data, since the relationship may stem simply from either more capital involved in large scale mining or better natural resources.

In 1969, the federal government passed new legislation establishing significantly stricter safety requirements for mining. Many of these requirements became effective for underground mines early in 1970. Due to the additional labor required in underground mines, productivity declined in all size categories but the smallest. The apparent growth in productivity for the smallest size category probably stems from the disappearance of the smallest and most inefficient mines. The number of mines producing less than 25,000 tons per year declined from 2715 in 1968 to 1955 in 1970—a decline of 28 percent in two years. As Table B-2 shows, the number of mines declined in all size categories below 100,000 tons per year while growing in the intermediate size categories between 100,000 tons and 500,000 tons per year. Interestingly enough, the largest size categories, which have been combined in Tables B-2 and B-3 into one, also declined in these two years. As can be seen from Tables B-2 and B-3, these declines appear in terms of relative share of the output, absolute level of output, and number of mines. As a general conclusion from this data it would appear that small size mines are no longer very economic, in part due to government safety requirements, and that the most economic portion of the

Table B-1. Labor Productivity in Coal Mines by Size and Type of Mine

| Size of Mine (thousands of tons per year) | Underground Mines | | | Strip Mines | | |
	Tons per Man Hour 1968	1970	Ratio of 1970/1968	Tons per Man Hour 1968	1970	Ratio of 1970/1968
Less than 25	1.18	1.28	1.08	1.93	2.35	1.22
25-50	1.73	1.72	0.99	2.73	3.34	1.22
50-100	1.94	1.84	0.95	2.95	3.43	1.16
100-250	2.06	1.71	0.83	3.54	4.01	1.13
250-500	2.01	1.65	0.82	4.41	5.03	1.14
500-750	2.09	1.82	0.87	5.77	4.92	0.85
750-1,000	2.15	1.94	0.90	4.88	6.08	1.25
1,000 & over	2.51	2.24	0.89	7.76	7.17	0.92

Source: Bureau of Mines.

Table B-2. Changes in Underground Mine Size From 1968 to 1970

Size of Mines (thousands of tons)	Number of Mines		Percent Change from 1968 to 1970	Output in Millions of Short Tons		Percent Change from 1968 to 1970
	1968	1970		1968	1970	
Less than 25	2715	1955	−28.0	18.2	13.6	−25.1
25-50	382	325	−14.9	13.3	11.6	−12.7
50-100	249	226	−9.2	17.3	15.5	−10.6
100-250	173	207	19.7	26.7	32.5	21.5
250-500	123	145	17.9	44.5	53.6	20.6
Over 500	208	191	−8.2	222.2	211.1	−5.0
Total	3850	3053		342.2	337.9	

Source: Bureau of Mines.

Table B-3. Relative Change in Number and Output by Underground Mine Size from 1968 to 1970

Size of Mines (thousands of tons)	Percentage of All Mines		Percentage of Total Output	
	1968	1970	1968	1970
Less than 25	70.5	64.0	5.3	4.0
25-50	9.9	10.6	3.9	3.4
50-100	6.5	7.4	5.1	4.6
100-250	4.5	6.8	7.8	9.6
250-500	3.2	4.7	13.0	15.9
Over 500	5.4	6.3	64.9	62.5

Source: Table B-2.

industry is for mine sizes larger than 100,000 tons per year. The largest size mines may also not be as efficient as intermediate size ones, although this may be simply a statistical accident.

There are good reasons to believe that the largest mines are economic and that they will increase their share of the market in the future. Sales of bituminous coal to power companies has increased from 19 percent in 1950 to 46 percent of total consumption in 1960, to 62 percent in 1970 [*Statistical Abstract*, p. 656]. A combination of factors have increased the economic importance of large-scale mining for supplying power plant needs. Power generating plants have grown considerably larger over time with a concomitant need for vast amounts of coal delivered to a single generating station. With the development of unit trains to deliver coal, the most economical operation is a long-term supply contract for coal to be delivered from a single mine to a single power plant. While data is lacking on the overall percentage of consumption sold on long term contracts—over one year—reports of individual large coal companies indicate that it must be substantial. For example, a 1951 survey of West Virginia coal marketing indicated that 60 percent of its coal was sold in the spot market and most of the rest on contract for less than a year, but in the mid-1960s Continental Oil reported that about 70 percent of Consolidation's tonnage sold to electric utilities was made under long-term contracts [*Hearings on Concentration by Competing Raw Fuel Industries*, p. A 505]. Peabody Coal reported that about 94 percent of its coal sales to electric utilities in 1966 were made under long-term contracts [*Hearings*, p. A 506].

Between 1968 and 1970 not only did productivity fall in underground mines, but output actually declined. As a result the postwar trend of an increasing share of output produced by strip mines accelerated as the percentage of production produced in underground mines declined from 63 percent to 56 percent, while the percentage coming from strip mines increased from 34 percent to nearly 41 percent. The new safety regulations had not affected strip mining by 1970, but the growing environmental agitation was leading to an increase in strip mining operations before any antistrip mining legislation was passed. As a result the number of strip mines increased by 44 percent. As Table B-4 shows, the number of small mines increased as a percent of all strip mining operations as well as increasing their share of the total output. On the other hand, strip mining operations larger than 100,000 tons a year lost share of market in the two years. The data therefore suggests that there are no real

Table B-4. Relative Change in Number and Output by Strip Mine Size from 1968 to 1970

Size of Mines in Thousands of Tons per Year	Percentage of All Mines		Percentage of Total Output	
	1968	*1970*	*1968*	*1970*
Less than 25	48.9	51.0	3.6	4.4
25-50	16.9	17.1	5.0	5.5
50-100	13.6	14.6	7.9	9.1
100-250	11.3	9.8	13.1	12.9
250-500	3.8	2.7	10.6	8.6
Over 500	5.5	4.8	59.8	59.6
Total	100	100	100	100

Source: Bureau of Mines.

economies of scale in strip mining even though productivity seems to increase with size.

In summary then there would seem to be no large economies of scale in coal mining, although there may be some small advantage to large size (over 500,000 tons a year) in meeting federal safety requirements and possibly in serving the increasingly important electrical utility industry. While this conclusion seems warranted for production units, that is, actual mines, it may be that multiestablishment companies experience savings over savings unit firms. It is to that question that we now turn, although the evidence here is less firm than on individual mine operation.

MULTIUNIT FIRMS

In 1967 census data indicates that large coal mining companies (that is, those with over 500 employees) produced 52 percent of the receipts of companies classified in coal mining, while in 1963 such large companies produced only 46 percent of the receipts. In 1967, there were 11 such companies while in 1963 there were only 7 [*Enterprise Statistics*, p. 424 and p. 129]. This growth in large companies took place at the same time that the number of smaller companies declined 44 percent. Thus, by the survivor technique, large companies appear to be the most economical.

Data from the 1958 and 1963 Census of Mineral Industries indicates that in 1958 only 6.7 percent of the total companies involved in coal mining were multiunit operations. But five years later, the percentage of multiunit operations had declined to 5.1 percent of the total. While there may have been a trend towards larger multiunit companies since, it seems implausible to argue that they are becoming the dominant mode, although they do produce a large portion of the coal. This suggests that multiplant economies are small in coal mining. A priori they would appear to be small. Common ownership of several mines would not produce the savings that might arise in some areas of manufacturing from scheduling of output. While there might be savings from common management, they would appear to be small.

Any gains from R&D would be greater for a large firm with several mines, but would still undervalue the gains for the industry. That is, a successful R&D program that found,for example, a cheap way of turning coal into gas or which found a cheap way to desulfur coal would have huge benefits for the industry as a whole but for a single company, unless the process could be patented, would bring much smaller gains. Thus, there would be some gains from doing R&D by large firms. As a matter of fact little R&D is done by coal mining firms.

A common selling organization may also have cost benefits for a firm but such savings are likely to be small in coal mining. Buyers are increasingly large utilities or large users who normally solicit bids for coal. A firm need only submit bids to compete. Thus, little is gained from having several mines with a single selling agency.

ECONOMIES OF INTEGRATION WITH NONCOAL MINING INTERESTS

In recent years there have been a number of mergers of coal mining firms and firms in other industries. Of the 16 major coal interests acquired by firms in other industries in recent years, nine were acquired by 5 major oil firms, and 7 were acquired by 6 other large industrials such as General Dynamics, Kennecott Copper, and Wheeling-Pittsburgh Steel [*Hearings*, p. 46]. This recent trend has aroused considerable comment and controversy. Some have seen this as an attempt by oil companies to monopolize the energy market. Others have seen it producing savings by creating energy companies. Still others believe that it has resulted solely from the tax advantages to purchasing coal mining operations. Because of special provisions in the tax law (these provisions were changed in 1969) dealing with mineral development, coal mines have been more profitable to a purchaser than to a firm which has owned the enterprise for a long period. The result naturally has been that many of the major coal mines have been bought by outside interests. It is noteworthy, however, that over half of the mines purchased have been acquired by petroleum firms. The question then arises as to whether there are any cost savings or improvements in economic efficiency that would explain oil company acquisitions.

In recent years, oil companies have become increasingly involved in coal mining. In 1970, 11 out of the largest 25 oil companies held about 24 percent of the leases on public coal land outstanding [*Hearings*, p. 433]. However, it should be noted that only about 2 percent of the public domain lands classified as coal lands are under lease [*Hearings*, p. 433] and the Bureau of Mines has estimated that about 65 percent of the 220 billion tons of reserves are not now under lease to any coal company [*Hearings*, p. 130].

Petroleum companies have explained their movement into coal as a need to build a secure base of supply for their research on methods of converting coal to pipeline gas and on the production of gasoline from coal. For example, Howard Hardesty, Senior Vice President, Continental Oil Company, testified:

> It is my firm belief that the association of Continental and Consol has resulted in increased supplies of coal at fair and reasonable prices and that the blending of talents and research facilities of the two companies had made and will make a substantial contribution toward meeting the nation's escalating demands for fuel. [*Hearings*, p. 98]

Hardesty claimed that the blending of Continental and Consol research and engineering facilities hastened the development of new technology. He said:

Our combined research efforts reach across the horizon and involve pollution-free synthetic fuels, mine health, and safety, environment, efficiency and transportation. [*Hearings,* p. 101]

Many of the oil companies are actively conducting research on the conversion of coal to synthetic liquids and gas. The oil companies claim that owning coal mining facilities will provide them with the raw materials for transforming coal into other fuels. They will then be in a position to market synthetic gas or petroleum through their regular channels.

This argument is not completely convincing. Do the oil companies need to have coal mining operations to do research on conversion of coal? Since patent protection may not be perfect, especially where much of the research is being done under government contract, it may pay a firm doing such research to be in a position to convert coal to a liquid or a gas in order to fully exploit the technology when developed and perhaps get a jump on its competitor. But the ownership of mines is not necessary for this, since the coal can be purchased at competitive prices on the open market for any conversion effort.

Oil companies have traditionally owned some proportion of their crude supplies necessary for their refineries. This approach seems to be carried over to the production of petroleum products and gases from coal. The question of whether there are economies of vertical integration in the petroleum industry will be discussed below. To the extent there are such economies, they would probably apply as well to the integration of coal and oil firms when conversion of coal becomes commercially economical.

If, however, coal were priced above competitive levels, then a firm planning to use considerable quantities of coal would find it worthwhile to enter the mining industry in order to secure coal at marginal cost. There is no evidence, however, that coal is being priced above competitive levels nor does it appear that it will be priced above in the future. Were coal to be priced above competitive levels, there would be strong incentives not only for users of coal to enter the industry but also for new producers to appear and reap above competitive returns on their investment. With such vast, untapped resources of coal available, such entry would appear to be likely in any case where coal is priced above long-run marginal costs.

One justification for petroleum companies entering the coal industry is that they can bring considerable additional capital to coal mining. If the capital market is rationing capital or capital is unavailable to coal mining for some other reason, then a capital rich oil firm can help bring needed resources to coal mining. There have been periods in recent years when acute capital shortages have developed and some types of customers found borrowing difficult. However, there is no evidence that such capital shortages affected coal mining more than other industries. If in fact the capital market is working well, then this argument for integration has no validity.

On net, therefore, there may be some economies for joint research in the general area of hydrocarbons which will have benefits for both the coal industry and the petroleum industry. These benefits may be more easily captured by a firm involved in both industries although the evidence is far from clear. Other potential savings from combining oil and coal production are even more speculative.

There is some evidence on this subject although it is not completely consistent. Rate of return data indicates that independent coal mining firms do not suffer any obvious handicap. Independent coal firms earned an average 14.5 percent rate of return in 1970 whereas Consolidated Coal, owned by Continental Oil, according to company executives earned only 10 percent that year [*Hearings*, p. 104]. Consolidated Coal appears to be typical of oil-owned coal facilities. In fact, it is by far the largest of such operations and accounted in 1968 for almost half of the coal produced by mines owned by oil firms [*Hearings*, p. 12].

It is also apparent that the major independent coal mining firms are expanding as rapidly as are the operations owned by oil firms or other noncoal mining interests. From 1962 to 1969 the output of the four large independent coal mining firms, which were among the 15 largest coal producing firms in the U.S., grew about 75 percent. Total industry output grew only 33 percent while output of the 11 other largest firms grew 73 percent. Thus the independent firms apparently are able to keep up with the nonindependents. It should be noted that much of the growth of all the firms in the top 15 was due to merger and the purchase of other coal mining facilities [*Hearings*, p. 47].

On the other hand, there is some evidence that the coal firms owned by petroleum firms do have some advantage over the rest of the industry. Five of the 10 largest coal companies in 1968 were owned by firms with other energy interest. In 1968 these firms opened two-thirds of the new mines [*Hearings*, p. 343], but provided for only about 26 percent of the total industry output [*Hearings*, p. 12]. They are growing considerably faster than the rest of the industry.

From 1968 to 1970, Consolidated Coal, which is the largest of the petroleum owned coal companies, increased its output 25 percent while industry output actually declined and the total number of mines in the industry grew only 5.1 percent. During that period, Consol developed 17 new mines, for a total of 56. For the four years before the merger of Consol and Continental Oil, Consol invested $76 million in new capacity while in the four years subsequent to the merger their investment for new capacity increased to $184 million [*Hearings*, p. 99].

The data presented above portrays a mixed picture. The profit data suggests that the independent coal firms do better than the petroleum company subsidiaries, but this evidence is weak. It is not clear that the figures are on a comparable basis. In any case, they are only for one year and could be distorted by a heavy investment program being carried on by Consol that had not yet begun to produce income. The data on the growth of large independent coal firms and the other large firms suggests that they grow equally fast, but most of the growth is accounted for by merger. In other words, the data on growth between 1962 and 1969 indicates that both groups were purchasing coal facilities at about the same rate. Given the tax advantages of purchasing coal mining property it is not surprising that these firms were expanding at a rapid rate.

The most convincing data is from the growth of Consol over the period 1968 to 1970 when the rest of the industry was declining. If this is considered typical of petroleum subsidiary behavior then there are obviously social gains from the integration of oil and coal. On the other hand, these figures are for only a short period and may be an aberration. At least they suggest caution in any policy designed to prevent petroleum companies from purchasing coal facilities or for any policy of divesting past acquisitions.

While the coal operations of major oil firms are small relative to their oil operations and unlikely to make much of an impact on their overall rate of return, it is interesting to note that two of the four major oil companies purchasing major coal enterprises in recent years—Continental Oil and Standard Oil of Ohio—earned less than the average of the top 20 oil firms (see Table B-6). One, Gulf Oil, earned about the average, while only Occidental Petroleum, which had purchased in 1968 the Island Creek Group with 6.8 percent of the coal industries production, made significantly more than the average. In fact, it was the industry leader for that year.

CRUDE OIL

Crude oil production takes place through the drilling and operating of wells. Most of the businesses involved in this enterprise are small although the giant oil companies dominate the field. There are clear economic advantages from operating oil fields as a unit and there has been an increasing tendency of states as well as the federal government, to require that new fields be operated on that basis. When different wells pumping from the same field are owned by different firms or individuals, unitized production is often hard to achieve even when large economies could result from it. Thus the natural size unit for production is the oil field size unit. Oil fields differ greatly in size, some being mammoth such as the ones in the Middle East and some are of insignificant size.

Economies of scale beyond the field size are hard to identify. Exploration is often carried on by small outfits hoping for a lucky strike. But to explore an area like the arctic or off-shore may take millions of dollars and be possible only for a gigantic firm. Increasingly oil exploration is taking place in more hostile territories and thus requiring larger firms to bear the risk.

Table B-5 shows the percent of total value added contributed by

Table B-5. Selected Statistics on the Crude Petroleum and Natural Gas by Size Category of Company

Operating Companies Ranked by Value of Shipments & Receipts	Avg. Number of Barrels of Oil Shipped in 1,000 Barrels (1967)	Percentage of Total Value Added			Value Added per Manhour 1967	Value Added per Dollar Capital Expenditure 1967
		1958	1963	1967		
First 8 companies	206	43.2	46.1	54.0	$152	4.3
9-16 companies	65	15.6	13.9	17.4	90	4.9
17-24 companies	23	8.9	7.5	6.4	95	5.1
25-32 companies	10	3.2	3.5	3.4	116	3.0
33-68 companies	10	7.6	7.2	6.5	59	5.0
69-100 companies	2	2.8	2.9	2.0	41	5.9
All others Less than	0.1	18.7	19.0	10.3	27	3.6
Total Less than	2	100	100	100	84	4.3

Source: U.S. Bureau of Census, *Census of Mineral Industries: 1963*, Vol. I, Summary and Industry Statistics (U.S. Government Printing Office, Washington, D.C., 1967), p. 13B98.
U.S. Bureau of Census, *Census of Mineral Industries: 1967*, Industry Series: Supplement MI667(1)-13A, "Crude Petroleum and Natural Gas," p. 4.

each size category of firms for the years 1958, 1963, and 1967. Using the survivor technique it is apparent that only the two largest size categories are clearly economic. Other size categories are losing share of output. Even among the two largest categories, only the largest eight firms gained as a share of the industry output from each census to the next census.

Table B-5 also shows the amount of value added per manhour and value added per dollar of investment. As can be seen, value added per manhour is inversely correlated with the size of the company. The largest companies have significantly higher value added per manhour than do smaller. This, of course, may be due to a higher level of investment per manhour for larger companies and may not permit them to have significantly lower costs than smaller operators. But as can be seen on the basis of one year's figures, value added per dollar of capital expenditures while lower for the biggest eight companies than for many of the smaller categories, is just equal to the industry average. It seems unlikely that the efficiency in terms of manhours is entirely offset by less value added per unit of capital.

PETROLEUM REFINING

A considerable number of estimates have been made of economies of scale in petroleum refining. Surprisingly enough, the estimates are reasonably consistent, which gives confidence in the figures. However, most of the data and estimates are for the refinery and not for companies.

One of the earliest estimates was made by Bain on the basis of engineering data gathered from published sources and from industry interviews. For the petroleum refining industry, he found that a plant of the minimum efficient size—120,000 barrels per day—would provide 1.75 percent of industry capacity and would have cost in 1951 between $225 and $250 million including transportation facilities [Bain, pp. 72, 158, and 233]. After adjusting for inflation, it would take about $400 million today to construct such a refinery. Bain also claimed that there were no economies from owning more than one refinery.

In a speech before the Economics Club of Detroit in March of 1973, Rawleigh Warner, Jr., Chairman of Mobil Oil Corporation, asserted that the minimum economic size for a new refinery was about 160,000 barrels a day and would cost in the neighborhood of a quarter of a billion dollars. This size refinery would be about 1.2 percent of industry capacity. These figures are reasonably consistent with Bain's estimates for two decades earlier.

The inventor of the survivor technique, George Stigler, applied it to petroleum refining. He found that the plant size between 0.5 and 2.5 of industry capacity had grown relatively rapidly between 1947 and 1954, and were therefore efficient [Stigler, p. 69]. These figures bracket Bain's estimates. For companies, Stigler found that the efficiency sizes ranged from 0.5 to 10 percent of industry capacity and since large plants (over 2.5 percent of industry capacity) have lost share of market, concluded that the growth of companies beyond 2.5 percent of capacity was due to the economies of multiple plant operation. But since the minimum efficient size plant and company were the same, it would appear that there are no significant savings in multiplant operations.

Saving applied the survivor technique to a large number of census industries for the years 1947 to 1954. He found that the minimum efficient size for petroleum plants was 100 employees or about 0.12 percent of industry output [Saving, p. 600].

Leonard Weiss, using the survivor technique for the years 1958 to 1961, found that there were two optimum size categories, one for small relatively specialized plants and one for large diversified refineries. The minimum optimum size he found for the diversified plants was a refinery with capacity of about 150,000 barrels per day [Weiss, p. 249].

William G. Shepherd also attempted to estimate the minimum efficient size by the survivor technique utilizing census of manufacturer's data for 1947-1958. Shepherd reported his data as unclear.

Census data for 1963 and 1967 shows that refining establishments with fewer than 20 employees have been growing (in terms of the percent of value added) more rapidly than the industry, that establishments greater than 100 employees but less than 1,000 have also been growing more rapidly than the industry. Thus the same bimodal result Weiss found appears in the more recent data. The growth of small refineries may be due to the allocation of crude oil import tickets to such refineries. It seems fair to conclude that the most efficient sizes for general refineries are plants with at least 100 employees. Refineries larger than 250 employees have been growing even more rapidly. While data for plants of 500 to 1,000 employees was not given separately in the 1967 census, it would appear that there has been some reduction in optimum size in terms of employment in recent years.

C. Pratten and R.M. Dean investigated economies of scale in petroleum refining for Great Britain. Their basic conclusion was that cost savings were evident on the basis of engineering data up to the largest refineries then available in the U.K. Their largest category was 10 million long tons per year capacity or about 200,000 barrels per day. As suggested above, engineering estimates of economies of scale are very likely to indicate decreasing costs for increasing size since they fail to take into account the factors that will lead to higher costs, such as management inefficiencies. In an update of these estimates, Pratten found that costs continued to decline up to a capacity level of 20 million tons (400,000 barrels per day) although the saving for a 20 million ton plant over a 10 million was small. Thus a plant of about 10 million tons would exhaust most of the savings, and if other factors such as the additional time necessary to construct a large refinery and any additional management inefficiencies are taken into account, it would seem that these British figures are consistent with the American based estimates.

OIL COMPANIES

As is well known, all the large oil firms are integrated from crude oil production through distribution. While normally they do not have sufficient crude to supply all their own needs, they often have substantial crude interests. There are, however, independent refiners, independent marketers, as well as independent crude producers. The important question, however, is whether integration brings savings and whether larger firms have cost advantages over smaller ones.

There are a number of reasons why vertical integration may produce savings over having independent enterprises operating at each level.[1] For integration to produce savings, internal coordination might result from situations where frequent and difficult bargaining are required, because rapid change, uncertainties, and complicated technologies are involved. Savings may result if securing accurate information is difficult or costs of policing contracts are high.

None of these advantages of integration are apparent in the integration of crude oil production and refining. The product is simple, reasonably standardized, takes little special investment that might become obsolete with change (except possibly for pipelines), and information needs are not so great that lack of complete and accurate information would impose any real handicap.

Large oil companies may be better able to bear the risk of prospecting for oil, especially in areas of high costs. The tax laws, however, make it more profitable for individuals with marginal tax rates over 48 percent (the corporate rate) to explore for oil than for firms. Such individuals have the federal government as a partner who bears a proportion of any loss equal to the individual's marginal tax rate. Thus exploration for oil should be dominated by wealthy individuals in high tax brackets and large oil companies that can spread the risk over a large number of ventures. If it were not for the tax laws, oil exploration might be even more concentrated in the hands of petroleum firms.

While the risk of exploring for oil may explain why most large petroleum companies are integrated from production through refining, it would not explain a movement into coal production. The risks in such an enterprise are obviously of a different magnitude than oil exploration.

There is some evidence on the question of economies of scale for petroleum enterprises. Stigler found that refining companies without crude pipelines survived as well as those with such pipelines [Stigler, p. 70]. John Moroney estimated the coefficients for Cobb-Douglas production functions for various industry groups, including SIC 29 (Petroleum and Coal Products)—a category dominated by petroleum refining. He found constant returns to scale using data from the 1958 census of manufactures [Moroney, p. 46].

If there are substantial economies of scale among integrated petroleum companies, they should be reflected in profit rates. No such relationship is evident. Table B-6 lists the 20 oil companies which have appeared on Fortune's 500 list for the last five years with their average rate of return on invested capital over this period. As can be seen, the rates of return are uncorrelated with the size of the firm. The most profitable oil company was Occidental Petroleum with an average rate of return of 16.8 percent, but in 1971 it was only the 11th largest. The next most profitable was Clark Oil and Refining Company, with a rate of return of 16.2 percent, yet it was in 19th place. Standard Oil of New Jersey earned 12.2 percent, slightly better than the average, while the second largest, Mobil, earned only 10.5 percent. These statistics therefore do not indicate any overall economies of large scale operation.

Table B-6. Profits as a Percent of Invested Capital, Twenty Largest U.S. Petroleum Firms (average of the years 1967-1971)

Firm *(ranked according to 1971 sales)*	*Average* *Profit Rate*
1. Standard Oil of New Jersey	12.2
2. Mobil Oil	10.5
3. Texaco	14.1
4. Gulf Oil	11.8
5. Standard Oil of California	10.4
6. Standard Oil of Indiana	9.7
7. Shell Oil	10.9
8. Atlantic Richfield	8.8
9. Continental Oil	10.0
10. Tenneco	11.2
11. Occidental Petroleum	16.8
12. Phillips Petroleum	8.4
13. Union Oil of California	9.5
14. Sun Oil	9.6
15. Ashland Oil	11.2
16. Standard Oil of Ohio	8.8
17. Getty Oil	9.1
18. Marathon Oil	11.6
19. Clark Oil	16.2
20. Commonwealth Oil Refinery	13.4
Average	11.2

Source: *Fortune*, May issues.

NOTE

1. See Oliver E. Williamson, "The Vertical Integration of Production: Market Failure Considerations," *American Economic Review*, May 1971, pp. 112-123.

DATA SOURCES

U.S. Bureau of the Census. *Statistical Abstract of the U.S.:* 1972 (U.S. Government Printing Office, Washington, D.C., 1972).

U.S. Bureau of Mines. "Injury Experience in Coal Mining, 1968," prepared by F.T. Moyer and Mary B. McNair (U.S. Department of Interior, 1972).

U.S. Bureau of the Census. *Enterprise Statistics:* 1967, Part 1—General Report on Industrial Organization (U.S. Government Printing Office, Washington, D.C., 1972).

U.S. Bureau of the Census. *Census of Mineral Industries:* 1958 and 1963, Vol. 1, Summary and Industry Statistics (U.S. Government Printing Office, Washington, D.C., 1961, 1967).

BIBLIOGRAPHY

Joe S. Bain. *Barriers to New Competition* (Cambridge: Harvard University Press, 1965).

John R. Moroney. "Cobb-Douglas Production Functions and Returns to Scale in U.S. Manufacturing Industry," *Western Economic Journal*, VI (December, 1967), pp. 39-51.

Reed Moyer. *Competition in the Midwestern Coal Industry* (Cambridge, Mass.: Harvard University Press, 1964).

C.F. Pratten. *Economies of Scale in Manufacturing Industry* (Cambridge: Cambridge University Press, 1971).

C. Pratten and R.M. Dean. *The Economies of Large-Scale Production in British Industry* (Cambridge: Cambridge University Press, 1965).

T.R. Saving. "Estimation of Optimum Size Plant by the Survivor Technique," *Quarterly Journal of Economics*, Vol. LXXV (November, 1961), pp. 569-607.

William G. Shepherd. "What Does the Survivor Technique Show About Economies of Scale?" *The Southern Economic Journal* (July 1967), pp. 113-22.

George J. Stigler. "The Economies of Scale," *The Journal of Law and Economics*, 1 (October 1958), pp. 54-71.

U.S. Congress, House Subcommittee on Special Small Business Problems of the Select Committee on Small Business, *Hearings on Concentration by Competing Raw Fuel Industries in the Energy Market and Its Impact on Small Business*, 92nd Congress, First session.

Leonard W. Weiss. "The Survival Technique and the Extent of Suboptimal Capacity," *The Journal of Political Economy*, LXXII (June 1964), pp. 246-61.

Appendix C

The Coal Industry

Reed Moyer

This chapter is divided into two parts which, in some respects, are interconnected. In the first part of the study I analyze certain aspects of entry conditions in coal mining. I determine capital costs associated with entry into conventional coal mining operations and pay particular attention to capital costs required for coal mines used to supply coal gasification plants. I determine the implications on the ease of entry into coal mining from these capital requirements. Also I study cost and productivity data to determine the extent to which scale economies exist both in underground and in strip mining. The extent to which they exist bears on entry conditions in the industry.

The second part of the study looks at some implications for the coal industry of a potential massive shift toward the use of low sulphur coal to meet possible restrictive air pollution standards. Since much of the country's low sulphur coal is likely to be used as feedstock for coal gasification plants, the second part of the study is intertwined with the first. Furthermore, since ownership of low sulphur western coal differs from ownership of eastern coal reserves (where most coal mining currently occurs), a shift toward western coal mining has interesting market structure implications which this study analyzes.

ENTRY CONDITIONS IN COAL MINING

In 1971 bituminous coal production in the United States totalled 552.2 million tons.[1] Underground mining accounted for 50 percent of the total, strip mining, 46.9 percent and auger mining 3.1 percent. As Table C-1 shows, the percentage represented by strip mining has steadily increased, especially in recent years. Auger mining accounts for a small but steady percentage of the total. This technique, combining features of underground and surface mining, will not

Table C-1. Bituminous Coal Mining Production, 1920-1971
(millions of tons)

| Year | Tonnage | Percent of Total Production | | |
		Underground	Strip	Auger
1920	568.7	98.5	1.5	—
1925	520.0	98.8	1.2	—
1930	467.5	95.7	4.3	—
1935	372.4	93.6	6.4	—
1940	460.8	90.6	9.4	—
1945	577.6	81.0	19.0	—
1950	516.3	76.1	23.9	—
1955	464.6	73.9	24.8	1.3
1960	415.5	68.6	29.5	1.9
1965	512.1	64.9	32.3	2.8
1967	552.6	63.1	33.9	3.0
1969	560.5	61.9	35.2	2.9
1971	552.2	50.0	46.9	3.1

Source: *Bituminous Coal Facts 1972*, National Coal Association, n.d.

concern us further in this study. Instead our attention centers on underground and strip mining.

Mining Techniques

To understand cost and entry conditions and capital requirements in coal mining requires first a brief explanation of strip and underground mining techniques.

Coal mining may be carried out either below or above ground. In underground mining, coal is brought to the surface in one of three ways: through use of a shaft, slope, or drift (horizontal) opening. Topographic conditions and equipment used determine which of the three systems is adopted. The openings are used to convey men, supplies, and equipment to and from the working areas and to transport raw coal from underground workings to the surface.

In a conventional, mechanized, underground mine, mining is divided into several distinct operations. The coal is first undercut to facilitate its breaking away from the solid face. The seam is then drilled, these drill holes are filled with explosives, and a section of the coal is blasted away from the seam. Mechanical loading machines gather in the shot-down coal with clawlike arms, pulling it onto a built-in conveyor. The coal discharges from the conveyor into a waiting shuttle car or a longer extensible conveyor which transports it to the main haulageway. From there, the coal moves via either another conveyor or an underground small-gauge railroad directly to the surface, or in the case of a shaft mine, to a skip hoist which subsequently lifts the coal above ground.

Though this system is far removed from former techniques in which coal was hand loaded and mules and horses provided the energy for underground transportation, still it is a laborious and costly method of extraction. In recent years, continuous mining machines have been perfected which combine into one operation the separate jobs of undercutting, drilling, blasting, and loading.

Whether a conventional or continuous mining system is used, most underground mining is done by the room-and-pillar method. The area underground is crisscrossed with haulageways and entries which divide the mine into separate segments. Each segment again is divided by entryways into workable areas, referred to as rooms. Barriers and pillars of coal are left as roof supports between rooms and around the main shaft and haulageways to prevent subsidence. The loss of coal in these supports reduces the coal recovery in most underground mines to 45 to 60 percent of the original reserve.

Each room is a separate production center, outfitted with a complete set of machines and connected with haulageways by its own minor transportation system. The older a mine becomes, the farther from the main shaft these production centers move. Underground transportation costs increase concomitantly. Diminishing returns in underground mining arise principally from the wavelike expansion of production farther and farther away from the shaft or slope opening.

The strip-mining production cycle differs sharply from the methods used in underground production. The consolidated and unconsolidated material (called overburden) lying above a coal seam is first removed by shovels or draglines whose bucket capacity ranges in size from a cubic yard or two to 220 cubic yards on the behemoths recently put into production. The coal having been laid bare by the excavating equipment, smaller shovels then load it into large trucks to be hauled via mine roads to the preparation plant for processing.

Strip mining offers decided advantages over underground in several respects. First, mining is conducted in the open, eliminating the need for ventilation and increasing the safety factor. Moreover, the use of large, mechanized equipment reduces labor requirements and cuts costs. Finally, removing the requirement for support pillars and barriers conserves coal by increasing coal recovery to 80-90 percent of the original reserve.[2]

The distinction between underground and strip-mined coal disappears when the coal enters a preparation plant. The raw coal is usually crushed and sized and, in many operations, washed to remove impurities, and then loaded onto conveyor belts or into railroad cars, trucks, or barges for shipment to market.

Capital Investment Requirements

Very little published data exists on the capital costs incurred in coal mining. Most of this information is maintained in the confidential records of the coal mining companies. Occasionally one finds published estimates of capital costs for strip and underground mines, but they fail to be supported by hard data. Estimates of capital costs for strip mines run from $4-10 a ton of annual capacity; for underground mines the estimates usually run from $10-20 a ton of annual capacity.

Several factors account for the wide ranging estimates. First is the size of the operation. Scale factors may either increase or decrease the capital costs per ton of output. Consider a strip mining example. The major item of capital expenditure is the large unit(s) of stripping equipment used to remove overburden and lay bare the coal seam. The trend until recently has been toward the use of draglines and shovels with increasingly large bucket capacities. From a maximum capacity several decades ago of 25 cubic yards, machines increased up to as much as 220 cubic yard capacity. As mines grew larger, the scale of most of the other major components in the mining process also increased—especially in strip mining. Loading shovels, haulage trucks, preparation plants, and drills, among other things, became available in larger sizes. In underground mining, larger size occurred more from the concurrent use of more mining modules than from an increase in the size of the mining equipment (except for preparation plants which, as in strip mining, also increased in size over time).

The use of larger equipment, however, does not necessarily spell lower capital costs per ton of output. An unpublished study by a strip mine operator reveals that capital costs per cubic yard of bucket capacity for stripping equipment decline up to around 60 yards, remain fairly flat for machines with buckets of 60-100 yards capacity, but increase for machines with bucket sizes above 100 yards.

Other factors contribute to varying capital costs. Mining conditions may differ substantially from one mine to another. Among factors affecting capital costs are: thickness of seam, depth of overburden, number of seams being mined, extent to which the coal is processed (washed, crushed, sized), mode of entry for underground mines (shaft, slope or drift), and roof conditions over the coal. Also, capital costs vary considerably depending on whether or not coal acquisition costs are included in the figures. Finally how one calculates the costs may influence the estimates. Whether one adds an allowance for funds used during construction, and includes start-up costs and working capital will affect the total. One must also decide whether to measure capital costs for initial mining operations or whether to include later capital costs required to *maintain* a given level of output.

The National Petroleum Council's energy study throws some light on capital costs in coal mining. Table C-2 summarizes their findings. Capital costs include initial mine investment, land acquisition and exploration costs, and working capital. They are based on historical values for the ten years prior to the estimate.

These figures require a word of caution. The estimates for 1970 represent the average original capital investment for mines in operation at that

Table C-2. Estimated Capital Investment per Annual Ton of Production at U.S. Coal Mines

Oper. year	*Underground Mines*				*Surface Mines*			
	1970	1975	1980	1985	1970	1975	1980	1985
Orig. cap. invest.	$ 7.15	$ 8.46	$ 9.20	$ 9.84	$ 6.39	$ 7.33	$ 8.07	$ 8.78
Tot. cap. invest. over life of mine[a]	19.66	23.17	25.03	26.64	10.59	12.15	13.79	14.44

[a]Less salvage value. N.B.: 30 year life (constant 1970 dollars).
Source: National Petroleum Council, *U.S. Energy Outlook*, p. 145.

time. N.P.C.'s model projects a growth in coal demand of 3 percent a year, so the figures for years beyond 1970 reflect *average* capital costs for each year, assuming that new mines are added to meet the demand growth and replace worked-out mines. Since the model projects a 30-year life for each mine, capital costs should presumably stabilize toward the end of the century at something over $10 a ton of annual capacity for the original investment in underground mines and over $9 a ton for surface mines. Since values are expressed in *constant 1970 dollars* these two figures should give rough estimates of capital investment requirements per ton of annual output in that year. These figures are deficient in several respects. An important deficiency is the failure to specify the hypothetical mine sizes used in constructing the model's investment figures.

We have other fragmented data on capital investments required for entry into coal mining. An unpublished Bureau of Mines study develops data for hypothetical eastern underground mines operating in a 72-inch seam. The estimate assumes the investments will be made to expand existing mines rather than to develop mines *de novo*. Hence the figures omit capital costs involved in digging a shaft or slope. Also since the model assumes the coal will be used as raw fuel for a coal gasification plant, the data exclude costs of erecting a cleaning plant. Based on 1972 costs, a mine with an annual output of 1.13 million tons would cost $6.56 a ton of annual output to erect and an additional $6.56 a ton for deferred capital costs. At 3.12 million tons of annual capacity, the total investment would be $15.3 million or $4.91 a ton of annual capacity with an additional $15.7 million required for deferred capital investment. Including investment to construct a cleaning plant and to dig a shaft or slope would approximately double the capital cost.

Taken alone, the capital investment required for different sized mines means little. It takes on added meaning, however, in light of the Bureau of Mines' finding that the optimum scale plant (lowest average mining cost per ton) was an underground mine with an annual output rating of 3.8 million tons. This output would require a capital investment of approximately $18 million on the stripped down basis referred to above, or $36 million including a cleaning plant and shaft or slope excavation costs. Later we elaborate the competitive implications of this investment level required to achieve optimum costs.

As previously indicated, the capital investment required for entry into coal mining varies substantially. But it probably fluctuates more for strip than for underground mining. This is because of the importance of stripping equipment in the initial capital cost and the variability of stripping ratios (roughly, overburden depth: coal thickness). The greater the stripping ratio, the more equipment required to uncover the coal deposit.

Again we have hypothetical investment requirements for model mines—this time for surface mines—in a Bureau of Mines study. Table C-3 summarizes the capital costs for twelve hypothetical strip mines based on costs effective in the latter part of 1969. None of the models includes a cleaning plant, although each operation provides for screening and crushing the raw coal.

The capital costs for the twelve hypothetical strip mines fluctuate substantially—from a low of $2.78 per ton of annual capacity to $16.00 for the Oklahoma mine operating in a very thin seam. Table C-3 demonstrates the significant impact on investment requirements of stripping ratios. The Southwestern United States mine (Arizona, Colorado, New Mexico or Utah) with an annual capacity of 5 million tons and an 8.8:1 stripping ratio requires a capital

Table C-3. Estimated Investment Requirements for Model Strip Mines, 1969

Location	Annual Production (mil. of tons)	Estimated Cap. Invest. (mil. of dol.)	Investment per Ton of Ann. Output	Stripping Ratio
Northern W.Va.	1	$12.7	$12.70	18:1
Northern W. Va.	3	28.0	9.33	18:1
West Kentucky	1	13.7	13.70	18.2:1
West Kentucky (2 seams)	1	8.3	8.30	10:1
West Kentucky	3	24.9	8.30	18.2:1
Oklahoma	1	16.0	16.00	24:1
Southwest U.S.	1	7.9	7.90	7.5:1
Southwest U.S.	5	28.7	5.74	8.8:1
Montana	5	13.9	2.78	3:1
Wyoming	5	13.9	2.78	3:1
N. Dakota	1	6.4	6.40	4:1
N. Dakota	5	20.8	4.16	5:1

Source: U.S. Bureau of Mines, "Cost Analyses of Model Mines for Strip Mining of Coal in the United States," Information Circular 8535.

outlay of $28.7 million, twice the cost of a similarly-sized Montana mine with a 3:1 stripping ratio.

In addition to the estimated investment cost of these hypothetical strip mines we have data on a projected strip mine designed to supply coal for a large coal gasification plant. This information is especially useful in view of the possibly expanded role that coal gasification may play in the energy picture. A strip mine in New Mexico to be developed by El Paso Natural Gas Co. to supply the coal feed for its proposed gasification plant would cost $45.7 million, (excluding leasehold costs) or $5.17 per ton for a mine with a capacity of 8.84 million tons.[3] This total includes $4.5 million as an allowance for funds used during construction. This operation has incurred substantial leasehold costs in excess of normal. When these are included the original capital investment total rises to $64.6 million, or $7.31 per ton of annual capacity. Costs are those in effect in 1972. These figures are perhaps more useful than the Bureau of Mines' model mine investment data in Table C-3, since they represent figures for an actual mine, whereas the Bureau of Mines' data represent estimates for hypothetical mines. Nonetheless, the $5.17 per ton capital investment figure for the El Paso operation comes close to the estimates per ton of capacity for the 5 million tons-per-year mines in Table C-3 for the Southwestern U.S. and North Dakota, after adjustment for the cost inflation between the two dates used for the estimates (1969 and 1972) and for the difference in the mines' stripping ratios. The El Paso mine will experience an average ratio of 5.53:1 compared with 8.8:1 for the Southwestern U.S. model mine and 5:1 for the North Dakota lignite mine.

From the standpoint of analyzing the level of entry barriers in coal mining, the size of the El Paso and other proposed coal gasification plants is significant. The initial coal gasification plants are scheduled to produce 250 million cubic feet a day (cfd) of synthetic natural gas. Both the proposed El Paso plant and a Texas Eastern Transmission Co. plant in New Mexico call for an initial capacity of 250 MM cfd. The Texas Eastern operation may be expanded later to a total of four plants of similar size. Coal requirements would similarly increase four-fold over the initial requirements. Northern Natural Gas Co. and Cities Service Co. have announced plans to study the possibility of a joint coal gasification and pipeline venture that would involve four coal gasification plants, each with a capacity of 250 MM cfd daily.[4] Peabody Coal Co. would supply the coal from its reserves in southeastern Montana. Other studies which analyze the feasibility and cost of producing synthetic natural gas, figure the minimum scale plant to be rated at 250 MM cfd. The only existing or proposed S. N. G. plants using coal as feedstock which are smaller than this size are experimental or pilot plants.

Plants rated at 250 MM cfd require different tonnages of coal depending on the coal's heating value. Plants of this size, producing at a 90 percent operating rate, have the following annual coal requirements for the coal feeds:[5]

	Coal Quality (BTU/lb)	Million Tons
Bituminous coal	11,500	5.3
Subbituminous coal	8,500	7.2
Lignite	6,750	9.1

In addition, the operations require the use of steam in the gasification process and as a power source, and this adds 20 percent to the coal requirements. Thus gasification plants such as those proposed in New Mexico and Montana, using subbituminous coal require complementary coal operations with annual capacity of approximately 9 million tons to support a 250 MM cfd gasification plant, and 36 million tons to support four modules of gasification plants each with that capacity. At a cost of $5.17 a ton of annual capacity (the El Paso figure), the initial investment requirement for the four coal mines necessary to supply the four gas plants would total $186 million. It is unlikely that this investment would occur at one time. The Texas Eastern Transmission Co. proposal (with Utah International, Inc. supplying the coal) calls for an initial investment in a single 250 MM cfd gasification plant with the possibility of three additional plants at a later date. The Northern Natural Gas-Cities Service-Peabody proposed operation would find Peabody supplying coal for two 250 MM cfd plants initially with two others to follow later.

In either case—with the development of single or multiple gasification plants—the capital requirements for the supportive coal mines are substantial and create a sizeable entry barrier. The barriers appear even greater when compared with capital requirements traditionally needed for the coal industry. Any of the proposed coal mines needed to supply feedstock for the gasification plants discussed above would have a capacity greater than the 6.7 million tons produced in 1971 by Utah International, Inc.'s Navajo mine, the largest U.S. coal mine. In the same year the average output of the 5149 bituminous coal mines was approximately 107,000 tons.[6] The fifty largest mines, accounting for 22.8 percent of total bituminous coal production in 1971, had an average output of 2,516,000 tons, 30-40 percent of the size required to supply a 250 MM cfd coal gasification plant.

There is an additional factor affecting entry into coal mining to support a gasification operation: the extensive coal reserves necessary to sustain the coal mine over the life of the operation. To amortize the gasification-coal complex requires a useful life of at least 25-30 years or coal reserves of 150-250 million tons, depending on coal quality. Utah International, Inc. has dedicated 249 million tons to the Texas Eastern Transmission Co. project; El Paso has set aside 225 million tons of its own coal reserves for its proposed plant. If the four Northern Natural Gas-Cities Service gasification plants materialize, Peabody Coal Co.—the coal supplier—will have to dedicate one billion tons of coal reserves to the project.

If coal gasification occurred in the middle west or in the east, both coal capacity (in tons of annual output) and coal reserve requirements would diminish because of the coal's higher BTU content per *ton*. An Illinois mine, for example, might need a 150-ton coal reserve to support a 250 MM cfd gasification plant, and slightly over six million tons of annual capacity. The capital investment costs, however, are liable to be higher than those required for western strip mines because the absence of large blocks of strip coal in the midwest (and east) requires the substitution of underground mines with their greater capital cost per ton of annual output.

One might ask whether entry barriers into mining operations supporting gasification plants are necessarily as high as here indicated. Is it not possible for several companies' mines to supply a single gasification plant, lowering the ticket of admission into the game for each participant? Several factors work against this development. First are scale economies in coal production which dictate the use of large units. A 250 MM cfd gasification plant would require coal from perhaps one and not more than two coal mines producing at optimum scale. (More on this when we discuss below the costs of production in coal mining.) Also, minimizing coal costs requires that costs of transporting coal from the mine to the gasification plant be eliminated or minimized. It is not feasible to incur the transportation charges involved in hauling coal from several mine locations to the gasification plant. Furthermore, reorganizing the enormous coal requirements for gasification plants has led coal producers (and others interested in acquiring coal reserves for conversion to synthetic fuels) to amass large contiguous coal blocks that can sustain a gasification plant for its useful life.

Although gasification plants represent a large potential market for coal, the electric utilities provide the largest existing coal market segment, accounting for 59 percent of U.S. coal production in 1971, up from 20 percent in 1951. In the same period, sales of coal to electric utilities increased 219 percent. Demand from this sector is estimated to more than double between 1970 and 1985.

Along with the growth in this sector have come several developments affecting entry conditions in coal mining, and as a result, the market structure in the industry. The growth of the utilities and their similarly growing need for large and dependable long-term supplies of coal have led to the utilities' increasing reliance on long-term coal contracts. These contracts call for the utility to purchase a large percentage of the requirements for given generating stations from several coal suppliers for periods of five to twenty years or more. The contracts protect the coal operators' profit margins over the life of the contract through price escalation provisions. The contracts effectively lock-in the coal operators to each submarket (the generating station) for the life of the contract and seal off these markets from outside competition during the contract's duration. Increasingly electric utilities contract with a single coal supplier for the entire coal requirements of new generating stations. As generating stations increase in size, so too must the coal mines necessary to supply those stations. We can throw light on the entry conditions in coal to serve this dominant market by studying the coal requirements of new and proposed electric generating stations. This we now do.

Table C-4 summarizes data on proposed new generating plants in the period 1972-1977 that are scheduled to burn coal exclusively. The data excludes extensions to existing generating plants, some of which involve additions to capacity that are as large as the units installed in new operations. Others involve the installation of smaller units. Many of the additions will be fired by coal supplied by a single coal operator. In other cases several suppliers may share the new business.

The focus of Table C-4 is on the larger, totally new plants that call for the commitment of new large coal supplies under long-term contracts. Some of the new operations will be mine-mouth power plants that have coal from a single large coal supplier operating at the site of the power plant providing the entire coal requirements for the generating station. Even in the case of generating plants located away from the coal fields, many, if not most, of them will be fueled by a single, essentially captive supplier. For each of the larger generating stations covered by Table C-4, coal requirements exceed three million tons a year; for the 27 intermediate-sized plants, the coal needs approach two million tons a year. Unlike the situation with coal gasification, many of the coal operations supplying electric utilities require coal cleaning facilities, which increases the capital investment requirements for the new mines. At a rate of $10-15 per ton of annual output, mines capable of supplying the entire coal requirements for the intermediate- and large-scale plants in Table C-4 would require capital investments of $19-46 million.

Whether this capital requirement sufficiently discourages entry to

Table C-4.   Proposed New Coal-Burning Steam-Electric Generating Plants, 1972-1977

Installed Capacity	No. of Plants	Total[a] Capacity	Coal[b] Tonnage	Avg. Annual Coal Consump. per Plant (tons)
−500 megawatts	11	3412	8809	801,000 tons
500-1000 megawatts	27	17691	50958	1,887,000 tons
+1000 megawatts	6	8360	18,400	3,066,000 tons

[a]Megawatts.

[b]Millions of tons. (Assumed annual coal consumption of 2.5 tons per kw of annual capacity for generating stations using bituminous coal, 3.5 tons for subbituminous coal and 5 tons for lignite.)

Source: *Steam Electric Plant Factors/1972 Edition*, National Coal Association, December 1972.

adversely affect competition depends upon the perspective from which the analysis is made. Although ownership in the coal industry is more concentrated than it was several decades ago, it is still highly fragmented. Of the more than 5000 coal companies in the U.S., all but a handful are financially incapable of mobilizing the capital necessary to develop the intermediate- and large-scale plants in Table C-4—and few others could finance even the smaller sized plants. In 1971, there were only 19 commercial (noncaptive) coal mining companies with annual output larger than the annual coal requirements of the eight scheduled large-scale plants in Table C-4, and 31 with output in excess of the requirements of the intermediate-sized plants.

There are a handful of other companies, outside of the coal industry, however, which are capable of entry, even on the basis of the largest power plants' needs. These are companies which control large blocks of uncommitted coal reserves but which, for good reasons, have failed to enter into active production, or produce at a fairly low level of output. Many of the potential entrants are large petroleum firms that have acquired substantial coal reserves in recent years and are either holding them for future use or are just beginning to develop the properties.[9] Humble, for example, operates a recently opened mine in Illinois (with 1971 tonnage of 1.2 million tons) and has coal reserves (and obviously the financial capability) sufficiently large to meet the needs of even the largest coal users. Atlantic Richfield has announced plans to develop a western coal mine to supply an Oklahoma utility with enough coal to warrant shipment by unit train of 10,000 tons daily.[10]

There is another related issue which affects market power in the coal industry and which indirectly bears upon the ability of potential entrants to penetrate the electric utility market. This is the fact that the coal industry is composed of a number of *regional* markets rather than a single national market. Electric utilities, similarly, operate in regional markets. Hence, there may be a much smaller number of coal companies competing for the sale of coal under long-term contract to supply a proposed new generating station than the information in the preceding two paragraphs might indicate. These few companies have developed strong ties with the electric utilities based on their long-standing supplier relationships with them. Outsiders lack this experience.

How concentrated are some of these regional markets? Two of the forty-four new plants included in Table C-4 are located in Indiana where three companies account for 95 percent of the production and of most of the coal reserves capable of sustaining large scale operations. Market conditions dictate that Indiana mines supply most Indiana electric utilities.

The situation is only slightly less restrictive in Illinois, where *three* new generating stations are scheduled to be introduced in the period 1972-77. There, eleven companies in 1971 (later reduced by acquisition to ten) accounted for 99 percent of the state's commercial production; for the top five, the figure was 77 percent. As in Indiana, market conditions call for the use of in-state production to supply the state's electric generating stations. Only nine commercial coal companies in Illinois had coal reserves in 1968 sufficiently large to sustain coal operations of the size called for by the generating stations underlying the data in Table C-4. And in two cases the companies' reserves are probably committed to existing operations; therefore, they do not permit the expansion of production necessary to fulfill new large commitments.

In other submarkets competitive conditions are less restrictive for one of two reasons: either more than one mining district or state (they are not always synonymous) can effectively compete in the submarket—thereby opening up competition to more producers—or concentration levels within given districts or states are lower. Nonetheless, market structures in other coal supply areas are not vastly different from those in Illinois and Indiana. Certain submarkets can be effectively supplied only from western Kentucky mines. In 1971 the six leading coal producers in that district accounted for 70 percent of the district's output. Three of the producers were also found on the list of Illinois' top producers. In fact 90 percent of the commercial coal market served by the Eastern Interior coal region (Illinois, Indiana and western Kentucky), which goes far beyond the three producing states, was served in 1971 by only thirteen producers. The top four producers accounted for 61 percent of the total commercial production.

Cost and Productivity Conditions

To complete the picture of entry conditions in coal mining, we need to investigate the shape of long-run cost curves. A study of long-run costs serves several purposes. It indicates the extent to which firms can achieve scale economies and helps us understand whether enlarging production units in the industry serves industrial efficiency. But more importantly for the analysis in this chapter, it aids in ascertaining if the optimum scale plants are large enough to penalize potential entrants which are incapable of achieving optimum scale, thereby contracting the number of sellers in the industry.

The industry standard for measuring productivity is tons produced per man/day. This figure is also inversely related to average mining cost. In a linear regression analysis relating stripping ratios and tons per man/day to the average value per ton for strip coal produced in the 15 major coal producing states, the partial correlation coefficient for the average value and tons per man/day figures was .59. I have attempted through regression analysis to determine the factors accounting for tons per man/day (TMD) in Illinois where data necessary for the analysis is readily available. Separate regressions were run for strip and underground mines, and in both cases a linear least squares model of the form $y = a + bx_1 + cx_2 + nx_n$ was used. The strip coal regression includes two independent variables—mine size and stripping ratio. The underground mine analysis has as independent variables, mine size, seam thickness, seam depth, age of mine, and a dummy variable representing whether the mine has a shaft or slope opening. Table C-5 summarizes the regression results.

In both cases mine size is a significant variable at the 0.05 level. Stripping ratio is also significant but not as significant as in the national analysis referred to above, where the partial correlation coefficient for TMD and stripping ratio was .72. For the strip coal analysis, the signs of the coefficient are what one would intuitively expect.

Mine size is the only statistically significant variable (at the 0.05 level) in the underground mine analysis, but the signs of each of the variables except x_3, seam depth, are what one would expect. The R^2 values for both equations are significant at the 0.05 level.

Table C-5. Regression Results of Coal Mining Productivity Analysis for Illinois Mines, 1971

		Strip Coefficient Standard	Error	t-value
Tons/Man/Day	Y			
Constant		44.717		
Mine size	X_1	0.00669	0.00255	2.62[a]
Stripping ratio	X_2	−895	0.421	2.12[a]

$n = 25$
$R^2 = .376$
F ratio = 6.65

[a]Significance at .05 level.

		Underground		
Tons/Man/Day	Y			
Constant		14.300		
Mine size	X_1	0.00266	0.00121	2.19[a]
Seam thickness	X_2	−0.167	0.859	0.19
Seam depth	X_3	−0.00557	0.00650	0.86
Slope or shaft	X_4	2.456	2.380	1.03
Age of mine	X_5	−0.116	0.854	0.14

$n = 23$
$R^2 = .579$
F ratio = 4.668

[a]Significant at .05 level.

Source: *1971 Annual Coal, Oil and Gas Report*, Department of Mines and Minerals, State of Illinois.

Although the regression results show the influence on productivity (hence, costs) of mine size, there is little data available which determines the optimum plant scale in coal mining. The unpublished Bureau of Mines study referred to in the discussion of capital investment costs reveals the following cost schedule for different sized hypothetical underground mines.

Yearly Output (millions of tons)	Selling Price/Ton (assuming 12% DCF)
1.13	6.61
2.27	6.11
3.12	6.05
4.25	6.29
5.10	6.27

These costs reflect conditions in 1972 for a West Virginia mine operating in a 72-inch seam and producing coal for a gasification plant. It lacks a cleaning plant and, since it is considered to be an extension of an existing mine, it does not include a shaft- or slope-digging cost in the capital investment requirement for the mine. Both of these factors reduce costs below normal for a new, full-scale underground mine. But our concern centers more on the *shape* of the cost function rather than its height. We find the traditional U-shaped curve, with an aberration at the 5.1 million ton output level. The optimum scale plant was one with an output of 3.8 million tons per year. To enlarge an underground mine involves adding production modules, but beyond a point this process evidently achieves diminishing returns. That is, up to a point there is divisibility in the underground mining process.

The same does not apply to strip mining. There, scale economies are restricted primarily by an operator's financial resources and his coal reserves position. Small strip mines use small-scale equipment—shovels, loaders, drills and trucks. As they increase in size these basic pieces of equipment grow larger with accompanying scale economies. The operation of a 220-yard dragline requires no more personnel than one with 18-yard capacity. To amortize a large-scale mine requires control of an equally large coal reserve which, with the corresponding steep financial requirements, may lie beyond the capacity of most coal operators.

Beyond a certain mine size scale economies cease for strip mines because, in effect, modularization begins. A mine may operate with several pits, each pit employing a large stripping tool (shovel or dragline), a loading shovel, drills, and large-sized trucks. It is doubtful, however, that a mine with a 6-million ton capacity will operate at a cost level much different from one producing 3 million tons a year, if the former operates two pits and the latter, one.

It is impossible to state categorically what size represents optimum scale in strip mining because much depends on mining conditions, especially the stripping ratio. Mining conditions influence the underground picture too, but not as significantly. We do know, however, that both underground and strip mines achieve optimum scale output at an annual output level of at least several million tons. This condition, along with the previously discussed need for large reserves and large scale output to serve the electric utility and coal gasification markets, strengthens entry barriers in the coal industry.

EFFECT OF SHIFT TO LOW SULPHUR WESTERN COAL

There are a number of cross-currents at work in the coal industry which may lead to profound changes in the industry's market structure. We have touched on some of them in the previous section dealing with entry conditions. Several others are related to the possibility of a massive shift in output from the high sulphur eastern coals to low sulphur coals found in the west.

This section studies the problem from two aspects: the possibility of a shift to the use of western coal to meet environmental standards on air pollution, and the shift to the same low sulphur western coal reserves to furnish feedstock for coal gasification. The motivation behind the move toward western coal in these two cases differs: in the first the search is for low sulphur coal; in the second low sulphur content of the coal is an immaterial consideration. Of primary concern is the low mining cost of the western coal reserves and their noncaking characteristics which encourage their use in coal gasification. The dual motivations lead to the same coal: that found in the Rocky Mountain and Northern Great Plains provinces, principally in Montana, Wyoming, North Dakota, and New Mexico. Should coal gasification flourish and the western states supply the coal, and should pressure continue to use low sulphur coal, we could witness a dramatic westward tilt in the location of the coal industry. This development would have not only locational significance, it might also influence ownership patterns in the industry.

Impact of Air Pollution Standards

Recently developed standards for stack gas emissions at the federal, state, and local levels make the continued use of much of the country's coal for power generation problematical. In many areas of the country, present and projected standards rule out the use of high sulphur coals which predominate in the eastern coal fields—those fields nearest the major coal markets. To render high sulphur coals acceptable, one of several developments must occur: either (1) stack gas emission control devices must be developed to reduce SO_2 levels to satisfactory amounts, (2) a way must be found to pretreat coal to reduce sulphur content to acceptable levels, (3) coal must be gasified (and become a low-sulphur-content fuel), or (4) emission standards must be relaxed to permit the use of high sulphur coal.

It is uncertain which of these alternatives will prevail. Presently, no methods exist satisfactorily accomplishing either of the first two alternatives, although research is pressing forward vigorously to develop acceptable stack gas scrubbing devices.[12] The long lead time necessary for developing coal gasification plants and their novelty in this country, rule out gasification as a near-term solution. Whether public policy shifts toward a relaxation of air pollution standards—the fourth alternative—remains to be seen.

What then can we expect if only coal with a low sulphur content can be burned? Coal resources in the United States total 3.2 trillion tons.[13] Economically-available reserves, however, total only 150 billion tons, 105 underground and 45 strip.[14] Undoubtedly this total would increase substantially if the need for more arose as "inferred" reserves became "measured" reserves, and as mining proceeded to less desirable (deeper and thinner seamed) coal beds. No data exists on the sulphur content of the 150 billion tons. Table C-6, however, shows the sulphur content for the 1½ trillion tons of coal reserves that have been mapped and explored. It indicates that over 90 percent of the low sulphur coal reserves are located in the west, away from the existing coal industry and from the centers of demand. Moreover, much of the eastern low sulphur coal is especially suited for use as metallurgical coke and is allocated to that market.

Table C-7 indicates the location of the economically feasible strip coal reserves. These reserves are especially significant in light of the cost data in the first section of this study which shows strip mining costs to be substantially lower than underground mining costs. Doubly significant is the fact that the low sulphur strip coal is located predominantly in the western states. Over one-half of the low sulphur tonnage in Table C-6 is found in Wyoming. Four states—Wyoming, Montana, New Mexico and North Dakota—account for 87 percent of the low sulphur coal and lignite reserves in Table C-7.[15]

Table C-6. Estimated U.S. Coal Reserves by Sulphur Content (millions of tons)[a]

Sulphur Content (%)	United States		East of Mississippi River	
	Tons	%	Tons	%
−0.7	720,060	46	50,062	11
0.7-1.0	303,573	19	45,219	9
1.0-3.0	238,374	15	177,281	37
+3.0	314,159	20	206,495	43
Total	1,576,166	100	479,057	100

[a]As of January 1, 1965.

Source: *U.S. Energy Outlook*, National Petroleum Council, December, 1972, p. 160.

Table C-7. Estimated Strippable Coal and Lignite Reserves in the United States, January 1, 1968, Classified by Sulphur Content, Coal Rank, and Coal Province[a] (millions of tons)

| Region | Strippable Reserves | | | |
	Low Sulphur	Medium Sulphur	High Sulphur	Total
	Bituminous Coal			
Eastern Province, Appalachian Region	1,862	1,433	1,876	5,171
Interior & Gulf Provinces	13	535	6,748	7,296
Rocky Mt. & N. Gt. Plains Provinces	482	160	8	650
Total Bituminous	2,357	2,128	8,632	13,117
	Subbituminous Coal			
Rocky Mt. & N. Gt. Plains Provinces	19,414	289	529	20,232
Pacific Coast Province	135	0	25	160
Total Subbituminous	19,549	289	554	20,392
	Lignite			
Interior & Gulf Provinces	650	684	0	1,334
Rocky Mt. & N. Gt. Plains Provinces	4,795	937	0	5,732
Total Lignite	5,445	1,621	0	7,066
Grand Total, U.S.	27,351	4,038	9,186	40,575

[a]Excludes Alaska.

Source: "Strippable Reserves of Bituminous Coal and Lignite in the United States," Bureau of Mines Information Circular 8531, 1971, pp. 14-16.

We need to analyze the extent to which western coal can supply the electric utility market. It is beyond the scope of this chapter to determine the relative share of the electric utility fuel market that coal might be expected to capture in the future under different cost and environmental standards assumptions. In 1971 electric utilities consumed 326 million tons of coal which gave coal approximately 47 percent of the electric utility fuel market.[16] The National Petroleum Council estimates that in the 1970-1985 period, coal consumed by electric utilities will increase approximately 78 percent.[17] This estimate is based on the assumption that the utility industry will not be "subjected to severe constraints on its decisions."[18] Presumably this implies a freedom to use high sulphur coals.

If high sulphur coals are banned from use in the electric utility market, growth of total coal demand would fall far short of the N.P.C.'s estimate. Markets in the eastern half of the country would suffer most due to relative scarcity of eastern low sulphur coal reserves that can be mined cheaply. But western coal markets ought to expand because of the abundant low cost reserves with low sulphur content available in the area. Western coal could supply not only its natural market areas but could also move into other markets traditionally supplied by Interior and Gulf Province mines or into markets now served by other fuels. Reports exist of western coal already moving into some midwestern markets.

I have derived hypothetical delivered prices for representative western and midwestern coals delivered to the Chicago electric utility market by unit train. A number of existing unit train rates provide the basis for estimating transportation charges on long and short haul unit train movements.[19] The N.P.C. and Bureau of Mines reports furnish estimates of FOB mine prices for a model Wyoming mine and for midwestern underground and strip mines. Table C-8 summarizes the results of the delivered cost analysis.

Table C-8 gives results of a high and a low estimate. Two sources provide estimates of coal prices (in 1970) from a hypothetical western strip mine (Wyoming in one case, "western" in the other, but both with 3:1 stripping

Table C-8. Comparison of Hypothetical Delivered Cost per Million BTU to Chicago Market from Wyoming and Midwestern Mines

	Wyoming	*Midwest*
Mine price (low estimate)	$1.83[a]	$4.60[b]
Mine price (high estimate)	2.75[b]	5.65[b]
Freight rate (low estimate)	4.16[c]	1.35[d]
Freight rate (high estimate)	4.16[c]	2.25[e]
BTU per pound	8500	11200
Delivered cost/BTU (low est.)	35.2¢	26.6¢
Delivered cost/BTU (high est.)	40.6¢	35.3¢

[a]Bureau of Mines estimate.

[b]N.P.C. estimate (read from graph).

[c]925 miles @ 4.5 mills/mile.

[d]150 miles @ 9.0 mills/mile.

[e]250 miles @ 9.0 mills/mile.

Sources: "Cost Analyses of Model Mines for Strip Mining of Coal in the United States," U.S. Bureau of Mines, Information Circular 8535, 1972, p. 97.

U.S. Energy Outlook, National Petroleum Council, December, 1972, pp. 148, 149, 162.

The Potential Market for Far Western Coal and Lignite, Vol. I, Robert R. Nathan Associates, Inc., December 27, 1965, pp. VI-22.

ratios) and from hypothetical underground and strip mines in the Illinois, Indiana, Ohio, Iowa region where in the N.P.C. model, mining conditions were presumed to be similar. The freight rates are based on hauls of 925 miles from northeastern Wyoming to Chicago, and 150-250 miles from Illinois mining districts to the same destination. If one uses the average price actually paid for Illinois coal in 1970 together with the low and high freight rate estimates, the delivered costs per million BTUs are 28.0¢ and 32.0¢.

Thus the hypothetical data shows that on an economic basis western coal probably cannot compete with most midwestern coal in the Chicago market. Markets to the east would find western coal even less competitive. Conversely, electric utility markets closer to the western coal fields would fall in the west's orbit. The break point is probably not far west of Chicago. Note that the BTU cost for Wyoming's low estimate is slightly lower than for the midwestern mine's high estimate. The freight rates imply large unit train shipments and would not apply to the rates for smaller tonnages. Furthermore, the cost estimates in Table C-8 are, in effect, averages and hide the possibility of delivered cost conditions varying from the averages for specific mining operations where conditions deviate from the average.

In any event, the rapid growth of western coal demand to supply the electric utility market in areas outside of its traditional market territory probably hinges on developments in the fight to maintain air quality. If SO_2 standards are relaxed or if satisfactory stack gas processes are developed, western coal may not supplant other coal regions' supplies to the electric utility markets. On the other hand, if those conditions do *not* prevail, western coal might well expand into midwestern markets. The comparatively narrow delivered cost differential between western coal and midwestern coal makes this supply substitution a fairly attractive solution to the air quality problem for electric utilities if stack gas desulphurization technology fails to develop as anticipated. It is unlikely that the electric utilities would incur coal costs much above existing levels, however, since nuclear power offers an alternative in many markets at reasonably competitive prices. The availability of nuclear power would certainly bar western coal from the eastern coal market areas.

This analysis has not recognized the possibility of western coal moving into electric utility markets formerly served by oil and natural gas, especially in the area west of the Mississippi River. The growing energy shortage makes this a real possibility; in fact several new western coal mines already serve such markets.

What are the possibilities for western coals to supply feedstock for coal gasification? As previously indicated, such a development might entail a shift to the use of western coals, not for their low sulphur content but to take advantage of their low mining cost.

The N.P.C. has estimated a potential market for coal for syngas plants of from 46.6 to 232.4 million tons a year by 1985, depending on the assumptions used in making the estimate. Several coal gasification projects have been announced and are scheduled for completion in 1976 and 1977 if present plans can be carried out. They and other first generation syngas plants will use the Lurgi process which has been used to gasify coal in Europe since the 1930s and in South Africa for two decades. The western coals to be used in the first U.S. gasification plants resemble the noncaking German coals which use the Lurgi process. Although the process favors the use of noncaking coals, progress has been made toward using the caking coals typical of the Appalachian coal fields after pretreatment. But this added process is estimated to add about 15 percent to the gasification plant capital investment and 6-7¢ per million BTU to the manufacturing costs for the syngas.[20]

Coal cost is obviously an important component of total gasification costs. Cost data supporting the El Paso and Texas Eastern gasification proposals shows that coal costs (for feedstock and steam and power generation) account for 27-28 percent of the plant-side price to be charged for the syngas. We need to determine, therefore, the competitive position of western coal vis-à-vis other coal regions to determine the prospects of gasification continuing to center in the west.

The first announced syngas plants, as we have seen, are scheduled to be located in western states—two in New Mexico, the third in Wyoming. The National Petroleum Council's energy study sees up to 30 gasification plants by 1985 in the four western states of New Mexico, Wyoming, Montana, and North Dakota.[21] Doubtless much of this gas will be consumed in the western states. Some, however, may move eastward. The proposed Northern Natural Gas-Cities Service-Peabody venture would presumably move into the upper midwest which is Northern's market area. How far east might we expect western syngas to move? What are the prospects of using western coal in the production of syngas to supply markets east of the Mississippi River as some have suggested?

The analysis depends upon assumed coal prices. We use here the hypothetical "prices" for western strip coal and midwestern underground coal in the N.P.C.'s energy study.[22] At prices of $2.75 a ton and 8500 BTUs per pound for western coal and $5.65 a ton and 11200 BTUs for Illinois coal, the total coal bill for a 250 MM cfd syngas plant would be approximately $24.2 million and $37.7 million respectively. This difference should add around 16¢ per thousand cubic feet to the plant-side price of the syngas.

Based on 1971 costs, Wyoming syngas should be produced for around $1.20 per thousand cubic feet. Pipelining charges would add around 25¢ per MCF for delivery to Chicago. Adding 16¢ per MCF to the cost of gasifying Illinois coal for the higher coal feedstock cost and allowing for pipelining charges on the movement of gas from the Illinois coal fields to Chicago makes Illinois- and Wyoming-produced syngas almost a standoff in that market. However, Wyoming coal has an intangible advantage of having a low sulphur content which makes its use in creating process steam less objectionable from an air pollution standpoint. On the other hand, the Wyoming coal is disadvantaged by being strip-mined, with the environmental implications which that involves.[23]

The foregoing cost analysis leads to the tentative conclusion that syngas using western strip coal should have a delivered cost advantage in the area north of Chicago and west of the Mississippi River. To what extent the high delivered cost of syngas will deter its rapid development as a gas source in this country is beyond the scope of this study. As previously indicated, the N.P.C.'s estimate of coal use for gasification plants runs as high as 232 million tons by 1985.

Developments in the electric utility and the coal gasification markets raise the probability of a substantial relative shift in coal output away from the eastern coal fields and toward the low sulphur western coal fields. What are the market structure implications of this shift?

Tables C-9 and C-10 summarize pertinent data on the ownership of western coal land. Table C-9 shows the groups that own coal in the southwestern states. The ownership patterns in Montana and North Dakota are not available in the same detail but are probably similar in broad outline. Indian tribal ownership may be less, but, as in the southwestern states, the federal and state governments and railroads (especially the Burlington Northern) predominate as coal-land owners.[24] In Montana the federal government controls 55 percent of the mineral (much of the surface controlled by private parties) and, in general, the state owns Sections 16 and 36, or approximately 5 percent of the total.[25]

Table C-9 gives more detailed ownership information for Montana, North Dakota, and part of Wyoming (apparently southwestern Wyoming reserves have been omitted). It should be accepted as a rough estimate of ownership in these states.[26] Note that total reserves exceed the Bureau of Mines figure given in Table C-7 for this region. This may partly stem from the different overburden

Table C-9. Estimated Ownership of Coal Lands in Five Southwestern States[a] (ownership in percentages)

State	U.S. Gov't	State Gov't	Indian Tribes	West. RR's	Private	Total
Wyoming	48	12	–	20	20	100
Utah	67	15	1	–	18	100
Colorado	36	14	6	20	24	100
Arizona	–	–	100	–	–	100
New Mexico	34	18	40	5	3	100
Total	45	13	13	12	17	100

[a]Provisional estimates based on incomplete data.

Source: *Southwest Energy Study*, report prepared for the Secretary of the Interior, November, 1972, p. 3-4.

Table C-10. Estimated Strip Coal Reserve Ownership in
Montana, Wyoming and North Dakota[a]

Company	Estimated Reserve (millions of tons)
Burlington Northern RR	2423
Montana Power*	195
Montana-Dakota Utilities	273
Peter Kiewit*-Pacific Power & Light*	1189
Star Drilling Co. (optioned to Peoples Gas Co.)	1000
Reynolds Aluminum	1000
H.F.C. Oil Co.	810
Carter Oil-U.S. Steel*	465
Carter Oil-Amax Coal Co.*	1295
Carter Oil-Consolidation Coal Co.*	1253
N. Am. Coal Co.*-Consol. Coal Co.*	380
North American Coal Co.*	855
Consolidation Coal Co.*	300
Westmoreland Coal Co.*-Amax Coal Co.*	1500
Gulf Oil Co.*	800
Peabody Coal Co.*-Montana Power*	1440
Peabody Coal Co.*	5207
Federal, State, Burlington Northern	677
Federal Government	5434
Carter, Amax*, Kerr-McGee Oil, Sun Oil, Mobil Oil, Atlantic-Richfield Oil, Peabody*, Federal	18000
Total	39496

*Coal-producing companies in 1972.

[a]It is not known whether tonnages for two or more companies are jointly owned or represent individual ownerships grouped together. The 18000 million ton figure, however, presumably represents individual ownerships.

Source: Coal mining company's confidential field report.

depths used in calculating reserves, but also probably stems from incomplete government mapping of western coal reserves.[27]

Several important market structure implications flow from the data in Table C-10. First, the total output in 1972 of coal companies shown in the table to be controlling reserves was 175 million tons, or about one-third of the national coal output. Or, put the other way, two-thirds of the U.S. coal industry is unrepresented in this potentially expanding region. This becomes more significant in light of the prevailing view in the industry that the principal available (and desirable) strip coal reserves in this area (federal coal lands aside) are pretty well tied up by the companies shown in Table C-10, plus the few others which are unaccounted for there. There are still large uncommitted federal coal lands potentially available for lease, but for the time being a moratorium exists on the leasing of most of these lands following an order of the Secretary of the Interior to delay further coal leasing, pending a review of the government's coal land leasing policy.[28]

Another interesting implication of the Table C-10 data is the heavy concentration of ownership among firms which are not now coal producers. Many of these companies are major petroleum producers. The substitutability of oil, gas, and coal in several major markets raises interesting questions concerning the tightening of market power in the energy sector resulting from oil company control of many of these reserves. Aside from this issue, ownership of these reserves is comparatively concentrated. Excluding firms in the table which presumably control reserves for captive mining operations, only sixteen companies (plus the federal government) control the entire reserve total.

NOTES

1. Hereafter this study will use the term "coal" to refer to the bituminous coal industry. This study focuses on the production of bituminous and subbituminous coal and lignite (which collectively will be referred to as "coal"); it ignores anthracite coal which is currently a nominal factor in the coal industry and will continue to play an insignificant role.

2. Reed Moyer, *Competition in the Midwestern Coal Industry*, (Cambridge: Harvard University Press, 1964), pp. 91-3.

3. Application of El Paso Natural Gas Co. at Docket #CP 73-131 for a Certificate of Public Convenience and Necessity, November 7, 1972, Exhibit Z-2, Schedule 2.

4. *Wall Street Journal*, March 24, 1973, p. 6. Subsequent to this announcement of the companies' plans, the Northern Cheyenne Tribal Council announced their intention to abrogate the coal leases which permit the mining of coal for the gasification projects.

5. *U.S. Energy Outlook*, National Petroleum Council, p. 163.

6. *Bituminous Coal Facts 1972*, National Coal Association, p. 52.

7. *Ibid.*, p. 84.

8. *U.S. Energy Outlook: An Initial Appraisal, 1971-1985*, Vol. 2, National Petroleum Council, November, 1971, p. 127.

9. Not included in these companies are the major petroleum firms which have acquired active coal operations and currently operate them as subsidiaries.

10. *Wall Street Journal*, March 12, 1973, p. 15.

11. *United States vs. General Dynamics*, U.S. District Court, Northern District of Illinois; Eastern Division, #67C 1632, filed April 13, 1972.

12. These devices exist today but not in a form which reduces SO_2 to acceptable standards under stringent regulations.

13. Paul Averitt, *Coal Resources of the United States*, USGS Bulletin 1275, (January 1, 1967).

14. Reported in *U.S. Energy Outlook*, National Petroleum Council, December, 1972, p. 140.

15. On a BTU basis the percentage would be lower because of the western coal's lower heat content.

16. Includes hydro as a "fuel."

17. *U.S. Energy Outlook*, p. 20.

18. *Ibid.*, p. 22.

19. *The Potential Market for Far Western Coal and Lignite*, Vol. 1, Robert R. Nathan Associates, Inc., December 27, 1965, pp. VI-22.

20. George Skaperdas, "Manufacture of Pipeline Gas from Coal," unpublished paper prepared for the Coal Task Group of the Other Energy Resources Subcommittee of the National Petroleum Council's Committee on U.S. Energy Outlook, April 5, 1971.

21. *U.S. Energy Outlook*, p. 168.

22. *Ibid.*, p. 148 and p. 162.

23. This analysis also ignores the important consideration of water supply which looms large in the gasification process.

24. In many cases, the mineral may be owned by, e.g., the federal government, but the surface by ranchers, requiring that prospective strip mine operators acquire effective control of the coal by dealing with both parties.

25. These are estimates of officials in Montana's Bureau of Mines & Geology, and the Bureau of Land Management.

26. Humble Oil Co., believed to control extensive reserves in the area, is not represented in the data.

27. The 18000-million-ton figure in Table C-10 includes coal at depths of 200 feet whereas the Bureau of Mines limit is 150 feet.

28. There are exceptions to the moratorium, but they are relatively insignificant.

Appendix D

Price-Output Behavior in the Coal Industry

Reed Moyer*

This chapter evaluates the conduct of the coal industry in recent years with respect especially to its pricing behavior, both in domestic and export markets, its output policies and behavior, and the relation between price and output decisions. It specifically analyzes the effect, if any, on output decisions of coal companies that have been acquired by petroleum and other large noncoal companies. We look at the factors allegedly affecting the recent sharp increases in coal prices to determine whether collusive forces were at work or whether the price increases can be attributed to other market and nonmarket factors. Some specific charges of price conspiracy will be analyzed to determine their validity.

OUTPUT BEHAVIOR OF NONCOAL CO.-CONTROLLED FIRMS

The structure of the coal industry has been characterized in recent years by increasing concentration of output and the acquisition of many large coal producers by petroleum and mining companies.[1] The expansion of the petroleum companies, especially, into other energy fields has created concern and has, in fact, led to an investigation by the Federal Trade Commission of some of the coal-oil mergers.

Interfuel acquisitions pose several potential threats. One of these is the danger of their creating sufficient market power to affect competition adversely. In a market where petroleum products and coal are competitive, the acquisition of coal companies might reduce the number of sellers of *energy* in that market with a concomitant potential increase of market power that may flow from increased concentration. We say that the number of sellers "might"

*Michigan State University.

rather than "will" be reduced since the petroleum company and acquired coal company may or may not market their products in the same geographic markets. The possibility for such interfuel competition exists principally in the electric utility market where many plants are equipped to burn either coal or oil.

If interfuel mergers increase market power we need to analyze the possible effects of that increase. Producers in imperfectly competitive markets may restrict output to maintain prices at levels permitting above-normal profits. This being true, we need to look at the output behavior of coal companies acquired by noncoal companies to find whether evidence of output restriction exists. We include the nonpetroleum acquiring companies because their large size gives them the potential for market power not available to the coal companies that they acquired.

Tables D-1, D-2, and D-3 summarize the output performance from 1966-1972 for the major coal companies acquired by oil and nonoil companies (henceforth referred to as coal-oil and coal-conglomerate companies). The data reveals no clear-cut pattern of output restriction. From 1968, the first year when the four petroleum companies controlled their respective coal operations, until 1972, annual tonnage of the four coal companies rose from 104.9 to 106.2 million. Their share of total coal[2] output declined slightly from 19.3 percent to 17.9 percent. Their share of the output from the fifty largest coal companies also fell moderately from 28 percent to 26.6 percent.

Output behavior of individual firms varied; there is no apparent pattern among them and certainly no superficial evidence of a coordinated output policy. Consolidation's (Continental Oil) market share from the year of its acquisition to 1972 rose from 9.6 percent to 10.9 percent but only after an erratic performance in the intervening years. Island Creek's (Occidental Petroleum) share declined as did Pittsburgh and Midway's (Gulf Oil). Old Ben's (Standard Oil of Ohio) remained fairly constant. Output increased in two cases and declined in the other two during a period when total tonnage increased modestly.

In this analysis we treat Arch Mineral as a separate case, unlike the other four coal-oil companies. Begun by coal operators, ownership has since shifted principally to Ashland Oil Company and the H.L. Hunt oil interests with management maintaining a small minority interest. Arch's early coal operations began from scratch, but approximately half of its 1972 tonnage resulted from the acquisition of Southwestern Illinois Coal Company.

This analysis of output behavior dating from the coal companies' acquisitions suffers from a defect. A decision to restrict output could be effected immediately following the company's takeover, but much of the output expansion could have a several year time lag due to the time required to develop new mines.

The absence of strong evidence supporting a move toward output restriction by the coal-oil companies is not surprising. Such a policy—whether

Table D-1. Output Behavior of Major Acquired Coal Companies 1966-1972 (million tons)

Coal Company	Parent Company	Year						
		1972	1971	1970	1969	1968	1967	1966
Peabody	Kennecott Copper	71.6	56.0	63.6	59.7	59.8[a]	59.4	54.0
Consolidation	Continental Oil	64.9	54.8	64.1	60.9	59.9	56.5	51.4[a]
Island Creek	Occidental Petroleum	22.6	22.9	29.7	30.4	25.9[a]	25.9	23.7
Amax	Amer. Metal Climax	16.4	13.3	14.4	11.3[a]	9.3	8.6	8.5
Old Ben	St. Oil of Ohio	11.2	10.5	11.7	12.0	9.9[a]	10.3	9.9
Arch Mineral[b]	Ashland Oil-H.L. Hunt	11.2	2.1	–	–	–	–	–
Freeman-Unit. Electric	Gen'l. Dynamics	10.0	11.5	14.1	14.0	13.0	14.1	13.6[a,c]
Pittsburgh & Midway	Gulf Oil	7.5	7.1	7.8	7.6	9.2	9.0	8.9[a,d]
Total U.S. Bituminous Coal Tonnage		595.4	552.2	602.9	560.5	545.2	552.6	533.9
Tonnage of Top Fifty Coal Producers		399.4	358.4	407.0	383.6	374.2	364.8	341.1

[a]Year company was acquired.

[b]Much of Arch Mineral's increased output came in 1972 from its acquisition of Southwestern Illinois Coal Co.

[c]Year when General Dynamics achieved complete ownership. Large stock interest previously held.

[d]Control of Gulf Oil dates to 1963.

Source: *Coal Mine Directory, Keystone Coal Industry Manual*, various years.

Table D-2.   Major Acquired Coal Companies' Output as Percentage of Total Coal Output, 1966-1972

Coal Company	Parent Company	Year						
		1972	*1971*	*1970*	*1969*	*1968*	*1967*	*1966*
Peabody	Kennecott Copper	12.0	10.1	10.5	10.7	11.0	10.7	10.1
Consolidation	Continental Oil	10.9	9.9	10.6	10.9	11.0	10.2	9.6
Island Creek	Occidental Petrol.	3.8	4.1	4.9	5.4	4.8	4.7	4.4
Amax	Amer. Metal Climax	2.8	2.4	2.4	2.0	1.7	1.6	1.6
Old Ben	Std. Oil of Ohio	1.9	1.9	1.9	2.1	1.8	1.9	1.9
Arch Mineral	Ashland Oil-H.L. Hunt	1.9	0.4	–	–	–	–	–
Freeman-Unit. Electric	General Dynamics	1.7	2.1	2.3	2.5	2.4	2.6	2.5
Pittsburgh & Midway	Gulf Oil	1.3	1.3	1.3	1.4	1.7	1.6	1.7

Source: Table D-1.

Table D-3. Index of Acquired Coal Companies' Output, 1966-1972 (1966 = 100)

Coal Company	Parent Company	Year 1966	1967	1968	1969	1970	1971	1972
Peabody	Kennecott Copper	100	110	110	110	118	104	133
Consolidation	Continental Oil	100	110	116	118	125	107	126
Island Creek	Occidental Petrol.	100	109	109	128	125	97	95
Amax	Amer. Metal Climax	100	101	110	133	169	156	193
Old Ben	Std. Oil of Ohio	100	104	100	121	118	106	113
Arch Mineral	Ashland Oil-H.L. Hunt	N.A.	N.A.	N.A.	N.A.	N.A.	N.A.	N.A.
Freeman-Unit. Electric	General Dynamics	100	104	96	103	104	85	74
Pittsburgh & Midway	Gulf Oil	100	101	103	85	88	80	84
Total U.S. Bituminous Coal Output		100	104	102	105	113	103	112

N.A.–Not applicable.
Source: Table D-1.

derived collusively or otherwise—might be effective if the firms controlled a substantial share of industry output. But as in this case, when their output collectively was less than 20 percent of total production, restricting output would simply permit non-oil-dominated firms to absorb the slack.

It might be more productive to pursue a different line of inquiry concerning the output behavior of the major non-independent coal companies. One might argue that their size, market power and access to large capital resources might strengthen their position vis-à-vis their independent competitors.[3] Let us compare market shares for the coal-oil and coal-conglomerate firms with those of the major independents. We compare output data for 1972 with 1968, the year in which all but one of the major nonindependents had been acquired. Tonnage for the seven major nonindependents listed in Table D-1 increased from 187.0 to 204.2 million tons. Their market share remained constant at 34.3 percent during the four year period. For the nine independents producing more than three million tons a year, output between 1968 and 1972 increased from 77.7 to 82.6 million tons.[4] Their market share too, remained quite steady, falling slightly from 14.3 percent to 13.9 percent.

Companies with output from one to three million tons increased their market share from 12.0 percent to 15.5 percent for the four year period. Most of these firms are independent although a few are captive. Moderately large independent coal firms, therefore, appear capable of withstanding competition from the larger independent and coal-oil and coal-conglomerate firms. Some independent firms have suffered and have lost market position. The following comments from the head of a major independent reflect their concern.

> The independent coal company, they say, is the model for the industry, the champion of free fuel competition, the first line of defense against the domination of fuel conglomerates.
>
> This is one part nostalgia and four parts baloney. . . . [T]oday few companies have the 10 or 20 or 30 million dollars it costs to put in a new mine. . . . [W]here could we expect to raise 30 million dollars for a big new coal mine when that amount would at least equal our total investment in the mines we operate today?[5]

In view of the market share data above, one must wonder whether the plight of the large independents reflects more the firms' management capabilities than it does their independent position. Some of the independents (as well as the coal-oil and coal-conglomerate firms) have been more aggressive than others in shifting resources into the growing western coal fields and into the export metallurgical coal markets, relying relatively less on the troubled eastern steam coal markets.

Production Payment Financing

A financing feature of several of the coal company acquisitions in the 1960s may have influenced coal output to some extent. Until 1969 special provisions in the tax law permitted third party financing of mineral company acquisitions which, in effect, allowed purchase payments to be treated as an operating expense called a production payment. This made the coal mines more valuable to the buyer than to the coal operator and permitted the payment for several coal companies of amounts apparently in excess of the companies' market value. This tax advantage (so-called ABC transactions) may account for some of the coal company mergers, but it would not account for the kind of buyer making the acquisition. Two of the leading mergers made with production payment financing were the Consolidation-Continental and Peabody-Kennecott transactions. Since the acquiring company is obligated under the terms of the purchase agreement to make large and continuing payments based on coal output, it has an incentive to *expand* rather than contract output. The Consolidation-Continental purchase agreement, in fact, spells out the extent of its output expansion requirements. The merger agreement calls for Consolidation "to increase the productive capacity of Consolidation's mines each year to a maximum of 55.7 million tons annually by 1971.[6] In 1965, prior to the merger, Consolidation produced 40.2 million tons of coal.

There is another aspect of production payment financing that bears scrutiny. "Do acquisitions of coal companies which involve production payment financing direct coal away from the 'spot market' and thus inflict competitive injury on the small utilities who are largely dependent on the spot market?"[7] This proposition was advanced in a study of competition in the energy sector conducted by the American Public Power Association (APPA) and the National Rural Electric Cooperative Association (NRECA) which were concerned about coal shortages in the spot coal market. These associations' members relied considerably on spot coal purchases.

The "spot market effect" theory lacks support on two grounds. First it ignores the fact that the same result (shortages in the spot coal market) could occur *in the absence* of an acquisition. There is nothing to prevent a going firm from raising capital through the use of production payment financing; it need not occur only in connection with a merger. Second, the theory fails to account for the existence of long term contracts which predated production payment financing by a number of years. In other words, if the spot coal market's supplies were inadequate one might rather blame the widespread use of long term contracts which underlie the production payment financing rather than the mode of financing itself.

Mergers Circumventing
Independent Entry

If noncoal companies acquire coal companies, one might argue that the structure of the industry does not change appreciably. Ownership may change, but the number of firms in the industry remains constant; concentration ratios stay the same. This is especially true in the case of coal-conglomerate mergers. The competitive effect of coal-oil mergers may be different, depending on the extent of, and importance one attaches to, interfuel competition.

Though structure may remain essentially unaltered, competition in the coal industry may nonetheless, be adversely affected by acquisitions by noncoal companies. Supply conditions in the absence of mergers may differ from those prevailing as a result of mergers. Therefore, we need to look a little more closely at conditions surrounding the entry of noncoal companies into the coal industry.

Unfortunately the motives affecting decisions of many noncoal companies to enter the coal industry are locked in corporate files or in the minds of corporate executives. In at least one case—Pittsburgh & Midway-Gulf—a coal-oil combination was a byproduct of another merger. Gulf Oil acquired Spencer Chemical Company whose subsidiary was Pittsburgh & Midway. Gulf apparently had no motivation to broaden its participation in the energy sector, although after the merger it viewed the coal operation as a desirable adjunct to its other operations and decided to retain ownership of Pittsburgh & Midway.

Other petroleum companies have expressed interest in coal as a way to diversify into alternative energy sources and to acquire a raw material for possible future oil liquefaction or gasification. Whether their purchase of coal properties is an attempt—as some critics charge—to monopolize the energy field is difficult to determine.

Oil companies have entered the coal industry in different ways. Continental, Gulf, Sohio and Occidental—among others—have entered through mergers. These acquisitions have been criticized as being potentially anticompetitive. They have led to demands for expost investigations of the mergers and possible divestiture. Other oil companies have acquired coal reserves—rather than operating companies—which they are now holding for future development or are beginning to develop. Humble, Atlantic-Richfield and Kerr-McGee, for example, either have operating coal properties or have announced plans to develop some of their coal reserves. Arch Mineral (Ashland-H.L. Hunt) combines features of these two coal-oil modes of operation, since it has both developed its own mines and has acquired existing coal properties.

From a competitive standpoint *de novo* entry of oil companies into the coal industry is preferable to entry through acquisition. In the latter case ownership shifts from one firm to another. With *de novo* entry, the number of firms in the industry expands. This is the conventional view and, to some extent, is valid in this instance. But the minerals sector differs from industrial and service sectors in one important respect. Market power derives not only from the control of output but from mineral reserve holdings as well. Which energy company possesses greater market power: Standard Oil of Ohio through its ownership of Old Ben with annual coal output of eleven million tons and coal reserves of eight hundred million tons or Humble Oil Co. (Exxon) with one to two million tons of coal output and 7-8 *billion* tons of reserves?

Motivation for the coal-conglomerate mergers is less clear than is the case with coal-oil mergers. In most instances the acquisitions apparently aim at diversification and is indicated by firms that already operate in natural resource fields. Coal mining is seen as being related from an operational standpoint to the firm's existing activities, in addition coal is viewed by many as a growth area benefitting from the expanding worldwide demand for energy. Hence we see American Metal Climax, Kennecott Copper, American Smelting and Refining, St. Joe Minerals, and Houston Natural Gas entering the coal industry through acquisition.[8]

Both in the coal-oil and coal-conglomerate merger cases, we need to ask the same question to assess the effect on competition of market entry through merger: would the outsiders have entered the coal industry in the absence of the mergers? We have both facts and conjecture to help answer the question. It is an important question to ask since its answer bears on the structure of the coal industry and the nature of the industry's supply function.

Existing independent coal companies often are approached by firms seeking to acquire them. The same was true of coal companies that have already been acquired. Sometimes the suitors were other coal companies; often they were companies outside the industry. When Peabody Coal Company was ordered to divest itself of six million tons of annual output following an anti-trust case settlement, it received inquiries from over thirty companies interested in possible acquisition of the mines. Some were coal companies; others were not. The same situation applied when the owners of Southwestern Illinois Coal Company decided to sell the company's mines. They received bids from both coal and noncoal companies.

The clearest record we have of entry of a noncoal company into the coal industry through merger involves Kennecott's purchase of Peabody Coal Company.[9] The Federal Trade Commission attacked the acquisition, and testimony in the case revealed Kennecott's motives and actions prior to the purchase. The F.T.C. used the "potential competition" theory to rule against Kennecott. They held that, in the absence of the merger, Kennecott would have become a coal producer on its own, which would have been procompetitive. In fact, Kennecott had acquired coal reserves from Knight-Ideal Coal Company, a small Utah coal operator, for $750,000 in 1965 and shortly thereafter tried to acquire three other coal tracts in the west. The company claimed it acquired reserves as a bargaining device and as a hedge to protect itself in its purchases of fuel for its western smelters. It never mined the acquired coal.

Kennecott also retained a consulting firm to seek other ways to enter the coal industry. It eventually entered active operations in coal by acquiring Peabody in 1968.

The strongest evidence against Kennecott appeared to be a letter from the consulting company to Kennecott confirming their assignment which read, "Kennecott has acquired some coal properties and it is anticipated you will enter the coal industry with a major effort. . . . Your general goal is to build a business with $100-$200 million sales a year." Kennecott argued that this statement referred to its intention to acquire an existing coal company rather than to start from scratch with the Knight-Ideal property.

In any event the F.T.C. ruled against Kennecott based on the belief that it was a potential competitor in an industry which was *trending* toward high concentration. The Supreme Court upheld the decision, and in early 1975 Kennecott was arranging for disposition of the Peabody properties.

Whether other companies—like American Metal Climax, American Smelting and Refining, etc.—offered "potential competition" is a matter of conjecture. Would they have entered the industry through their acquisition and development of reserves if they had been thwarted in their effort to acquire an existing coal company?

There is no evidence available to indicate whether or not the F.T.C. or Justice Department scrutinized the competitive potential in any coal-oil or coal-conglomerate merger aside from the Kennecott-Peabody case.

Collusion?

To this point we have looked at output behavior. The next major section of this study analyzes pricing behavior. The present section serves as a bridge between an analysis of output behavior and pricing behavior by studying an issue that has elements of both in it. We investigate here inferences and charges of collusion and conspiracy to limit coal supply and artificially raise prices.

The conspiracy charges stem principally from the Tennessee Valley Public Power Association, the American Public Power Association, and the National Rural Electric Coop Association which represent small electric utilities. Their charges refer specifically to price-output behavior in the coal industry in the period 1969-1971. In testimony before the House Select Committee on Small Business they marshall their "evidence."[10]

- A 60 percent increase in the price of coal between 1969 (average for the year) and December, 1970 despite an increase in consumption of coal by electric utilities of 4 percent for 1970 vs. 1969.
- The contrast between this sharp increase and the years of relative price stability in the immediately preceding period.
- A change in bidding patterns on T.V.A. and other public power company coal bids. Prior to the period in question the utilities enjoyed active bidding on their coal tenders. Suddenly, bidding activity dried up, and many of the firms that continued to bid did so on partial rather than on full coal requirements, as they previously had done.
- The failure of suppliers to deliver on long term contract requirements.

- The use by suppliers of what the utilities considered to be "false and misleading explanations" for coal supply shortages.[11] These explanations included the alleged existence of car shortages, increased export demand, reduced output and productivity stemming from enactment of the 1969 mine safety legislation, and environmental concerns. Let's analyze these statements and charges.

It is true that coal prices increased sharply in the period 1969-1971. Unfortunately, the 60 percent increase figure cited above is misleading, unrepresentative of coal prices in general and perhaps even unrepresentative of spot coal prices that it purports to cover. A.P.P.A., et al. use the steam coal component of the Wholesale Price Index to make their case. That index covers so-called list prices quoted by a fairly small sample of major coal producers. The B.L.S., which computes the index, views it as inadequate. Furthermore, only a small percentage of coal is shipped in the spot market; most is covered by long term contracts. Nonetheless, the impression which a coal buyer receives of price volatility would most likely come from spot coal prices. There's no doubt that prices on short term offerings increased substantially during the 1969-1971 period; however, the W.P.I. figures do not necessarily reflect the extent of those increases correctly. The average realization on *all* coal shipments rose 25 percent from 1969 to 1970.

A.P.P.A., et al. declare that coal prices remained stable from 1960 to 1965 despite a total output increase of 23 percent and a 40 percent increase in electric utility coal consumption. They contrast this performance with the period of sharply rising prices in the late 1960s and early 1970s. Prices in fact, declined during the first half of the 1960s from an average of $4.69 in 1960 to $4.44 in 1965. Indeed one can trace the declining price trend back to 1958 when average realization equalled $4.86 a ton.

But several factors differentiate conditions during the period of declining prices from the recent years when coal prices have increased substantially. In 1958 coal tonnage had fallen to 410 million tons; the mines worked an average of 184 days a year, scarcely 2/3 of capacity. From the period 1958 to 1965 the average number of days worked per year was 200. From 1966 through 1971 that figure rose to 219.

More important was the sharp and abrupt change in labor costs and productivity. From 1958 through 1965 average hourly earnings in coal mining increased 19 percent while productivity (measured in tons of output per man-day) increased 55 percent. From 1965 through 1971 hourly earnings rose 40 percent, tons per man-day rose a scant 3 percent.

Recall that one of the alleged "false and misleading explanations" for coal supply shortages in the 1970-71 period was the assertion that the 1969 Coal Mine Health and Safety legislation had reduced mine output. After a twenty-one year unbroken string of increases in output per man-day, the figure fell 9 percent from 1969 to 1971. In underground mines the decrease was even greater; there productivity declined 23 percent.

There is apparently no denying the assertion of A.P.P.A., et al. that coal supplies in the 1969-71 period were tight, and that suppliers often failed to deliver on long-term contracts. Coal producers and coal buyers alike attest to the inability of coal operators to meet their coal supply commitments. The key question is whether the coal producers engaged in a conspiracy to withhold supplies from the market in an effort to drive up prices. Those who allege the existence of such a conspiracy adduce no evidence to support the charge. We have already investigated the validity of one of the "false and misleading" explanations for the tight supply situation. What of the others?

Coal operators point to the existence of coal car shortages—a recurring problem in the industry. It is hard to find direct evidence to support or refute the assertion; however, actions of the railroads and the Interstate Commerce Commission provided circumstantial evidence for support. In 1970 the I.C.C. issued new rules accelerating the return of empty cars to railroads owning them, a move one associates with a car shortage. Additionally, railroads shipping coal to the ports for export instituted a permit system to free up cars for movement to domestic markets and to reduce demurrage at the ports. This action also points to the existence of a car shortage.

The increase in export demand may have contributed to the tight domestic coal demand in two ways. First, exports increased fairly sharply from 1968 to 1970, rising from a total of 50.6 million tons to 70.9 million. That increased tonnage sopped up 40 percent of the total coal tonnage expansion during the period. Furthermore, since much of the export tonnage is low sulphur metallurgical coal, export expansion made it harder to supply domestic markets with low sulphur coals in response to stiffer air pollution standards.

Another set of factors had an important bearing on coal supplies. In the late 1950s and through the first part of the 1960s coal operators faced twin perils—stagnant coal demand (with concomitant inadequate profits) and the threat of a massive move toward the use of nuclear reactors to produce electricity. Since coal's only expanding market was in the electric utility field, the nuclear threat was especially ominous. These parallel developments hardly encouraged the development of new mines, although some operators still continued to expand output despite the adverse conditions. Delays in the development of nuclear plants coupled with sharply rising nuclear construction costs, however, have forced the utilities to look to coal again as a long run fuel supplier in recent years. But new mines cannot be turned on and off like water from a spigot. Several years elapse between the conception and completion of a mine, and if demand outstrips the ability of the industry to develop new tonnage, temporary shortages may occur.

Some coal operators also report reluctance to build new coal mine capacity in fields with high sulphur coal in the face of tough antipollution laws and regulations restricting the use of that kind of coal. This condition may lead to a continuation of tight coal supplies especially if, as seems possible, pollution standards are relaxed during the current energy crisis.

The various pressures tightening the supply of coal in the 1969-71 period led to a reduction of coal stockpiles, especially for electric utilities. In February, 1970 electric utilities had an average of a fifty-four day stock of coal, 23 percent below the level two years earlier. T.V.A. found their stockpiles to be at "a very dangerous level" during this period. At one point some of its generating stations had only a few days supply of coal on hand.

Some of this squeeze on stockpile reserves stemmed from the tight coal supply which prevailed in the country, but in the T.V.A. case, at least, the condition resulted partly from a decision to resist buying coal at inflated prices. Aubrey J. Wagner, T.V.A.'s Chairman testified before a House Subcommittee that T.V.A. drew down its stockpiles rather than to pay 27-28¢ per million BTU for coal when it had been used to coal at 20¢.[12] Eventually, T.V.A. yielded and bought coal at the higher prices to assure continuity of its coal supplies.

By September, 1971 coal stockpiles at the nation's electric utilities had increased to a 108-day supply. Doubtless this buildup to abnormally high levels, partly in anticipation of an industry strike, contributed to the supply and price pressures objected to by the A.P.P.A., et al.

The conspiracy charge is buttressed by evidence of higher profits from coal mining in 1970 vs. 1969 offered by S. Robert Mitchell who prepared a report for the A.P.P.A., et al. to support their position. Most of the major coal companies, as we have seen, are controlled by noncoal producers hence, we lack information concerning profits from their coal operations. However, some major independent coal producers showed fairly sharp profit increases for 1970 vs. 1969. Pittston increased profits as a percentage of sales to 6.9 percent from 4.1 percent; for Westmoreland Coal Company the figures were 5.2 percent vs. 1.5 percent; for North American Coal Co., 3.4 percent vs. 2.9 percent; for Eastern Associated Coal Co., 7.7 percent vs. 5.8 percent.[13] *Statistics of Income* data for reporting coal companies (excluding the diversified companies) shows industry net income after tax rising from $−5.5 million in 1969 to $160.3 million in 1970. The 1970 performance stood in sharp contrast to the lackluster profit record of the coal industry throughout the 1960s when profits for the independents never exceeded $69 million.[14]

Representatives of the several public power associations alleging the existence of a price conspiracy in the coal industry took their case to the Justice Department to seek a grand jury investigation. The Justice Department found insufficient evidence to take such action. That posture found support from T.V.A. which suffered from short coal supplies and elevated prices. An investigation of the concentration in the energy field by a House Subcommittee developed the following two colloquies between Congressman Joe L. Evins and Aubrey J. Wagner, Chairman of T.V.A.

Mr. Evins: "Is this competitive bid process really injecting any competition into your purchases?"

Mr. Wagner: "Well, in competitive bidding we have not gotten generally uniform prices on coal. We have gotten a good spread of prices on coal and even in the negotiating prices there is not uniform pricing."
and,

Mr. Evins: "You have not seen any evidence of conspiracy or combination or uniformity in prices?"

Mr. Wagner: "I have not seen any evidence that I can put my hands on and say that is it."[15]

PRICE BEHAVIOR

We turn now to a closer investigation of pricing behavior in the coal industry, relating it to output behavior and to changes in wage costs and productivity rates. We analyze pricing behavior before and after 1968 to determine whether mergers with noncoal companies have had an anticompetitive impact. This section also investigates recent pricing behavior of mines in specific states to assess the effect of different regional market structures in pricing patterns. Finally we separate coal exports (excluding exports to Canada) from domestic shipments to calculate the effect of export prices on average coal realization and to determine price-output behavior in that special submarket.

Price-Output Relationships

Table D-4 summarizes measures of productivity, wage costs, prices, output, and capacity utilization in the coal industry from 1958-1972. The "modified wage cost" data in column 3 requires some explanation: It shows figures which relate tons per man-day (the standard industry measure for productivity) to average hourly earnings in the industry. If tons per man-day were converted to a figure representing output per hour and if that number were divided into average hourly earnings, the result would be a figure closely approximating labor costs per ton related to wages paid. It would still exclude such nonwage costs as contribution to the union's welfare fund, social security contributions, etc. The "modified wage costs" figure is thus a kind of proxy for wage costs, and is calculated to show the effect on wage costs of changes in productivity and earnings' rates.

The decade from 1958 to 1968 was marked by a fairly sharp reduction in wage costs. Prices declined in the first half of the decade in the face of stagnant tonnage and fairly low operating rates. From 1963 to 1968, average realization reversed its downward trend and recovered moderately as output and operating rates increased. Wage costs during this period, however, leveled off.

Steadily declining prices from 1958 to 1963 in the face of a moderate increase in output and operating rate shows an evident lack of market

Table D-4. Productivity, Wage Costs, Prices, Production, and Capacity Utilization, Bituminous Coal Industry, 1958-1972

Year	(1) Avg. Hourly Earnings	(2) Tons/Man-Day	(3) Modified Wage Costs (1÷2)	(4) Index of Wage Costs (1968=100)	(5) Average Realization	(6) Realization Index (1968=100)	(7) Average Number Days Worked Per Year[c]	(8) Annual Coal Output (MM tons)	(9) Output Index (1968=100)
1958	$2.93	11.33	$0.259	130.2	$4.86	104.1	184	410.4	75.3
1959	3.11	12.22	0.255	128.1	4.77	102.1	188	412.0	75.6
1960	3.14	12.83	0.245	123.1	4.69	100.4	191	415.5	76.2
1961	3.12	13.87	0.225	113.1	4.58	98.1	193	403.0	73.9
1962	3.12	14.72	0.225	113.1	4.48	95.9	199	422.1	77.5
1963	3.15	15.83	0.200	100.5	4.39	94.0	205	458.9	84.2
1964	3.30	16.84	0.196	98.5	4.45	95.3	225	487.0	89.4
1965	3.49	17.52	0.199	100.0	4.44	95.1	219	512.1	93.9
1966	3.66	18.52	0.198	99.5	4.54	97.2	219	533.9	97.9
1967	3.75	19.17	0.196	98.5	4.62	98.9	219	552.6	101.4
1968	3.86	19.37	0.199	100.0	4.67	100.0	220	545.2	100.0
1969	4.24	19.90	0.213	107.0	4.99	106.9	226	560.5	102.8
1970	4.58	18.84	0.243	122.1	6.26	134.0	228	602.9	110.5
1971	4.93[a]	18.03	0.273	137.2	7.07	151.4	210	552.2	101.3
1972	5.84[a]	17.50[b]	0.334	167.8	7.66	164.0	N.A.	595.4	109.2

N.A.–Not Available.

[a] Adjusted to include pro-rated value per hour of increase in welfare fund payment to 60¢ a ton effective Nov. 12, 1971; increased to 65¢ a ton effective Nov. 12, 1972.

[b] Preliminary estimate.

[c] Capacity is calculated by the Bureau of Mines to be 280 operating days a year, but this measure is of doubtful validity. It is probably overstated.

Sources: *Bituminous Coal Annual, 1972,* National Coal Association, Bureau of Mines.

power among coal producers. During the period tonnage increased 11.8 percent while average realization fell steadily from $4.86 to $4.39 a ton, a drop of 9.7 percent. Closer examination, however, modifies the conclusion on market power. First, operating rates throughout the period were at low levels. Tonnage in 1961 was at the second lowest level in twenty-two years. Most of the tonnage increase during the five year period came in the last year; output remained almost unchanged during the earlier years. Second, sharply reduced wage costs permitted coal to be sold at lower prices without necessarily adversely affecting profit margins. During this period average hourly earnings increased only 7.5 percent while tons per man-day rose 39.7 percent, creating a 22.8 percent fall in wage costs. Doubtless part of the average realization reduction resulted from producers being able to offer coal at lower prices as a result of lower labor costs.

In the period 1963-1968, coal demand increased steadily from 458.9 to 545.2 million tons, an 18.8 percent increase, and operating rates similarly rose to more satisfactory levels. The average price level increased only 6.6 percent for the same period. Again the record is inconclusive with respect to the industry's possession of market power. Just as price reductions were damped down during the earlier period, so too were price increases restrained when demand burgeoned.

Undoubtedly, an important factor moderating price changes throughout the period was the growing use of long term contracts. These contracts provide for the shipment of coal at prices that change only in response to certain cost increases (most importantly, from wage rate changes).[16]

From 1968 to 1972 average coal prices moved up substantially, increasing 64 percent during a period when tonnage rose only 9.2 percent and operating rates remained fairly constant. It was evidence of this sort of an explosive price rise that led critics to suggest the existence of a conspiracy to raise prices artificially. But note that while coal prices elevated sharply, wages costs increased even more. The higher wage costs resulted from wage and welfare fund payment increases, but more importantly from a big reduction in productivity, caused principally from enforcement of the 1969 Coal Mine Health and Safety legislation.

What does this price-output behavior indicate? First, it puts the elevated prices in a different perspective by relating them to changes in the industry's most important cost determinant. In 1960, a sample of District 8 mines found mine labor costs to account for 52 percent of total production costs.[17] But what of remaining costs? Unfortunately, little reliable industry data exists on the breakdown of coal production costs by major components. Doubtless they increased less than did labor costs and more than likely approximated price increases in the rest of the economy.

Thus profit margins undoubtedly increased during the period 1968-1972, in response to the increased price level.

State Price Behavior

To get a better insight into the influence of structural conditions on pricing behavior, we need to look at conditions in various submarkets.

The data in Table D-5 points to several striking conclusions. The first is that the companies producing in states with the highest seller concentration and highest operating rates tended to increase prices *less* during the inflationary 1968-1972 period than did companies in the other states. We usually associate high concentration with market power; similarly, as mines operating rates approach and exceed capacity their tight supply situation should reflect itself in inflated prices. In Illinois five companies accounted for 77 percent of the 1971 production; the top ten produced 99 percent of the state's output. In Indiana, three companies dominated output, producing 95 percent of the state's total. In Missouri, one producer sold 85 percent of the state's output, and in Kansas only two producers accounted for over 99 percent of total output. Output in the last two states is small—scarcely one percent of the national tonnage—but Illinois and Indiana account for 15 percent of total U.S. coal output. In Kentucky, where Table D-5 presents data for the entire state, results in the western half of the state serving one market area were quite different from those in the east, which serve separate markets. In western Kentucky eight producers account for 82 percent of total production; and average realization increased 43 percent between 1968 and 1971. In eastern Kentucky prices rose an average of 76 percent; yet, the top eight producers account for only 22 percent of output.

Underlying these data are two important market factors. First is that the concentrated ownership in the five producing areas referred to above is made up principally of the large coal-oil and coal-conglomerate companies listed in Table D-2. Thus we have additional evidence—admittedly less than conclusive—of the failure of the major producers to exert extraordinary upward pressure on prices. The second market factor of importance is the heavier reliance in these five producing areas on long-term utility contracts. These contracts undoubtedly tended to damp down price increases. Note also that smaller price increases in Illinois, Indiana, Missouri, and Kansas did not result from lower wage cost increases; the average wage cost increase for those states were similar to the increases for the other states.

Conversely, the existence of a more fragmented coal industry in the other coal districts listed in Table D-5 permitted higher price increases as operators were able to capitalize on the tight coal market which led to elevated spot coal prices.

A final factor accounting for the different price behavior in the two groups of mining states probably was the influence of the export market where prices rose substantially faster than they did in the domestic market. Districts 7 and 8, comprising mines in West Virginia, Virginia, Tennessee and eastern Kentucky, account for 91 percent of non-Canadian exports. These mining

Table D-5. Prices, Wage Costs and Capacity Utilization, Various States, 1968-1972

State	1968 Price Index	1968 Wage Cost Index	1968 No. Days Worked	1969 Price Index	1969 Wage Cost Index	1969 No. Days Worked	1970 Price Index	1970 Wage Cost Index	1970 No. Days Worked	1971 Price Index	1971 Wage Cost Index	1971 No. Days Worked
Kentucky	100.0	100.0	198	105.8	104.0	216	145.3	112.8	207	165.9	129.6	192
West Virginia	100.0	100.0	218	107.7	108.2	212	149.1	133.9	219	179.3	159.2	195
Pennsylvania	100.0	100.0	229	109.3	108.0	245	135.4	127.2	248	158.6	145.6	228
Illinois	100.0	100.0	250	107.7	112.3	267	122.7	134.6	269	136.2	155.3	249
Ohio	100.0	100.0	238	103.5	107.9	249	118.9	118.4	235	132.3	134.9	228
Virginia	100.0	100.0	210	112.0	110.3	206	163.8	130.5	219	171.9	145.9	204
Indiana	100.0	100.0	253	106.4	104.4	282	118.6	111.4	271	133.5	123.7	260
Tennessee	100.0	100.0	206	104.4	107.1	193	134.6	108.1	192	175.8	121.7	214
Missouri	100.0	100.0	310	103.1	118.6	301	104.5	121.7	318	115.9	131.8	280
Kansas	100.0	100.0	298	105.2	116.6	299	108.5	107.4	284	111.1	157.7	275

Source: *Minerals Yearbook*, U.S. Bureau of Mines, various years.

districts exhibited the sharpest rise in average realization during the period under review. The unusual pricing behavior in the export sector dictates a closer examination of that special submarket.

Export Pricing Behavior

Table D-6 presents some salient statistics on recent pricing behavior in the export market. The data first require some explanation. They cover only overseas exports and exclude Canada. The shipments thus account for around 70 percent of total exports and 83 percent of high valued metallurgical coal exports. The average realization figures for exports given in Table D-6 have been calculated by subtracting an estimate of the weighted average shipping and dumping charges per ton from the available statistics on export prices FOB east coast ports. No published data exist on the FOB mine price of overseas exports, but the estimates in Table D-6 are believed to be quite accurate. Canadian exports are excluded from the analysis due to the inability to separate transportation charges from the published delivered price data.

A peripheral item of interest in Table D-6 is a time series giving the average realization for coal *less* overseas exports. It shows a slightly lower rate of increase for domestically consumed coal than the data in Table D-4 indicates. With Canadian exports included the rate of increase would be even less.

One is struck by the enormous increase in export coal prices during the period covered by Table D-6 and especially since 1968. While domestic prices increased 60 percent, export prices were rising at more than double that rate—124 percent. The disparity between the domestic and export prices widened considerably. In 1972 the export price level stood 82 percent above the average domestic coal price; in 1968 the export price advantage amounted to only 33 percent. In 1960 a large sample of District 8 mines revealed overseas export prices *lagging* 11 percent behind the average price for all shipments and 19 percent behind domestic metallurgical shipments at a time when non-Canadian exports accounted for 6 percent of total coal sales.[18]

Much of the rapid increase in the overseas export price level can be traced to a fairly sharp bulge in demand for world coal exports between 1967 and 1970.[19] The major coal exporters (excluding Russia) increased coal exports 34 percent during the period; for the United States the increase amounted to 42 percent. For non-Canadian shipments the United States increase was even sharper—53 percent. Japan, almost single handedly, accounted for the United States' export expansion, increasing its purchases by 15.4 million tons between 1967 and 1970, all of it metallurgical coal.

This demand expansion helps partly to explain the increase in the export price level between 1967 and 1970. Note, however, that in 1971 overseas exports fell 25.3 percent, yet export prices *rose* 25.4 percent. In the domestic market prices also increased, but only 12.6 percent in the face of a 6.8 percent production decline. In 1972 overseas exports demand continued to decline,

Table D-6.   Export Coal Price-Output Behavior, 1967-1972[a]

Year	Export Tonnage (MM tons)	Exports as Percent of Total Production	Average FOB Mine Price, Export Shipments	Index of Export Prices (1967 = 100)	Average Nonexport Coal Price	Index of Nonexport Coal Price (1967 = 100)
1967	34.2	6.2	$ 5.92	100	$4.54	100
1968	33.9	6.2	6.06	102	4.58	101
1969	39.4	7.0	6.76	114	4.85	107
1970	52.3	8.7	9.67	163	5.93	131
1971	39.1	7.1	12.13	205	6.68	147
1972	37.8	6.4	13.28	224	7.28	160

[a]Excludes exports to Canada.

Sources: *Minerals Yearbook*, Volume I, U.S. Department of Interior.
World Coal Trade, Coal Exporters Association of the United States Inc., U.S. Bureau of Mines.

falling 3.3 percent from the 1971 level, but average realization rose another 9.5 percent. Between 1970 and 1972 domestic coal prices increased an average of $1.35 a ton as production for domestic consumption fell 1 percent; overseas export prices went up $3.61 a ton while demand fell 28 percent!

A measure of the profitability flowing from increased export prices is revealed in the annual statements of Pittston Company, a major coal exporter. Their profits from coal increased from $9.5 to $23.5 million, 1970 vs. 1969, on almost identical *total* coal sales (roughly 20.5 million tons). Pittston's export sales for the period, however, rose from 6.3 to 11.2 million tons.

Cost increases doubtless account for part of the sharp run-up in export prices, especially between 1970 and 1972. Recall that wage cost increases for the period amounted to 46 percent (see Table D-4). But domestic coal production was subject to roughly similar cost pressures and managed smaller price increases under favorable demand conditions. Since most of the export coal moves under long term contracts, the sharpness of the price increases is even more puzzling. Built-in escalation provisions in the contracts should permit price increases based on increased costs—as they do in domestic contracts—but they would not account for increases of the magnitude indicated in Table D-4.

It might be illuminating to know the initial prices on new export contracts each year to determine the effect of these prices on average realization. It is conceivable that the average for all export shipments is made up of a combination of existing contracts whose prices increase in response to cost increases and new contracts whose prices exceed those covered by older, existing contracts. Unfortunately, the data to make this kind of analysis is unavailable.

A significant aspect of the export market situation is the concentration of sellers in a few hands. Twelve companies which are members of the Coal Exporters Association of the United States, Inc., affiliated with the National Coal Association, account for 75-80 percent of U.S. coal exports. Two nonmembers, A.T. Massey Coal Company, and Island Creek Coal Company handle a large share of the remaining tonnage. Most of the export companies primarily handle sales of their parent producer companies' output, but they also act as brokers for other producers. Any deeper investigation of export coal pricing behavior might investigate the effect of the market structure and the impact, if any, of an industry association binding together most of the leading exporters.

POLICY ALTERNATIVES

In reviewing policy alternatives it might be useful to break the discussion into two parts: a review of policies that might be adopted to correct or reform existing situations, and a discussion of possible policy prescriptions designed to limit future activities that might reduce effective competition.

Divestiture

Evidence exists elsewhere in this study (Duchesneau's) of increasing concentration of sellers in the coal industry, a good deal of it resulting from mergers. For this discussion we study separately mergers between coal companies on the one hand and coal-oil and coal-conglomerate mergers on the other. The question we investigate specifically is whether divestiture in either case might be warranted to improve conduct and performance in the energy sector. Let's look at coal-coal mergers first.

There is ample case law to support overturning mergers violating Section 7 of the Clayton Act.[20] Action under Section 7 may prevent a merger or lead to divestiture where "the effect of such acquisition may be substantially to lessen competition or to tend to create a monopoly."

The courts have already called for divestiture in a coal-coal merger under Section 7. Peabody Coal Company, under a consent order, divested itself of six million tons of annual capacity and 120 million tons of coal reserves in a case growing out of Peabody's acquisition of Midland-Electric Coal Corporation and Stonefort Coal Mining Co.[21] All three companies operated in the state of Illinois where, as we have seen, seller concentration is high. The Justice Department also sought to overturn the acquisition by General Dynamics of United Electric Coal Co. several years after the fact. General Dynamics, through a subsidiary, also controls the Freeman Coal Company. Both Freeman and United Electric are large coal producers in Illinois. A government victory in this case would have led to divestiture of the United Electric properties; however, the court held in favor of General Dynamics. The case is before the Supreme Court on appeal and may be decided before the end of 1973.

Another possible divestiture is involved in the Kennecott Copper Company case in which the lower courts have sustained the Federal Trade Commission's allegation that Kennecott's acquisition of Peabody Coal Co. constituted a Section 7 violation.[22] Although Kennecott was not an active coal producer, it had acquired a coal mine (which it shut down) and, prior to the Peabody acquisition, had sought ways to enter the coal industry on a large scale. This case, also, is pending in the Supreme Court on appeal, and may be decided before the end of 1973. A judgment against Kennecott would lead to a divestiture of the Peabody properties.

The effectiveness of divestiture as a remedy in coal-coal merger cases depends, in part, on the outcome in the General Dynamics case, especially, and, to some extent, on the Kennecott case outcome. In *General Dynamics*, the District Court ruled against the government on the grounds that coal had not been shown to be a separate product market, that neither the Eastern Interior coal region nor the state of Illinois, in which the mines operated, was a separate geographical market, nor that the merger had lessened competition.

If the Supreme Court upholds this position, divestiture as a remedy for coal-coal acquisitions may be difficult to achieve.[23] It would be hard to find many other markets where structural conditions would better favor a Section 7 case than in the state of Illinois.

If the Supreme Court overrules the lower court in the General Dynamics case, divestiture of other coal properties will continue to be a feasible remedy to reduce seller concentration. But does it make sense to recommend a policy of divesture to unscramble other coal-coal mergers? A case might be made for the divestiture of Southwestern Illinois Coal Company from Arch Minerals Company since both sell large quantities of coal in the Illinois market area where we have noted high seller concentration. Consistency of antitrust policy would dictate such a policy since Arch purchased Southwestern after both the Midland Electric case had been settled and after litigation had begun in the General Dynamics case. Aside from this situation it is hard to find a basis for divestiture of coal properties based on previous coal-coal mergers unless one wants to recreate the highly fragmented industry that existed prior to the 1940s, with its attendant ruinously competitive behavior.

How about the coal-oil and the coal-conglomerate mergers? Does a policy of divestiture seem called for in these cases? As earlier noted, we may find Peabody divested from Kennecott if the Supreme Court upholds the lower court's finding in this case. Whether moves to divest in other similar situations will be, or should be, made remains to be seen. In some respects the Kennecott case is unique because of evidence indicating Kennecott's prior involvement in the coal industry and its intent to enter the industry on a massive basis.

If divestiture is called for in the coal-conglomerate mergers it might make sense if one of the following conditions prevails:

- Evidence exists of a lessening of competition in the coal industry following the mergers.
- The so-called "deep pocket" theory prevails—that is, the dominant position of the acquired firm (backed up by its acquirer's large resources) gives it a competitive edge over its smaller rivals. This condition, of course, is related to the first concerned with the lessening of competition.

Our investigation of both price and output behavior finds no evidence that, taken as a group, the coal-conglomerate companies either outperformed their rivals in gaining market share ("deep pocket" situation) or exerted undue influence over prices. Much of Amax's output expansion (see Table D-2) had occurred or was in the development stage prior to its acquisition by American Metal Climax.

A tougher policy question concerns the coal-oil mergers. Would divestiture in these instances promote competition? The problem is more difficult than either the coal-coal or coal-conglomerate situations, because, in addition to possessing great financial power ("deep pocket" threat), the acquiring oil companies control at least one, and in a number of cases, several potentially competitive fuels (petroleum, uranium, oil shale, natural gas). It might aid the analysis of the impact of coal-oil mergers on competition to look not only at past activity, to determine whether divestiture is called for, but at the future as well to recommend a policy concerning the role, if any, of oil companies in the coal industry.

A key factor in analyzing the competitive impact of coal-oil mergers is a determination of the relevant product market. If coal and oil are considered to be separate and unique products serving separate markets, it is harder to make a case against coal-oil mergers than would be the case if the cross elasticity of demand between the two products were high. Supreme Court concurrence of the lower court's interpretation of the relevant product market in the General Dynamics' case will strengthen the concept of *energy* markets encompassing several fields. In the short run—in certain end-use markets—coal and oil are substitutable in demand. In others they are not. Substitution possibilities, of course, increase in the long run.

With respect to the effect of coal-oil mergers in lessening price competition, one needs to analyze both the impact on current market conditions and the potential for future injury to competition. We have found in this study no evidence of collusion or conspiracy to raise prices or limit ouput artificially in the domestic coal market. Admittedly, the evidence for this assertion is weak. The potential for future injury to competition depends on several factors whose dimensions are unclear. For example, the Justice Department refused to block the Consolidation Coal-Continental Oil merger based on Continental's selling little or no petroleum products or natural gas in Consolidation's market territory. High transportation charges would have ruled out Consolidation's moving into Continental's markets.

But one must look beyond current conditions to assess the competitive impact of coal-oil mergers. The failure of an oil company and a coal company to operate in the same market area does not rule out the future possibility of their selling in the same geographical market. In fact since the Continental-Consolidation merger, Consolidation has acquired large western coal reserves in a prominent Continental market area. Furthermore, a critic of this merger asserts that to maximize competition in regional petroleum markets, impetus will have to come from existing oil companies entering new regional markets.[24] Thus the argument that two energy companies do not compete in the same geographical market at the time of a proposed merger needs modifying if one is looking at the *potentially* adverse impact of the merger.

In addition one must take into account the potential for future competition from gasified coal and from petroleum products produced by liquefying coal. Though coal, oil, and gas may not compete in certain markets today, the ability to liquefy and gasify coal may create new competition tomorrow.

There are other problems associated with coal-oil mergers. The planning horizon for the use of coal reserves may differ for coal companies serving traditional coal markets and oil companies acquiring coal properties for future liquefaction or gasification. Thus we might witness oil companies keeping large coal reserves off the market pending future development for synthetic fuels while potential entrants into the coal market—to serve, e.g., the electric utility market—are unable to acquire desirable coal reserves.[25] This situation may exist to some extent in Illinois today where huge coal reserves have been acquired in recent years by large oil and coal-oil companies. The problem here expressed applies as much to the situation where oil companies acquire coal reserves as it does to coal-oil mergers, but both conditions contribute to the problem.

A question relating to the foregoing is whether coal-oil mergers (and control through merger or otherwise of large coal reserves) will reduce incentives to develop oil liquefaction technology. The oil companies have a big stake in existing petroleum production facilities and may be reluctant to obsolesce them by developing processes for producing economical synthetic substitutes. So runs the argument. The argument has some appeal. On the other hand, the coal industry has hardly been noted for being progressive in research. But the growth in the size of the larger coal companies adds a new dimension to the ability of coal producers to undertake significant research programs. Until recent years even the largest coal companies were too small to afford more than a token research effort. An exception was Consolidation Coal Company which, before its acquisition by Continental, sustained a fairly extensive research program aimed at developing gasoline synthetically from coal. That program was jointly underwritten by Standard Oil of New Jersey (now Exxon). This arrangement raises the question whether, in the absence of coal-oil mergers, that pattern of joint sponsorship of liquefaction and other oil-related research could not be conducted by independent coal and independent petroleum companies. This possibility is enhanced by the growth of the largest coal producers which makes a respectable research effort more feasible today than was formerly the case.

There is a final aspect of the problem of the entry of oil companies into the coal industry, and the competitive effect of these moves. This is the matter of oil company ownership of coal lands, *independent of* their acquisition of existing coal companies. In Illinois, Montana, North Dakota, Wyoming, and, to a lesser extent, in other states, major oil companies have acquired leases or fee ownership of thousands of acres of coal lands containing billions of tons of coal. A review of government coal land leases in April, 1971 showed that oil companies held 77 leases on 187,250 acres.[26] The oil companies have acquired large coal reserves from nongovernment sources as well. Humble, for example, reported control of 3-1/4 billion tons of coal in Illinois in 1968.

Several oil companies with newly acquired coal reserves have opened coal mines. This development plus the drive for coal reserves by a large number of other oil companies raises the question whether oil companies which entered the coal industry via the merger route might not have entered independently if antitrust enforcement had barred their coal company acquisitions.

From a public policy point of view it is hard to justify coal-oil mergers—either those that have occurred or future mergers—if we intend to maintain vigorous competition in the energy sector. The benefits from such mergers have not been demonstrated; the potential injury to competition from a concentration of substitutable energy sources *could* be substantial. Note that the threat to the conduct of vigorous competition is more potential than actual. If government policy permits reasonably free access to remaining coal reserves, divestiture of coal properties as a result of *previous* coal-oil mergers may be unnecessary, and in any event, difficult to accomplish. To promote competition in energy markets government policy ought to discourage major oil companies entering the coal industry through future acquisition of existing coal companies; rather they should be encouraged to make independent entry into the industry.

As previously indicated, the control of vast coal reserves by oil companies could pose a grave competition threat, which ought to be dealt with. The federal government, through its ownership of forty million acres of western coal lands, holds the key to the coal reserve control problem. Currently, the Bureau of Land Management, which manages the government's mineral lands, has imposed a moratorium on the leasing of most of the federal lands. Before opening up remaining government coal lands to competitive bidding, the Bureau intends to develop a policy that will achieve a fair monetary return to the government for the coal it owns, that will match coal supplies with market demands in a rational manner, and that will insure protection of the environment. To further promote public welfare the government's leasing policy should provide for maximizing competition in the energy sector. This may require establishing procedures to encourage new entrants into the coal industry (mining government lands) and restricting coal acreage allotted to individual firms—especially companies in other energy industries. Present B.L.M. policy limits the control of government coal leases to 46,080 acres per state for a single firm. Consideration should be given to reducing the limit and placing it on a *tonnage* rather than acreage basis if reasonable tonnage estimates can be made. In addition to limiting reserve tonnage in this way, the government's leasing program could call for a *competitive* impact analysis in addition to an environmental impact statement.

Restricting the control of large blocks of undeveloped coal reserves (on government and nongovernment land) for speculation and delayed development (e.g., as a hedge against the need for future liquefaction or gasification) may also serve the public interest by permitting easier entry into the industry. One way to accomplish this might involve a federal, progressive property tax on coal reserves that an individual company (and affiliates) controls in excess of a "reasonable" limit. The firm's scale of operations and expansion plans could dictate what would be considered "reasonable."

Finally, we need to return to a consideration of a public policy position with respect to *future* coal-coal mergers. There is a trend toward larger capital requirements both for mines serving the electric utility market and those scheduled to produce coal for gasification projects. Since market demands, industry structure, and scale economies all point to the need for fairly large producing units in coal, public policy ought not to be overly restrictive with respect to coal-coal mergers. One can make a case for improving the viability of the industry and the level of competition therein if producing units are larger. Such a policy, however, should not provide for blanket approval of coal-coal mergers. Each case will require a careful examination of the structure of buyers and sellers in the relevant geographical market to determine whether the potential advantages to the parties merging are outweighed by the injury to competition that such a merger might entail.

NOTES

1. In the following analysis only General Dynamics Co. which owns the Freeman-United Electric Coal Company's coal operations is neither a petroleum or mining company. Freeman and United Electric were originally acquired by Materials Service Company, producer and distributor of aggregate materials and cement products. Materials Service was later merged into General Dynamics.
2. Hereafter we use "coal" and "bituminous coal" synonymously.
3. We exclude captive mines—those controlled by steel and electric utility companies.
4. *Coal Mine Directory, Keystone Coal Industry Manual.*
5. "Coal: Prospects and Problems," speech by Herbert S. Richey at the National Energy Forum, Washington, D.C., September 23, 1971.
6. Continental Oil Co., Listing application to the New York Stock Exchange, #A-23648, August 30, 1966, p. 4.
7. "Artificial Restraints on Basic Energy Sources," p. A445, in *Concentration by Competing Raw Fuel Industries in the Energy Market and Its Impact on Small Business*, Hearings before the Subcommittee on Special Small Business Problems of the House Select Committee on Small Business, 92nd Congress, 1st Session, July 12, 13, 14, 15, 20, 22, 1971.
8. The St. Joe Minerals acquisition of A.T. Massey Coal Co. and Houston Natural Gas Company's purchase of Zeigler Coal Co. are pending at the time this is written.
9. The Kennecott case is summarized in *Fortune*, September, 1971, pp. 98-101, 140-42.
10. Hearings, *op. cit.*, pp. 37-45.
11. *Ibid.*, p. 38.
12. Hearings, *op. cit.*, Vol. 2, p. 11.

13. Hearings, *op. cit.*, Vol. 1, p. 41.

14. This is based on *Statistics of Income* data with the same data coverage limitations mentioned in the text.

15. Hearings, *op. cit.*, Vol. 2, p. 19.

16. Ordinarily prices move only upward although some contracts, providing for sharing of productivity improvements, may result in lower prices.

17. "District #8 Report of Average Cost and Realization—Calendar Year 1960," confidential report issued by Southern Coal Producers' Association.

18. *Ibid.*

19. See *World Coal Trade*, Coal Exporters Association of the United States, Inc.

20. E.g., *United States v. Pabst, United States v. Philadelphia National Bank.*

21. *United States v. Peabody Coal Co., et al.*, U.S. District Court, Northern District of Illinois, Eastern Division, September 21, 1967.

22. Since the writing of this study, the Supreme Court has decided both the General Dynamics and Kennecott cases, the former against the government, the latter in its favor.

23. The Peabody-Midland-Electric case sets no precedent since it involves a consent order.

24. Hearings, *op. cit.*, Vol. 1, p. 73. Testimony of Beverly C. Moore.

25. The same result could stem from a natural gas distributor, having a current need for coal for gasification purposes, but being denied the most desirable reserves because they are locked up by oil companies whose coal needs for *liquefaction* are not as urgent.

26. Hearings, *op. cit.*, Vol. 1, p. 133.

Appendix E

Potential Competition in Uranium Enriching

Thomas Gale Moore

The steps in the production of nuclear fuel for power reactors are: mining, milling of uranium into "yellow cake" (U_3O_8), conversion into UF_6 for feed, enrichment, and fuel fabrication. Except for the enrichment step, which is a Government monopoly, the industry is in private hands.

Uranium enriching is the process of increasing the proportion of the isotope U_{235} relative to the isotope U_{238} in the uranium. Natural uranium consists of about 0.0711 percent of U_{235}, while light water reactors need uranium enriched to between 2 and 4 percent U_{235}. Of the cost of fuel fabrication, including mining, approximately half is attributable to enriching at current prices.

All major enrichment facilities in the free world currently use the gaseous diffusion method of enriching uranium. This method pumps the gas UF_6 through permeable barriers. A slightly higher portion of U_{235} goes through the barrier than the heavier U_{238}. The enriched gas is then sent through another stage where again it filters through a barrier becoming slightly more enriched, while the depleted gas is recycled through another stage. This process can continue until the desired proportion of U_{235} is achieved.

The work done in separating the isotopes U_{235} from U_{238} is called separative work. It takes approximately one kilogram of separative work applied to 2.35 kilograms of natural uranium to produce one kilogram of uranium containing 1.4 percent of U_{235} while stripping the uranium tails to 0.2 percent U_{235}. The separation factor by gaseous diffusion is 1.00429 which means that 600 metric tons of UF_6 must be pumped so that about 300 metric tons permeates through the barrier in order to accomplish one kilogram of separative work. This huge pumping requirement is responsible for the vast power needs of the gaseous diffusion process.

The U.S. Government owns three gaseous diffusion plants, built at an original cost of approximately $800 million each. Located at Oak Ridge, Tennessee, Paducah, Kentucky, and Portsmouth, Ohio, they have been operated as a unit by private contractors for the AEC. In 1970 they produced about 6,900 metric tons of separative work while consuming 2,000 mw of electrical power. The AEC reports that power accounts for about three-fourths of the out-of-pocket cost of operating these plants. The plants as they are have a maximum capacity of about 17,000 metric tons per year.

According to AEC projections, annual consumption of enriching services on a worldwide basis is expected to grow rapidly and will exceed the current capacity of these plants by 1978. While there exists small enriching facilities in Great Britain and France the U.S. has the only substantial capacity for meeting free world needs. To satisfy those needs the AEC has embarked on a "preproduction" program of stockpiling enriched uranium to satisfy demand after it exceeds capacity. In addition it has launched two major improvement programs for the existing plants that will increase capacity by 10510 metric tons per year by 1981. As a further step to postpone the inevitable day when new facilities will be needed, the AEC has determined to operate the plants with a 0.3 percent tails assey. This means that less separative work will be necessary (although more natural uranium will be needed) to produce a given amount of enriched uranium. Together these steps, according to the AEC, will satisfy demand through fiscal 1983.

A committee of the Atomic Industrial Forum has forecast that sometime during calendar year 1982, world demand will exceed world availability of enriched uranium. With the adoption by the AEC of 0.3 percent tails assey, the date when demand will exceed supply has been pushed back about half a year. Thus, both the Atomic Industrial Forum committee and the AEC agree that sometime during the period 1982-1983 new supplies from new plants will be needed if shortages are to be avoided.

In December of 1972, the AEC announced that any new plants should and can be built by private industry. Then Chairman Schlesinger of the AEC announced that the Government "does not intend to build additional enrichment plants. . . ." At the same time the AEC established ground rules for facilitating private industry in its quest for classified data on enrichment technology. The stage therefore has been set for the private construction of enriching facilities. At this time, however, there are no announced plans for the disposal of the government owned plants, nor has the AEC established any rules as to the manner in which the plants would be operated in competition with private facilities. Schlesinger has said though that while a private operator's charges might be higher than the AEC's, "he did not believe this would necessarily be a serious problem."

No matter who builds the next addition to enriching capacity a crucial decision will be the type of facility constructed. Besides the gaseous diffusion technology, there is the possibility of using centrifuges to separate U_{235}. Centrifuge technology has been under AEC development for years. A consortium of Germany, Britain, and the Netherlands has been building a pilot centrifuge plant and hopes to start constructing a production plant this year which will come on stream in 1976. On the other hand, the French have been trying to interest other nations in building a gaseous diffusion plant using French technology.

There are two advantages centrifuges have over gaseous diffusion: they are less subject to economies of scale and can, therefore, be built on a much smaller scale, and they use less power per unit of separative work. For example, the Tripartite Centrifuge project mentioned above plans a 200-300 metric ton facility while the minimum efficient size for a gaseous diffusion plant is about 8000 metric tons of separative work. Reynolds Metals proposed to the AEC in 1972 to build a 8750 metric ton unit for about $2.2 billion to come on stream as early as 1978. The Dutch plant that the Tripartite group plans on the other hand is forecast by the group to cost no more than $60 million for a 300 metric ton unit.

The lower needs for electricity are a significant factor in favor of centrifuge. At past AEC rates of less than 4 mills per kwh, electricity has accounted for about half of the total cost of separative work and about three-quarters of out-of-pocket costs. Electricity rates are rising and no one believes that the AEC or any group in the U.S. can secure such low rates again. As a consequence, the attractiveness of the centrifuge process is growing. Two pieces of indirect evidence suggest that centrifuge may be the process used in the next plant. In December of 1972, Chairman Schlesinger repeatedly made it clear at a news conference that he expected centrifuge to supplant gaseous diffusion before long. Seven companies have requested and have been given access to AEC classified enrichment technology and six of the seven have requested and been authorized to conduct R&D in centrifuge while only one is planning work on gaseous diffusion.

The attached table based on the Atomic Industrial Form's projections show the available enriched uranium for each year through 1985 and the cumulated demand. In the final column the excess demand that new plants will have to satisfy each year is shown. As can be seen in 1982, there would be need for one diffusion plant or many centrifuge plants. In the next year another diffusion plant or many additional centrifuge plants will be needed. In other words, the table shows that at least several new plants will be needed in the early 1980s and that if centrifuge is the economical way, which it looks like it will be, then there is the possibility of a multitude of plants, which may cost less than $100 million each.

If the government facilitates entry by private industry and if gaseous diffusion is the proven technology, the enriching business would go from a small number oligopoly in the first half of the 1980s to a more competitive structure by the end. On the other hand, if centrifuge is chosen entry will be easy, scale small, and the industry competitive. Once private industry is doing much of the enriching business, it should be possible for the government either to sell its enriching plants, which President Nixon had said would be desirable, or for the government to lease out the plants to private interests to operate as they wish in a competitive environment.

The real question therefore is not the economics of the enriching business but government policy. Will the government permit and encourage private industry to move into the area? The AEC has said that it wants private industry to build the next plant. Reynolds has already requested authority to do so. A combination of Bechtel, Union Carbide, and Westinghouse have requested classified data on enriching from the AEC in order to enter the enriching business. In addition the Japanese have expressed interest in participating in an enrichment enterprise in this country.

To make possible private enriching ventures the AEC announced a plan to provide access to its classified data for firms intending to investigate the economics of gaseous diffusion and centrifuge. It is also prepared to provide sufficient data and information to enable a firm to design, build, and operate facilities. This offer, however, is restricted in many ways and has been subject to vigorous criticism from industry. One requirement would preclude the permittee from giving outsiders any information, classified or unclassified which is "made available to the permittee by the Commission or . . . developed by the permittee, its employees or others . . . under the access permit or as a result of data or information made available by the Commission." As several companies pointed out, this would limit the normal exchange of business data between the permittee and prospective investors and customers. Another provision according to industry would give the AEC almost dictatorial powers over a private firm's organizational structure, the charges it establishes for separative work, other contract provisions, and even the structure of the industry.

Without going into excessive detail it is clear that industry is not convinced that the AEC means to facilitate private enriching ventures. Given the long lead time necessary to construct a plant—about 6 years for gaseous diffusion—if the AEC does not move rapidly to facilitate entry, it will have to construct the plant itself to avoid shortages in the 1980s. There are no doubt many at the AEC and on the Joint Committee on Atomic Energy who would prefer to see the business remain in government hands. In the long run, the possibility of private enriching will depend on the willingness of the AEC to permit considerably wider access to its restricted data than it has shown to date. In fact, to make the industry truly private with relatively free entry, would require declassifying enrichment technology. Unless the Administration strongly pushes the AEC it seems unlikely that they will ease access sufficiently to provide for a private enrichment industry.

POLICY ALTERNATIVES

There are three major policy alternatives in the nuclear enriching area: maintain AEC control and operation, establish a government corporation, or attempt to promote a competitive private industry. In weighing each of these alternatives, the impact on national security, on costs of enriching, on progressivity of the industry, and on prices charged must be evaluated.

Projected
Enrichment Supply and Demand

Calendar Year	Total World Availability	Total World Demand	Excess Demand To Be Met by New Plants
	(cumulative metric tons separative work)		(annual increment)
1973	33,760	14,475	
1974	47,860	24,067	
1975	63,780	35,365	
1976	81,710	51,031	
1977	101,760	70,444	
1978	124,360	94,763	
1979	150,560	124,540	
1980	180,320	159,319	
1981	211,970	207,719	
1982	244,620	252,622	8,002
1983	278,520	304,001	17,479
1984	313,920	363,010	23,609
1985	350,820	428,579	28,669

Source: *Nuclear Industry*, October 1972, p. 7.

From a national security point of view, it is obvious that government control and operation whether under a government corporation or by continuing AEC operation would be the safest. Fewest people would need be given access to sensitive material. In fact if a viable competitive private industry is to be established, most if not all data on enrichment technology must be declassified. True free entry cannot be provided if AEC clearance is necessary to receive access to the technology.

Assuming a private industry would imply complete declassification of enrichment technology would this imply a significant reduction in national security? It is not obvious that it would. Even though centrifuges can be constructed on a considerably smaller scale than diffusion plants and even though they take significantly less power than diffusion plants, declassification of the technology does not imply that every country in the world has the resources including skilled labor and equipment to build a viable centrifuge system. It would take considerable industrial sophistication to build a working system even if all data had been published. Thus the nightmare of a small or medium size dictatorship constructing enriching facilities in a barn and then blackmailing neighboring countries is an unreal characterization of the world. Publication of the data would mean that sophisticated industrial countries could more easily build their own enriching facilities on a competitive basis, but these countries are able to build such facilities now, albeit probably not as well as knowledge about U.S. technology would permit.

Economic theory suggests and experience teaches that a competitive private industry would in the long run have lower costs than would a government operation. Competition would force producers to search for the efficient solution. Continued government control would mean that pressure would be restricted to the much weaker force of foreign enrichment competition. Since a government corporation should be less subject to political control, it seems likely that it could be more forceful in taking steps to reduce costs and improve efficiency. Continued AEC control would mean continued influence by Congress through the Joint Atomic Energy Committee. While the AEC has been reasonably successful, as far as anyone can tell, in controlling costs, in the future political interference may generate higher costs.

A private industry will have a strong incentive to continue to innovate to reduce costs. The impact of government operations on progressivitiy is more speculative. Governmental budgetary restrictions may reduce R&D by the AEC below optimum levels, although this is far from certain. A government corporation, however, being a monopoly may in fact under invest in research since it is already assured of the market. R&D costs money and its results are uncertain, therefore the bureaucratic response may be to minimize efforts in this area.

Unless government operation is subsidized by the taxpayer it is unlikely that private operations would price higher than public. Such subsidization is quite likely under continued AEC operation but less likely with a government corporation. If centrifuge technology becomes economic, a highly competitive industry is likely and prices are likely to be lower under private operations than under a government corporation. On the other hand, if centrifuge does not turn out to be viable, a private enriching industry could—if the AEC sold its three gaseous diffusion plants—consist of four firms in 1982, five firms in 1983, six firms in 1984, and seven firms by 1986. This scenario would be the most optimistic under the assumption that gaseous diffusion remains the dominant technology. Thus the industry would be a small firm oligopoly much like the aluminum industry, at least until about 1984. However, this implies that all new capacity is constructed by new firms and it seems almost inevitable that existing firms would build part of the capacity. Thus even by the mid-1980s without legislation to guarantee that new capacity was built by new firms, it seems unlikely that there would be more than four or five firms in the industry. Their pricing would be limited, however, to the lowest cost of producing enriched uranium by centrifuge. Thus even if diffusion manages to keep its economic advantage a private enrichment industry could perform in a workably competitive manner, although not as well as it would were centrifuge to become the low cost method. If diffusion continues to be the dominant technology, legislation limiting the size of the facilities constructed by one firm for a period of about five years would help guarantee that by the late 1980s a reasonably competitive industry would be operating.

At the present time, it seems most likely that centrifuge will be the technology in the 1980s and a competitive industry will be possible. If a viable private sector is to be created the following steps are desirable: First, the enrichment technology must be made more freely available now; by 1980 it should be completely declassified in order to permit relatively free entry. Second, the existing diffusion plants must be sold and ownership transferred by 1982 to three different companies or combinations of companies. Plans for such a sale must be made well in advance, preferably by 1975, in order to permit private industry to plan new plants as well as the purchase of the old. Since by 1982, all of the preproduction will have been sold, the best time for private industry to take over the industry would be in that year.

Appendix F

The Outlook for Independent Domestic Refiners to the Early 1980s

John H. Lichtblau*

(Author's Note: This essay was completed in June 1974. Subsequent developments have affected some of the facts but not the general conclusions of the essay. 2/10/1975)

From the late 1950s to the beginning of the 1970s four of the principal features underlying the structure of the U.S. oil industry were (1) restrictions on the importation of crude oil and refined products; (2) the rapid growth in the supply of relatively low-cost foreign crude oil controlled by U.S. and other international oil companies; (3) the existence of excess domestic crude oil producing capacity; (4) the existence of excess domestic refining capacity.

Since 1971 all four of these features have either disappeared altogether or are in the process of doing so. In 1971 for the first time since World War II U.S. crude oil production, in the face of rising demand, was no longer able to register an increase over the previous year, despite the absence of any economic production restrictions. Each of the two subsequent years (and probably also the year 1974) have registered an actual decline in production. In 1973 U.S. refining capacity had to be utilized at virtually full operable capacity to meet demand. From the beginning of 1973 on foreign crude oil operations owned or controlled by private oil companies have been progressively transferred to the control and ownership of state agencies of the major oil producing countries; meanwhile, foreign oil costs have risen well above domestic costs. Finally, on May 1, 1973 the U.S. Import Control Program on crude oil and refined products was converted into an open-ended import system.

The following analysis inquires into the impact of these recent structural changes in the U.S. oil industry on the future of independent oil refiners. As a first step in this analysis we must consider how both these segments of the industry fared under the structure in existence before the changes that have occurred since 1971.

---

*Executive Director, Petroleum Industry Research Foundation, Inc., New York.

295

DEFINITIONS AND CHARACTERISTICS

The independent oil refiners are not a well defined group. The term "independent" is sometimes used to describe any refiner other than those who produce regionally or nationally advertised brands of gasoline. It is also often considered synonymous with small refiners. A third use of the term is to describe any refiner who is not integrated backward into crude oil supplies or forward into marketing. For our purpose we are classifying all those refiners as independent who have little or no integrated crude oil production. The reason why we are defining independence only in terms of *backward* integration is that forward integration into marketing can be achieved much more easily and changes the structure of the firm much less than backward integration into production. The latter assures control over a scarce natural resource and, under U.S. law, gives tax advantages not available to any other sector in the oil and gas industry.

We have arbitrarily set the cut-off point for integrated companies at a domestic self-sufficiency ratio of 25 percent. Thus, any refiner who must purchase 75 percent or more of his domestic crude oil requirement from nonaffiliated sources is considered independent.[a] There are altogether 130 refining companies in the U.S. We have assumed that the 18 refiners listed in the following table, all of which have self-sufficiency ratios in excess of 25 percent, represent the integrated refining sector. Their composite U.S. self-sufficiency ratio (net domestic crude oil production/domestic refinery runs) was approximately 62 percent in 1972. While the remaining segment of the refining industry may include a few small companies with a higher self-sufficiency ratio than our cut-off point for independents, by and large, this group consists overwhelmingly of firms which fit our definition of independent refiners.

Another characteristic of the independent refining segment is its relatively small size. The average size of the 18 integrated refineries is 100,000 b/d.[1] The average size of all other plants is 20,000 b/d. In terms of total company refining capacity the difference is even larger since most integrated refiners are multiplant companies while most Independents are single plant companies. Thus, the average refining capacity of the 18 integrated firms is 540,000 b/d, compared with 29,000 b/d for the average independent.

IMPACT OF IMPORT CONTROL PROGRAM

The Mandatory Oil Import Control Program which existed from March 1959 to May 1973 has generally worked in favor of the independent refiners, primarily because it was designed to do so. During the period 1959-1971 foreign oil had a cost advantage of $1-$1.50/bbl over domestic crude at the U.S. East Coast. An import ticket to a U.S. refiner had an approximate equivalent value.[b]

---

[a]The Emergency Petroleum Act of 1973 sets the independent refiner self-sufficiency ratio at 30 percent.

[b]Not considering any additional value it may have had for an internationally integrated refiner who could draw on his own foreign production.

Table F-1.   18 Major Integrated U.S. Oil Refiners, 1972 (000 b/d)

	Net Domestic Crude & Natural Gas Liquids Production	*Domestic Refinery Runs*
Standard of Indiana	487	1,037
Exxon	970	1,029
Texaco	792	1,012
Shell	638	1,001
Mobil	394	856
Standard of California	462	815
Gulf	561	767
Atlantic Richfield	400	713
Phillips	267	545
Sun	286	526
Union	299	416
Continental	210	339
Cities Service	216	269
Marathon	181	231
Getty	316	194
Amerada Hess	95	72
Kerr-McGee	42	37
Pennzoil	43	33
Total Crude & NGL (18 Co's.)	6,659	9,892
Total U.S.	11,215	12,698
18 Co's Share of the U.S.	59.4%	77.9%

Source: Company Annual Reports; U.S. Bureau of Mines Mineral Industry Surveys, *Petroleum Statement, 1972 (final)*.

Since the program allocated import tickets on a sliding scale in reverse relation to size, the smaller a company's total refinery runs the larger its relative import ticket allocation. The result was that the average integrated company received 0.15 b/d per b/d of refinery runs, while for the Independents the ratio was 0.24 b/d. On the West Coast (District V) which had a somewhat different import program from the rest of the country, ratios were 0.31 b/d and 0.71 b/d respectively.[c]

Another aspect of the Mandatory Oil Import Control Program was the trading of import tickets for domestic crude oil. This permitted the utilization of all overseas imports in District I-IV at the U.S. East Coast where, for logistic reasons, they had the maximum advantage over domestic crude oil. Since most independent oil refineries are not located at the East Coast, the trading of import tickets for domestic crude oil helped to maximize the value of their import tickets.

---

[c]Based on allocations issued for the first quarter of 1973, the last period for which all crude oil imports were based on quota allocations.

In the absence of a Mandatory Oil Import Program all U.S. refiners would have paid less for their crude oil in the period before 1973, both because more foreign crude would have been imported and because the price of domestic crude would have had to adjust to that of imported oil. While this would have helped all refiners it would not have given the Independents the *relative* advantage which the government granted them under the Mandatory Import Control Program. Furthermore, under free imports, competition from integrated companies with excess foreign producing capacity would have been more severe. Many of these companies would have built additional refineries in the U.S. to run on controlled foreign crude oil. In fact, several of the "newcomers" to foreign oil production, particularly the companies which found oil in Venezuela and Libya in the late 1950s and early 1960s, had originally undertaken the search for this oil in order to supply their expected U.S. requirements. Also, U.S. companies with foreign refineries could have been expected to bring products to the U.S. where they would have competed with domestically refined products.

Altogether, then, it is reasonable to conclude that none of the provisions of the Mandatory Oil Import Control Program have worked to the direct disadvantage of the domestic independent refiners while at least one provision—the sliding scale of allocating import tickets—has given them a relative advantage over their integrated competitors throughout the nearly 13 years of the effective functioning of the program.

The breakdown of mandatory oil import allocations came when foreign crude oil prices began to equate with domestic prices at the U.S. East Coast and then overtake them. When this occurred, the relative advantage of the Independents evaporated. The official abolition of the Oil Import Restriction Program in May 1973 was only a recognition of an already existing situation.

THE INDEPENDENTS AND U.S.
CRUDE OIL PRODUCTION

The end of the period of import restrictions coincided approximately with the end of domestic excess crude oil producing capacity. Until 1970 the U.S. oil producing capacity had consistently been above actual domestic crude oil requirements. The reasons for this excess producing capacity are complex and an analysis falls outside the scope of this study. However, one principal cause was the legal restriction of crude oil production in several major producing states to the level of Maximum Efficient Recovery (MER) or market demand whichever was lower. The restriction was accomplished through a system of prorationing. Since in every year prior to 1971 market demand was below MER, an excess producing capacity, immunized by law from exerting pressure on prices, existed. The prorationing system undoubtedly kept domestic crude oil prices at a higher level than they would have been otherwise. While this had a negative effect on all refiners, it affected the independent segment relatively more adversely, since the integrated refiners benefitted from the high crude oil prices to the extent of their integrated production. Their benefits were further enhanced by the depletion allowance, a tax deduction on crude oil profits whose size is a direct function of the wellhead value of crude oil.

A recent Treasury study asserts that "pro-rationing actually aided the independent refineries, in that they were assured of a crude oil supply by merely having the allowable production limitation raised. It was not until recently when all U.S. wells were producing at 100% of the maximum efficiency rate that pro-rationing became ineffective in being able to provide more crude oil for refinery use."[3]

While it is of course true that the market demand factor in prorationing assured every refiner of his required supplies, as long as market demand was below MER, it does not follow that the independent refiner would necessarily have fared less well regarding crude oil supplies under a free market system. As pointed out before, he probably would have done better in regards to crude oil prices. On the other hand, the high domestic crude oil prices were precisely the reason for the high value of import tickets, since the tickets in essence reflected the difference between foreign and domestic crude oil values. Hence, in the absence of prorationing, import ticket values would clearly have been less. This would have affected the Independents more than the integrated refiners because of the method of distributing the tickets.

In this connection we must also consider the availability of foreign crude oil supplies under the Mandatory Oil Import Program. Most foreign oil production during that period was controlled by the seven major international oil companies. In addition a small group of other integrated companies—the newcomers—who had ventured abroad in the late 1950s and early 1960s and had discovered significant quantities of oil in places like Venezuela and Libya also supplied the U.S. Independents. For a variety of reasons competition in the international oil market was very keen throughout this period so that oil was readily available on a spot or contract basis at declining real prices. This, too, added to the value of import tickets.

MARKET SHARE AND PROFITABILITY
OF THE INDEPENDENTS

Altogether, under the circumstances described independent refiners managed to maintain their share of the U.S. refining market throughout the 1960s, as Table F-2 makes clear. From the beginning of 1966 to the end of 1972 the Independents built a total of 18 new refineries with a combined capacity of 146,000 b/d and expanded the capacity of existing plants by 260,000 b/d.[2]

**Table F-2. Percentage of Total Certified Refinery Inputs
for Total U.S. Refineries (Grouped by Independents & Majors)**

	1960 %	*1966* %	*1970* %	
Districts I-IV				
Independents	16.2	16.5	17.1	
Majors[a]	83.8	83.5	82.9	
	1964 %	*1966* %	*1968* %	*1970* %
District V				
Independents	11.1	11.5	11.6	13.6
Majors	88.9	88.5	88.4	86.4

[a]Includes major integrated oil companies plus Ashland Oil, Inc. If Ashland were included in *Independents*, figures would show higher concentration in the industry by the independents.

Source: Office of Oil & Gas, Department of the Interior (formerly Oil Import Administration).

Another approximate measure of the relative shares of major and independent refiners in the U.S. can be found in Table F-5. It shows that the share of the 18 major integrated oil companies stood at 79.7 percent of U.S. refinery runs in 1969 and at 77.9 percent in 1972. The balance—2.8 million b/d in 1972—consisted mostly of independent refiners whose share presumably increased as that of the Majors declined. The Independents represent therefore a significant and dynamic element of the U.S. refining industry. Their most important function is that of principal suppliers to independent wholesale and retail marketers. According to a staff report by the Federal Trade Commission, in 1971 independent East and Gulf Coast refiners sold 70 percent of their total gasoline output to independent marketers for whom this represented 52 percent of total gasoline supplies.[4] These figures suggest that the economic viability of independent refiners and independent marketers are closely interrelated.

The profitability of independent refiners in recent years may be illustrated by an analysis of the earnings of 5 publicly owned independents— Clark Oil, American Petrofina, Ashland Oil, Standard Oil of Ohio and Crown Central Petroleum. While the refineries of these firms are considerably larger than those of the average Independents in our classification, they had to compete throughout this period with integrated refiners who had the advantage of controlled domestic and, in most cases, foreign crude oil production. The data indicate that while the Independents had consistently a lower rate of return on stockholders equity than the major integrated companies, in two of the four years shown the difference was relatively modest. Furthermore, three of the five Independents shown in the table, registered annual rates of return at least equal with those of the 18 Majors throughout the period. It should also be pointed out that the five Independents' composite rates of return were not below the average for U.S. manufacturing.

THE NEW OIL IMPORT PROGRAM

The new Oil Import Program which went into effect on May 1, 1973 replaced the imports quota system with an import fee system. Crude oil and refined products can now be imported freely by anyone provided a fee is paid. For crude oil that fee will be 21¢/bbl from 1975 on. For new refineries—both grass root and expansion of existing plants—75 percent of the crude oil fee will be waived for the first five years of operation. The new system also provides for a scheduled phasing out of import quotas between 1973 and 1980. However, during this period quota imports can be brought in without payment of the import fee. This gives quota oil a maximum 21¢/bbl advantage over nonquota oil imports. This advantage is very small, compared to both the $1-$1.50 value which import tickets commanded between 1960 and 1971 and to total crude oil imports which are rising while the level of fee-free quota imports is reduced annually until it is phased out in 1980. Thus, for practical purposes, the relative advantage of most independent refiners over their Major competitors because of the method with which imports were allocated has ceased to exist. As will be discussed later, this does not necessarily reflect a decline in interest on the part of the government in the viability of independent and small refiners.

Table F-3. Rate of Return on Common Equity[a]

	1970	1971	1972	1973
American Petrofina	16.7	9.6	12.3	13.4
Ashland Oil	12.8	9.6	14.9	16.6
Clark Oil	14.1	8.8	10.1	27.8
Crown Central Petroleum	10.8	1.5	2.7	14.8
Standard Oil (Ohio)	6.4	5.7	5.6	6.6
Composite 5 independent refiners	9.2	6.9	8.6	11.3
Composite 18 integrated refiners	10.8	10.9	10.3	14.5

[a]Net Income as a ratio of stockholders' equity (common stock and surplus).

Source: Carl H. Pforzheimer & Co., Comparative Oil Company Financial Statements, 1970, 1971, 1972, 1973, New York, N.Y.

In establishing the new policy the government obviously took into consideration the expectation of a growing shortage in refining capacity over the next several years which would have meant that no U.S. refinery would have experienced difficulties in disposing of its products for the next 5 to 6 years.

The expectation proved correct for 1973, as is evidenced by the fact that even though U.S. refineries were running at virtually full capacity throughout most of the year, imports of gasoline, middle distillates, and residual fuel oil rose to a record high, as per table below, until they were affected by the Arab oil embargo towards the end of the year.

The massive price increases in foreign and domestic crude oil in the last quarter of 1973 and the first quarter of 1974 have somewhat changed this outlook. Public response to the higher prices together with government imposed or encouraged energy conservation measures are likely to reduce oil demand in 1974 below the level of the previous year and will cause the growth rate over the next several years to be quite modest compared to both the historic rate and the previously projected rates. The Federal Energy Administration has estimated that 604,500 b/d in refining capacity will be added in 1974, 875,000 b/d in 1975 and 1,192,400 b/d in 1976.[5] Even if we assume realistically that only three quarters of all the additional capacity will actually be ready by the end of the latter year, it would increase U.S. refining capacity by 14 percent over the end-1973 level of 14.2 million b/d, considerably more than the now expected increase in demand during this period.

However, one of the stated purposes of the new oil import policy is to reduce imports of refined products. If new or expanded U.S. refineries will have a yield pattern which enables them to cut into the existing import markets for distillate fuel oil, residual fuel oil and the expected import requirements for naphtha (as a feedstock for petrochemicals and synthetic natural gas), it is probably still correct that they will have an assured market for their products for the next 5-6 years.

How does the government expect to achieve this reduction in imports? Under the new Import Program all refined products imports have to pay an import fee which is scheduled to rise to 63¢/bbl by 1975 (except for

Table F-4.   Selected U.S. Oil Imports (000 b/d)

	Gasoline	Distillate Fuel Oil	Resid. Fuel Oil
1970	67	147	1,528
1971	59	153	1,583
1972	68	181	1,742
1973	132	380	1,827

Source: Bureau of Mines Mineral Industry Surveys, Petroleum Statement 1971 (Annual) (December) 1973.

imports from Canada for which the 63¢ fee becomes effective in 1980). The fee is a less effective impediment to imports of gasoline and distillate fuel oil than the volumetric restrictions which prior to 1973 were virtually absolute on gasoline imports and permitted only a very small amount of distillate fuel oil to come in.

However, in the case of residual fuel oil, the one major oil product which previously was not subject to import controls, the fee—if sufficiently high—could generate significant import substitution from new domestic refineries designed to achieve higher yields of this product than existing U.S. plants. Import substitution for distillate fuel oil might run on the order of 10-12 percent of U.S. demand. The potential for import substitution of gasoline is likely to be very small under the assumed new reduced growth rate in the demand for this principal product of U.S. refiners. However the import demand for naphtha—a primary refined product most of which is currently converted into gasoline—is expected to grow rapidly to meet petrochemical feedstock and synthetic gas requirements, providing significant opportunities for import substitutions.

The effectiveness of the import substitution under the oil import program of 1973 depends of course on the size of the import fee. The relation of the fee to the cost of oil imports products has greatly changed since April 1973. Prices of imported products are now (end June 1974) about three times as high as in April 1973 when the fees were established. This has obviously eroded their effectiveness to some extent. Since the Administration continues to regard import fees as an instrument to encourage domestic refinery construction—FEA Administrator John C. Sawhill stated at a Senate hearing on May 29, 1974 that "these fees are needed to encourage the planning and construction of refining capacity domestically"—it may decide to increase the fees to bring them more in line with current prices.

The reduction of import competition through import fees or other measures could help the domestic Independents relative to those U.S. Majors with foreign refining capacity (13 out of the 18 companies in that category) because access to controlled foreign refining capacity gives the latter added flexibility in supplying the U.S. market.

In the absence of effective products import control, such as a sufficiently high fee, import competition may increase, since other countries, too, are reacting to the price rises and the Arab use of oil as a political weapon by reducing consumption and moving towards substitute energy forms. This could create additional excess refining capacity abroad over the next several years, particularly if the oil exporting countries should decide, as seems likely, on large-scale refinery construction of their own. On the other hand, the competitiveness of independent U.S. fuel oil and gasoline marketers who import directly could be adversely affected by high import fees on these products, especially in the transition period before the achievement of U.S. refining self-sufficiency. This has been recognized in the proposed Oil and Gas Energy Tax Act of 1974 (H.R. 14462), passed by the House Ways and Means Committee in May 1974. The Act would allow exceptions to the import fee for independent marketers.[6]

PROFIT MARGINS AND REFINERY SIZE

It may be assumed that existing domestic price controls on refined products will either be abolished in 1975 or the government will permit price levels which give refiners a reasonably attractive rate of return. Both actions would be consistent with the official policy of encouraging the construction of domestic refineries in preference to a rising level of imports.

Under this assumption the profit margin will be determined by the most efficient plant. Given the generally accepted significant economy of scale of refinery operations,[7] these are likely to be the largest plants.

Historically, economy of scale has favored the integrated majors. As pointed out, as of January 1, 1973 the 18 largest integrated oil companies in the U.S. owned a total of 110 plants with an average plant size of 100,000 b/d. By comparison, the average size of the 110 other U.S. refineries was just below 20,000 b/d. However, a number of the small refineries are specialty plants, making primarily lubricants or asphalt, which are not fully comparable with full-range refineries.

A recent National Petroleum Council study predicted that "as energy demand increases, single-train refineries will undoubtedly increase in size to the 200,000 to 250,000 b/d range. . . ." Most new integrated refineries can be expected to be in that range as will be some of the plants built by the larger independent refiners. Most of the other independent refineries can be expected to maintain their market share through expansion of existing facilities. Thus, the average plant size of all refiners will continue to rise, as it has in the past. Whether a consequence of this trend will be an increase in the existing differential in plant size between integrated majors and full-range independent refiners depends largely on the ability of the Independents to procure the necessary capital and crude oil supplies.

One important factor in considering capital availability and ability to expand will be the impact on small refiners of the change in gasoline quality required under the Clean Air Act of 1970. The reason appears to be primarily the economy of scale of refinery operations. According to a study by the consulting firm Bonner and Moore Associates, Inc., "The proposed lead removal would impact more severely on small refiners than on refiners with larger scale operations."[8] The study concluded that for three small inland refiners in California (on which the study was based) "the investment impact could be five times more costly on a gallon-of-gasoline-manufactured basis than it would be for larger refiners . . . the three small refiners . . . are faced with manufacturing cost increases ranging from 1.38¢/gallon to 1.79¢/gallon in order to meet the proposed lead requirements. A large refiner, however, would operate at no increased manufacturing cost due to sufficient credit gains and would, in fact, show a small saving."

The study points out that the small refiners could meet the proposed lead requirements by exchanging or purchasing "finished or intermediate products rather than construct small, uneconomical units." However, refiners in inland locations not serviced by products pipelines or navigable waterways may not be able to engage in such transactions. These refiners may therefore find it difficult to survive under the projected standards of lead removal from gasoline. Other small refiners may have to accept dependency on their larger competitors in the form of exchange agreements in order to meet the new lead standards.

All of this is likely to hurt the *small* independent refining segment of the oil industry and may force plants in this category to either turn to the manufacture of specialty products or face the threat of an eventual shutdown.[d] On the other hand, independent refiners with large single plants will be no more affected by the required lead removal from gasoline than integrated companies. Thus, the overall effect of lead removal on refinery size is likely to accelerate the trend towards large independent refineries and the decline in the number of small gasoline producing refiners.

UPCOMING CHANGES IN INTEGRATED REFINERY OPERATIONS

One historic advantage of integrated refiners over their independent competitors likely to decline in the years ahead is the former's ability to subsidize refining and other downstream operations out of crude oil earnings. Overseas this form of financing downstream activities has been standard practice for the major oil companies. In the five years 1967-72, for instance, U.S. oil company earnings in the major consuming areas such as Western Europe and Japan returned on the average (−) 1.0 percent and 1.4 percent respectively on net book investment while in the major producing countries the returns averaged in excess of 39 percent.[9] Clearly, the downstream investment would have been economically unjustifiable for these corporations except when viewed together with their returns in the producing countries. In fact, the Majors operating in the Eastern Hemisphere have often answered the producing countries' criticism of the high corporate rate of return on crude oil production by pointing to the very low rate of return (or actual losses) on downstream operations and insisting that all integrated operations had to be taken into account in judging a company's profit position, since all are interrelated.

The original reason for this particular distribution of earnings

[d]The majority of these plants are already specializing in such products as lubricating oils and asphalts.

between upstream and downstream operations abroad was the tax system. Until 1950 Middle East governments levied only royalties on the production of foreign oil companies. After that an income tax was instituted but the rate was generally significantly lower than that levied by most oil importing nations on corporate earnings. U.S. companies in producing countries had the additional advantage of reducing their U.S. tax liability on foreign earnings by the depletion allowance.

The tax advantage of maximizing integrated company earnings in the producing countries disappeared in the mid-1960s as these countries raised their tax rates and determined the prices at which oil had to be exported for purposes of calculating income tax liabilities. Nevertheless, for a variety of institutional and other reasons earnings of the integrated international companies have continued to be centered in the producing countries. One reason is that many importing countries have formal or informal price controls on oil products which have prevented local refiners from passing on the full increase in the tax and royalty cost of world crude oil.

In the future this situation is likely to change. The trend towards nationalization and host-country participation since 1971 is escalating at such a rapid rate that it is fairly safe to predict that within a very few years the Major Internationals will have lost much of their controlled foreign crude oil production. This means not only that they will no longer be able to supply independent refiners around the world out of their crude oil surplus, but the Majors themselves will have to purchase at least their incremental crude oil requirements in the open market under the same conditions as the Independents. Over a period of time such a development could greatly reduce any existing economic disparity between Majors and independent refiners on a global basis. A likely further consequence of the decline of the Majors' role in the production of crude oil will be an attempt on their part to increase the profitability of their downstream investments.

We may therefore see refinery profit margins rise in most importing countries, relative to their long-term pre-1973 levels. The governments of these countries are unlikely to oppose such increases as long as they would result only in putting the return on refining and marketing investment approximately on a par with that on other capital intensive manufacturing industries. This would of course benefit the independent refiners abroad who could operate under the profit umbrella provided by the Majors instead of having to compete with operations subsidized out of upstream earnings. It is therefore unlikely that over the next six to seven years independent refiners abroad will be squeezed out of the market by Majors operating their refineries at submarginal rates of return.

The above scenario describes the changing *international* role of the Majors. However, it is also likely to apply to their future role in the United States. For one thing, the steady decline in domestic crude oil and natural gas liquids production since 1970 is causing the domestic self-sufficiency ratio of the integrated companies to drop, as is shown in Table F-5. For the 18 largest integrated refiners shown in the table the composite self-sufficiency ratio dropped from 69.6 percent in 1969 to 67.3 percent in 1972.[e] With a further decline in domestic oil production and an increase in refinery runs in 1973, it is likely that the ratio dropped again last year.

Thus, except for West Coast refineries which can expect to receive

Table F-5. Ratios of Net Crude and NGL Production to Refinery Runs for 18 Major U.S. Refiners and Total U.S., 1972 and 1969

	1972			*1969*		
	Net U.S. Prod'n (crude & NGL)	*Refinery Runs*	*Ratio of Prod'n to Runs*	*Net U.S. Prod'n (crude & NGL)*	*Refinery Runs*	*Ratio of Prod'n to Runs*
	(000 b/d)		*%*	*(000 b/d)*		*%*
Standard of Indiana	487	1,037	47.0	452	971	46.5
Exxon	970	1,029	94.3	867	992	87.4
Texaco	792	1,012	78.3	744	929	80.1
Shell	638	1,001	63.7	548	901	60.8
Mobil	394	856	46.0	350	829	42.2
Standard of Calif.	462	815	56.7	547	687	79.6
Gulf	561	767	73.1	602	688	87.5
Atlantic Richfield	400	713	56.1	420	671	62.6
Phillips	267	545	49.0	268	517	51.8
Sun	286	526	54.4	254	467	54.4
Union	299	416	71.9	289	393	73.5
Continental	210	339	61.9	196	269	72.9
Cities Service	216	269	80.3	193	254	76.0
Marathon	181	231	78.4	157	165	95.2
Getty	316	194	162.9	299	195	153.3
Amerada Hess	95	72	131.9	92	120	76.7
Kerr-McGee	42	37	113.5	34	47	72.3
Pennzoil	43	33	130.3	39	32	121.9
Composite 18 Co's.	6,659	9,892	67.3	6,351	9,127	69.6
Total U.S.	11,215	12,698	88.3	10,827	11,448	94.6
18 Co's Share of Total U.S.	59.4%	77.9%		58.7%	79.7%	

Source: Company Annual Reports; U.S. Bureau of Mines Mineral Industry Surveys, *Petroleum Statements (Annual)*, 1971 and 1972.

---

[e]The true self-sufficiency ratio was somewhat lower in both years, since most NGLs are not used as refinery feedstock.

Alaskan oil in large quantities from 1977 on, incremental crude oil requirements of the Majors will have to come primarily from imports which will probably have to be purchased under conditions similar to those described above for importing affiliates in other countries.

The declining domestic self-sufficiency ratio will also reduce the value of the depletion allowance for integrated companies. Furthermore, there is a growing sentiment in Congress and the Executive Branch and even within the domestic oil industry to reduce tax benefits to oil producers in return for a free market price of crude oil and natural gas or at least to tie such benefits directly to expenditures for exploration. This would eliminate the opportunity of integrated companies using these benefits to subsidize downstream operation, as has been charged by independent refiners and marketers.[f]

All of these developments suggest that integrated domestic refiners will in the future put greater emphasis on the profitability of refining and other downstream investments than has been the case in the past. The declining margins in excess domestic refining capacity and the official policy aimed at reducing the levels of refined products imports will support this trend. If refineries and other downstream facilities are to be built at the required rates, the necessary investments will have to be increasingly justified on their own profitability rather than as an adjunct to crude oil production which is declining domestically and is being nationalized abroad.

FUTURE PROFITABILITY OF THE INDEPENDENTS

Independent refiners should benefit from this policy of their Major competitors, provided they can get the capital and the crude oil to expand their operations. These two conditions are of course interrelated. As pointed out, the government's stated policy of discouraging the growth of refined products imports, should give any competitive refiner with the right yield pattern a ready market for the next 5-6 years. He may also benefit from the growing shortage of natural gas which will raise middle distillate oil demand. Thus, new investments in independent refining capacity can be expected through the 1970s, provided the refiner can obtain an assured long term supply of crude oil at competitive cost.

This condition will be the key to the future of independent refiners. To the extent to which it will depend on crude imports one may envision a

---

[f]No conclusive evidence exists that integrated oil companies in the U.S. are actually engaging in this practice. In refuting the charge the Majors point to the fact that they regularly invest considerably more capital in their production sector than the savings realized as a result of the depletion allowance. Consequently, they say, these savings are not available for other purposes.

scenario under which the Independents will fare reasonably well in the future, although it is fraught with uncertainties.

As control of foreign crude oil is increasingly transferred from the private international companies to the state companies of the producing countries, the latter may as a matter of policy favor the independent refiners over the Majors in disposing of their uncommitted crude oil. Such a policy would increase the number of refiners competing in the world oil market, thereby reducing the possibility of the formation of a monopsony or oligopsony among major crude oil refiners with its potential downward effect on crude prices. Additionally, some of the national oil companies may feel they can obtain better terms from the Independents than from the Majors whose much larger economic resources give them generally more bargaining strength.

The system of auctioning off part of the governments' crude oil share, adopted in 1973, is a significant step in this direction. Independent refiners have been substantial buyers at these auctions. Some countries have actually excluded the Big Seven international majors[g] from participating in the auctions. While the auction system is likely to decline as a vehicle to sell oil directly, following Kuwait's recent failure to secure bids at its announced minimum prices, the volume of oil sold directly by producer countries to Independents is on the rise.

However, so far, most of these transactions are spot or short-term deals which, by definition, makes them relatively insecure. Yet, bankers or other lending institutions to whom most Independents must turn for the financing of new plant capacity are understandably reluctant to lend very large sums for the construction of an industrial complex whose raw material supply is not assured.

The attempts of independent refiners to obtain assured supplies through partnership agreements with national oil companies have so far met with very little tangible success. However, this situation could change fairly rapidly. As world crude oil surpluses begin to mount in response to the price-engendered demand curtailments and oil import substitutions, some of the producing countries may become as interested in assured long-term outlets for their crude as the refiners in the importing countries are in obtaining assured long-term supplies.

The Independents are likely to benefit from such an attitude. But this will not necessarily give them an edge over, or even equality with, the Majors in obtaining foreign crude oil supplies. For one thing, the Majors, by virtue of a combination of old and new agreements with the producing countries in which they operate, continue to have access to some oil below the officially established export market values. For another, the Majors are more actively, and probably also more successfully, searching for new crude oil supplies both in the traditional exporting countries (often under radically new types of agreements) and in new areas than the Independents.

---

[g]Exxon, Texaco, Mobil, Standard of Calif., Gulf, British Petroleum and Royal Dutch Shell.

In other words, while the Majors are clearly and rapidly losing their historic control over world crude oil supplies, for some time at least they will still continue to have better access to these supplies than independent U.S. refiners. Nevertheless, if we assume the ready availability of foreign crude oil supplies at quoted prices—a condition which will not be determined by the Majors but by members of the Organization of Petroleum Exporting Countries (OPEC) and by that organization's cohesiveness—the Independents should be able to obtain foreign crude oil on relatively more favorable terms than in the past. Unfortunately for the Independents, unless they can gain access to firm long-term commitments this may not be enough to convince money lenders to invest in refinery projects to be fueled by imported crude oil. In this respect, too, the Majors have an advantage because of their greater ability to self-finance new projects.

DOMESTIC CRUDE OIL AVAILABILITY

What about domestic crude oil which accounts for about 70 percent of U.S. crude oil requirements?[h] For most Independents access to domestic crude may prove more important for their survival than foreign crude. The reason is the complete reversal in foreign and domestic price relations since October 1973. Virtually all U.S. crude oil can now be produced for much less than the landed cost of any foreign crude oil and about 60 percent of U.S. crude oil output is limited by federal control to a price which is less than half that of delivered foreign crude. The other 40 percent is not subject to price control and sells roughly at parity with foreign oil. Since all new production above the 1972 base period is not subject to price control, controlled crude oil has become an increasingly attractive commodity to a U.S. refinery. Not surprisingly, the distribution of this commodity is not left to market forces but is carried out by the Government which has endowed the commodity with this special attraction. To the extent to which independent refiners need low cost domestic crude to remain competitive, domestically and internationally, their future will therefore be determined by governmental policies. What these policies might be is discussed below.

U.S. GOVERNMENT POLICY AND
THE INDEPENDENTS

There is a distinct possibility that domestic crude oil will continue to be allocated to U.S. independents, as well as crude short integrated refiners, by government decree both for existing and new or expanded plants. Under the Emergency Petroleum Allocation Act of 1973 which went into effect on

---

[h]Based on domestic crude oil production of 9 million barrels per day and crude oil refinery runs of approximately 12.5 million barrels per day.

January 15, 1974 every U.S. refiner is given equal access to the domestic and imported crude oil volume available in 1973, based on a nationwide ratio of crude oil runs to refinery capacity. The Act specifically calls for "the preservation (of) the competitive viability of independent refiners, small refiners, non-branded independent marketers and branded independent marketers."[i] It also provides for the expansion of independent and small refining capacity by specifically authorizing the President ". . . in the case of crude oil (i) to take into consideration market entry by independent refiners and small refiners during or subsequent to calendar year 1972, or (ii) to take into consideration expansion or reduction of refining facilities of such refiners during or subsequent to calendar year 1972."

While the Act is scheduled to expire on February 28, 1975, the intent of Congress to preserve the independent refining as well as other independent segments of the oil industry has been clearly stated. This intent precedes the Arab oil embargo of October 16, 1973 which gave rise to the passage of the Emergency Petroleum Allocation Act. Thus, in May 1973 the Senate Committee on Interior and Insular Affairs expressed in its report on an earlier version of the Emergency Petroleum Allocation Act the Committee's "concern for the non-integrated sectors of the petroleum industry and for the small integrated firms. These enterprises have an importance to the U.S. economy far out of proportion for their market shares, because they continually spur the major integrated firms to improve their own efficiency in production, refining, transportation and marketing."[10]

One may question the economic assumptions in this statement but not its sentiment. Congress has gone on record to express its desire to maintain the independent segment of the oil industry in the face of the revolutionary changes which have taken place in the foreign and domestic oil situation. Given the present strong anti-Major company sentiment in the country at large, it is likely that existing legislation to assist the Independents will be extended and that additional legislation will be adopted.

This does not mean that the existing crude oil allocation program will be, or should be, continued. If foreign crude oil ceases to be in short supply the need to allocate it may also cease. For, as we have seen, the Independents would then be able to obtain foreign supplies under conditions that are probably somewhat more favorable than in previous years. But domestic oil has now become a far more desirable commodity than foreign oil, if for no other reason than that it represents a considerably more secure and stable supply source, as has been dramatically demonstrated in recent months. But most important is the fact that the price-controlled share of domestic crude represents currently the

---

[i]The regulations for the Act define an Independent Refiner as one who obtains at least 70 percent of his crude oil requirement from uncontrolled producers and who markets a substantial volume of his refined gasoline through independent marketers. A small refiner is defined as one whose total refining capacity does not exceed 175,000 b/d.

lowest priced oil in the world. This particular domestic crude plays now the same role as low cost imported crude did in the period prior to 1973. Refiners—primarily the less integrated or independents—who would not have had access to imported crude in the years 1959-1972 while their competitors did would have had great difficulty surviving. The same can now be said about controlled domestic crude oil.

The prospect of another long-term bout with government allocation of crude oil supplies may not be a pleasing one. The value of this oil has been determined by foreign government action to raise prices and by domestic government action to keep them low, both without consideration of market forces. What would happen if the domestic price controls were removed? The principal beneficiaries would of course be the crude oil producers. Refiners would be negatively affected in inverse relation to their integration ratio: The more integrated the less they would feel the impact of the price increase. Independent refiners who by definition have little or no integration would therefore be the principal victims of the removal of crude oil price controls.

The need for mandatory allocation would of course be lessened by the removal of price controls. However, domestic crude, even at parity with imported crude oil, would still have the aforementioned intrinsic advantage of being more secure, economically and politically. Hence, it may still be a preferred crude and since, for the next decade at least it will not be available in sufficient quantities to meet all demand, some Independents may want it allocated even in the absence of a price advantage.

OUTLOOK FOR THE SMALL INDEPENDENTS

Of course, even the most sympathetic legislation may not be of much help to those small independent refiners in inland areas where local crude oil production is declining and the construction of a pipeline to deliver crude oil from abroad or from other U.S. regions to offset the decline would be uneconomical because of the small volumes required, at least initially. A number of smaller Independents in inland oil producing states in the Southwest and the Midwest are increasingly finding themselves in this position. Some of them might be helped by the government's new program to make royalty oil from production on federal lands available to such plants. However, the volume of federal royalty oil in inland areas is relatively small.

The future for some of these refineries is therefore bleak. They located where they did because of the availability of crude oil. If their supply goes into a terminal decline their economic viability might do likewise. This would appear to apply particularly to small refiners in areas which can be economically supplied from larger plants with access to adequate crude oil supply.

The new crude oil allocation program might keep such refiners in business a while longer. But if they cannot get supplies for logistical reasons the allocation program cannot help them indefinitely. Their principal hope lies therefore in the discovery of new local supplies or the reversal of the decline in existing supplies through secondary or tertiary recovery methods. Current domestic crude oil prices for new oil should be high enough to bring forth additional supplies where they exist. The maximization of secondary recovery may require the removal of the old oil price ceiling when such production is initiated.

NEW INDEPENDENT REFINERIES

It is significant that despite their current difficulties in obtaining secure, competitive crude oil at home or abroad, proposed refinery constructions or expansions by Independents appear to be at an all-time high. Of 46 scheduled projects recently tabulated by the Federal Energy Office for the years 1974-76 only 10 will be built by the 18 major companies. These 10 account for about 50 percent of all identified projects. Much of the rest will be built by Independents or newcomers. It should be kept in mind that these projects are announced but not under actual construction. Some will undoubtedly be scrapped or postponed for a variety of reasons. As discussed earlier, projects by Independents are more likely to become casualties than those of the Majors with their more assured supplies and financial resources. Still, many of these plants will probably be put on stream within the time indicated by the FEO.

There are also indications that a number of independent terminal operators located at the East Coast of the U.S. plan to integrate backward into refining in order to have a more secure source of supply for their products requirements. These operators have their own docking facilities for ocean going tankers and own or control sufficient terminal storage capacity to buy in cargo (tanker) volumes. They account for approximately 25 percent of all residual fuel oil and distillate fuel oil and a much smaller share of all gasoline supplied to the U.S. East Coast. Most of their residual fuel oil is brought in from abroad, since domestic refineries for economic reasons have long minimized the output of this particular product. On the other hand, most of the cargo buyers' distillate fuel oil and gasoline requirements have traditionally come from U.S. Gulf Coast refineries. In fact, up to 1972 independent cargo buyers fulfilled an important function within the U.S. oil industry in relieving Gulf Coast refineries of excess products—particularly distillate fuel oil during the off season. However, with the decline in excess refining capacity since 1972 much of their Gulf Coast supply began to dry up. It is in response to this situation which became acute in 1973 that some of the independent cargo buyers are considering the construction of their own refineries. At the same time they are also looking to lessen their traditional dependence on the Majors for their residual fuel oil imports. Hence, the planned refineries of these companies are designed to yield a much larger volume of residual fuel oil per barrel of crude oil processed than most existing U.S. plants. Northeast Petroleum, (Boston, Mass.), Sprague Oil Co., Steuart Petroleum Co., (Washington, D.C.) and Belcher Oil Co. (Miami, Fla.) are among companies that have announced tentative plans to move in this direction. These projects would of course increase the number of independent refineries in the U.S.

BEYOND 1980

The period beyond 1980 is extremely speculative at this stage. We do know that the quantum jumps in oil prices, both foreign and domestic, will have two predictable consequences. The growth rate in demand will be slower than had been projected and new oil and other energy supplies will become available at a more rapid rate than had been projected. The cumulative effect of both these changes will take time and will probably not be fully felt until after 1980. It is quite conceivable that in the early 1980s domestic refining capacity will again be in excess of market requirements, following a period of progressive import substitution by new domestic refineries, and that crude oil will be freely available to all buyers at competitive costs. How the independent refiners will fare under those conditions cannot be forecast at this period in time. But it will probably be different both from the period of the next five to six years which will be characterized mainly by the quest to obtain secure and competitive crude oil supplies and the period from 1960 to 1971 when the key to refining profitability was access to low cost foreign oil.

NOTES

1. Bureau of Mines, Mineral Industries Survey, *Petroleum Refineries, January 1, 1973.*
2. Department of the Treasury, Staff Analysis of the Preliminary Federal Trade Commission Staff Report on its Investigation of the Petroleum Industry, July 2, 1973, p. 42.
3. Department of the Treasury, Staff Analysis of the Preliminary Federal Trade Commission Staff Report on Its Investigation of the Petroleum Industry, July 2, 1973, pp. 47-48.
4. Preliminary Federal Trade Commission Staff Report on its Investigation of the Petroleum Industry, 1973, p. 11.
5. Statement of John C. Sawhill, Administrator, Federal Energy Office, before the Senate Committee on Interior and Insular Affairs, May 29, 1974.
6. Report No. 93-1028, House of Representatives, *Oil and Gas Energy Tax Act of 1974*; pp. 62-4, May 4, 1974.
7. "The petroleum refining industry, in its continued post World War II effort to reduce unit capital and operating costs, continues to build operating units and complete grass roots refineries of ever increasing

capacity, utilizing larger single train equipment, based on the principle that it costs less to build and operate one unit double the size of two smaller units." Source: National Petroleum Council—*Factors Affecting U.S. Petroleum Refining*, September, 1973, p. 71.

8. *An Economic Study for Three Inland California Refineries (Including Analysis of Proposed Lead Removal Schedules*—Bonner & Moore Associates, Inc., Houston, Texas.

9. Survey of Current Business, September, 1973, *U.S. Direct Investments Abroad*, pp. 24-5.

10. Senate Committee on Interior and Insular Affairs, Report to a company S1570. May 17, 1973, pp. 24-5.

Appendix G

Firm Size and Technological Change in the Petroleum and Bituminous Coal Industries

Edwin Mansfield*

The purpose of this chapter, commissioned by the Energy Policy Project of the Ford Foundation, is to summarize what is known about the relationship between firm size, on the one hand, and invention, innovation, and diffusion, on the other hand, in the petroleum and bituminous coal industries. In addition, an attempt is made to describe the implications of these findings for merger and divestiture policies regarding these two industries. Further, a brief description of important data gaps and needed research in these areas is included. The importance of these topics to any discussion of competition in the energy industries is obvious: Clearly, one of the most significant aspects of market performance is technological progressiveness, and it is essential, in considering the effects of changes in an industry's market structure, that proper account be taken of the effects on the rate of technological change.

More specifically, this chapter takes up the following topics: the size of firm and expenditures on research and development (R and D) in the petroleum industry; the size of firm and the productivity of petroleum R and D; and the costs and returns from petroleum R and D, as well as some recent changes in the petroleum industry's attitude toward R and D. We also discuss R and D in the bituminous coal industry; the relationship between size of firm and innovation in petroleum; this relationship in bituminous coal; and the results in other industries, such as chemicals, pharmaceuticals, and steel. The latter part of the chapter deals with the diffusion of new techniques in petroleum, the diffusion of new techniques in bituminous coal, and the relevant evidence concerning the diffusion process in other industries. We then summarize the implications of the findings for merger and divestiture policy, and point out some important gaps and limitations in the available data, as well as needed future research.

*Wharton School, University of Pennsylvania.

SIZE OF FIRM AND R AND D EXPENDITURES
IN THE PETROLEUM INDUSTRY

"Research" is original investigation directed toward the discovery of new scientific knowledge, and "development" is technical activity concerned with non-routine problems encountered in translating research findings into new and improved products and processes.[1] The petroleum industry is one of the nation's leading spenders on research and development, its total expenditure for this purpose being about $500 million in 1971.[2] The R and D carried out by the petroleum industry is directed at a wide variety of fields, not just petroleum processes and production; for example, the petroleum firms have done considerable R and D in the chemical, insecticide, and other such areas. The importance of chemical R and D by the petroleum industry is indicated by the fact that in 1970 the petroleum industry spent $128 million, or almost one-quarter of its R and D funds, on chemical R and D.[3]

What is the relationship between size of firm and the size of R and D expenditures in the petroleum industry? No one would deny that some minimum size is needed before a firm can maintain a profitable and effective R and D program, but are giants like Exxon required for this purpose? One relevant question bearing on this issue is whether the largest petroleum firms spend more on R and D, relative to their size, than somewhat smaller ones. Based on data for 9 major petroleum firms for 1945-59, the evidence seems to indicate that the answer is no: In the upper reaches of the size range, a one-percent increase in firm sales seemed to be associated with a 0.86 percent increase in R and D expenditures.[4] Similar results have been obtained by Scherer and others for most industries.[5] Although there is a certain threshold size (which varies from industry to industry) that must be exceeded if many kinds of development projects can be undertaken effectively, a firm's R and D expenditures generally do not increase in proportion to its size in the range much above this threshold size.

Since "research and development" is a somewhat ambiguous concept, it is important to look carefully at the kinds of activities that the major petroleum firms carry out under the heading of R and D. Table G-1 shows some of the relevant characteristics of the R and D programs of eight major petroleum firms in 1964. As you can see, the bulk of the R and D projects carried out by these firms is regarded as being relatively safe from a technical point of view, the median probability of technical success being at least 75 percent in most of the firms. As you can also see, only a small percent of the money goes for basic research, and most of the R and D projects are expected to be finished and have an effect on profits in 5 years or less.[6]

There is some evidence that the laboratories of the major petroleum firms were not responsible for the radical inventions that occurred prior to World War II. For example, John Enos concludes that: "The most novel

Table G-1. Characteristics of the R and D Programs of Eight Major Petroleum Firms, 1964

Characteristics	1	2	3[a]	Firm 4	5	6	7	8
Percent of total R and D expenditure devoted to:								
Basic research	2	0	11	6	15	9	7	24
Applied research	54	N.A.	87	65	45	39	N.A.	33
Development	44	N.A.	2	29	40	52	N.A.	43
Percentage distribution of projects by expected rate of return, if successful								
0-19 percent	N.A.	N.A.	N.A.	20	0	N.A.	60	13
20-29 percent	N.A.	N.A.	N.A.	20	10	N.A.	5	5
30-39 percent	N.A.	N.A.	N.A.	10	10	N.A.	5	5
40-100 percent	N.A.	N.A.	N.A.	20	30	N.A.	5	33
Not known	N.A.	N.A.	N.A.	30	50	N.A.	25	44
Percentage distribution of projects by estimated probability of technical success								
0-24 percent	11	N.A.	N.A.	25	0	N.A.	10	5
25-49 percent	11	N.A.	N.A.	5	10	N.A.	5	10
50-74 percent	22	N.A.	n.A.	10	10	N.A.	5	30
75-100 percent	36	N.A.	N.A.	50	30	N.A.	60	40
Not known	20	N.A.	N.A.	10	50	N.A.	20	15
Percentage distribution of projects by expected number of years to completion and an effect on profits								
Less than 2 years	40	50	N.A.	50	25	N.A.	65	54
2 to 5 years	40	25	N.A.	30	25	N.A.	25	11
More than 5 years	20	25	N.A.	20	50	N.A.	10	35

[a]There is evidence that in 1966 this firm devoted about 60 percent of its total R and D expenditures to applied research and about 30 percent to development. Thus, the figures shown below for the firm may be too high for applied research and too low for development.

N.A. Not available

Source: See note 6.

ideas—cracking by the application of heat and pressure, continuous processing, fractionation, catalysis, regeneration of catalysts, moving and fluidized beds— occurred to independent inventors, men like Dewar and Redwood, Ellis, Adams, Hondry, and Odell who occasionally have contributed their talents, but never their permanent employment or loyalties, to the oil companies."[7] In recent years, however, the role of the major oil firms in promoting radical invention seems to have increased, although, as suggested by Table G-1, the bulk of the R and D done by the big oil companies is directed at minor improvements, not radical advances.

At the same time, in fairness to the oil companies, it is important to add that the large, established firms in other industries seem to have much the same kind of record. That is, the bulk of their R and D seems to be directed at relatively safe, short-term objectives, the radical advances often come from outside their laboratories, and they seem to be better at adapting, developing, and improving the novel inventions of others than coming forth with their own. For example, even in the pharmaceutical industry, about one-half of the major innovations during 1935-62 were based on discoveries made outside the innovating firm. Table G-2 compares the average probability of technical completion of projects in a sample of 19 industrial laboratories in the 1960s: you can see that the average is about the same in petroleum as in the other industries that are included.

What is the relationship between firm size and the nature of the R and D carried out by a firm? Based on data for the 1960s, it appears that, among major petroleum firms, increases in size of firm seem to be associated with a greater leaning toward risky and long-term projects, but the relationship is not as clear-cut as might be expected. In this size range, there seems to be no statistically significant relationship between a firm's size and the percentage of its total R and D expenditures devoted to basic research. Moreover, the results seem to indicate that, although the differences between the largest firms and relatively small firms are sometimes considerable, the differences between the biggest firms in the sample and firms a fraction of their size are seldom large, if they exist at all. Thus, the available evidence seems to indicate that, although the big firms do proportionately more R and D of a more basic, technically risky, and long-term nature, than small firms, the differences between the biggest petroleum firms and merely big ones are not great, if they exist at all.[8]

Table G-2. Average Probability of Technical Completion, by Industry, Based on Data for 19 Laboratories

Industry	Probability of Technical Completion	
	Average	*Range Among Laboratories*
Petroleum	0.50	0.10-0.74
Electronics	0.73	0.20-0.99
Chemical	0.70	0.37-0.99
Pharmaceuticals	0.32	0.12-0.62
All Laboratories	0.56	0.10-0.99

Source: See note 6.

SIZE OF FIRM AND PRODUCTIVITY OF
R AND D IN THE PETROLEUM INDUSTRY

In the previous section, we were concerned entirely with inputs to the inventive process, not outputs. Despite the enormous problems in measuring the productivity of R and D, it is important that we try to determine, as best we can, whether the available evidence suggests that the largest firms in the petroleum industry have been able to get out more inventive output from a dollar of R and D than smaller firms. In other words, holding R and D expenditures constant, is "inventive output" higher in the largest firms than in somewhat smaller ones? To try to answer this question, we used Schmookler's list of important inventions in petroleum refining[9] and my list of important petrochemical innovations to construct an index of inventive output for eight major petroleum firms during 1946-56, and we regressed this index on the firm's size and the average of the firm's R and D expenditures in 1945 and 1950.

Specifically, we assumed that

$$\frac{I_i}{R_i} = a_0 + a_1 R_i + a_2 S_i + z_i,$$

where I_i is the weighted number of inventions carried out by the ith firm, R_i is its R and D expenditures, S_i is its size (measured in terms of sales), and z_i is a random error term. The statistical results provided no evidence that a_1 or a_2 is statistically significant. Moreover, the observed sign of a_2 is negative, not positive. Thus, there is no evidence, based on this very crude analysis, that the productivity of R and D expenditures is higher in the biggest firms than in somewhat smaller ones. Moreover, there is no evidence of economies of scale in R and D in this range of variation of R and D expenditures.[10]

Another way to try to obtain evidence concerning this question is to get subjective rankings by knowledgeable observers of the relative productivity of various firms' R and D establishments, in much the same way that Cartter and others have gotten them in connection with graduate schools. Needless to say, these results are extremely crude, but they are not uninteresting. Five R and D executives in the petroleum industry and seven professors of chemical engineering at major universities were asked to rank each of a number of major oil companies by their inventive output per dollar of R and D. There was close agreement among the rankings of the various experts. To see whether the average ranking of a firm was associated with its size or the size of its R and D expenditures, we assumed that

$$P_i = l_0 + l_1 R_i + l_2 S_i + z_i^1,$$

where P_i is the average rank of the ith firm, R_i is its R and D expenditures, S_i is its size (measured by 1963 sales), and z_i' is a random error term. Least-squares estimates of l_1 and l_2 are not statistically significant, and the results are strikingly similar to those described in the previous paragraph, despite the fact that they are based on entirely different and independent data. As before, there is no evidence that the productivity of R and D is higher in the biggest firms than in somewhat smaller ones, and there is no evidence of economies of scale in R and D in this range of variation of R and D expenditures.[11]

PETROLEUM RESEARCH AND DEVELOPMENT: COSTS, RETURNS, AND RECENT CHANGES

To carry out research and development requires considerable sums of money. Table G-3 shows the estimated cost of developing six major cracking processes. Clearly, the costs have increased enormously as time has gone on: for example, the cost of developing the Burton process was about $200,000, while the cost of developing fluid catalytic cracking was over $30 million. However, the time involved in development has not changed much, on the average: judging from the results in Table G-3, it has neither increased nor decreased sharply. Note that the costs of improving a new process are often greater than the cost of developing the initial version of it: for example, the improvements of the TCC and Houdriflow process cost almost $4 million, whereas the initial development work was about $1 million.

Looking at the whole process of research, development, and improvement, we find that about 20 years were taken to complete this process in 5 of the 6 cases in Table G-3: clearly, the process of development is long, as well as costly. For this reason, very small petroleum firms cannot do much in the way of process development. According to Enos, the oil companies have not established process-development departments (which Enos uses as an indicator that they are ready to do much R and D) before they reached outputs of 150,000 barrels per day. Besides the big oil companies, there are firms, like Universal Oil Products, Lummus, Kellogg, and Foster Wheeler, which specialize in process development. Finally, it is important to note that the cost of innovating includes the cost of constructing the first commercial unit incorporating the new process, as well as the R and D costs. Table G-4 shows that these costs, as well as the R and D costs, can be very substantial.

Given the extent of the costs of innovating, firms must expect commensurate benefits, if they find it profitable to invest in R and D. How big have the returns been from process development in the petroleum industry? Table G-5 contains Enos's estimates of the returns from the major cracking innovations, these estimates almost certainly being understatements. The results indicate that successful process innovation in petroleum refining was extremely profitable, the returns often being at least 10 times as much as the cost of the innovation. However, it is also noteworthy that the return on investment in process innovation seemed to decline over time, the returns from the Burton,

Table G-3. Estimated Expenditure of Time and Money in Developing New Cracking Processes

Process	Preliminary Activities		Development of the New Process		Major Improvements in the New Process		Total	
	Time Interval	Estimated Cost	Time Interval	Estimated Cost	Time Interval	Estimated Cost	Time Interval	Estimated Cost
Burton	None	None	1909-1913 (5 years)	$92,000	1914-1917 (4 years)	$144,000	1909-1917 (9 years)	$236,000
Dubbs	1909-1916 (8 years)	N.A.[a]	1917-1922 (6 years)	5,000,000	1923-1931 (9 years)	1,000,000+	1909-1931 (23 years)	7,000,000+
Tube and Tank	1913-1917 (5 years)	$275,000[b]	1918-1923 (6 years)	600,000[c]	1924-1931 (8 years)	2,612,000	1913-1931 (19 years)	3,487,000[d]
Houdry	1923-1924 (2 years)	N.A.[e]	1925-1936 (12 years)	11,000,000	1937-1942 (6 years)	N.A.	1923-1942 (20 years)	11,000,000+
Fluid	1928-1938 (10 years)	N.A.	1938-1941 (4 years)	15,000,000	1942-1952 (11 years)	15,000,000+[f]	1928-1952 (25 years)	30,000,000+
TCC and Houdriflow	g	g	1935-1943 (9 years)	1,150,000	1944-1950[h] (7 years)	3,850,000	1935-1950 (16 years)	5,000,000

aA sum of $25,000 was paid to Jesse A. Dubbs for his asphalt and emulsion-breaking patents. There is no record of the amount spent by Universal Oil Products in challenging the Burton process patents, but it was probably several times this amount.

bThis sum is the amount paid for the Ellis patents and for Rogers' claim against Adams and The Texas Company. Not included are the expenses incurred by E.M. Clark while he was working for the Standard Oil Company (Indiana).

cThis consists of $498,000 for development expenditures and $102,000 for legal expenses.

dThis amount is Jersey Standard's estimate of total expenses through 1931. Approximately $850,000 more was spent primarily on technical services from 1932 through 1957.

eThe amount which Houdry and his associates spent on the development of the process of obtaining motor fuel from lignite is not known. The $11,000,000 given in the column under the development of the new process is the total for both the lignite and the successful Houdry process.

fThe Model IV Fluid unit was introduced in 1952. This is assumed to be the last major improvement in the Fluid process. In 1942 the seven firms working on the Fluid development expected to spend $10,000,000 to $15,000,000 in the next three years. The higher figure is taken; this almost certainly understates the total for the entire ten-year period from 1942 to 1952 because expenditures for 1945 through 1952 were also substantial, Jersey Standard alone spent nearly $30,000,000 from 1935 through 1956.

gThe time and expense in developing the Houdriflow process might well be included here, as the TCC and Houdriflow processes were based upon Houdry's earlier work.

h1950 was chosen as the terminal year because it was then that the first air-lift TCC unit and the first Houdriflow unit were both installed.

Source: John Enos, *Petroleum Progress and Profits*, M.I.T. Press, 1962.

Table G-4.   Construction Costs for the First Commercial
Cracking Units Incorporating New Processes, 1913-1955

| | | | Construction Cost | |
| | | | At Time of | In 1939 |
Process	Date	Capacity of Unit (barrels per day)	Construction	Dollars
Burton	1913	89	$6,750	$12,950
Tube and Tank	1922	570	90,000	108,900
Houdry	1938	6,750	2,191,000	2,191,200
Fluid	1942	12,750	2,060,000	1,889,000

Source: John Enos, *op. cit.*

Table G-5.   Cost of and Returns from Cracking Process
Innovations, 1913-1957

| | Cost of Innovation | | Returns from Innovation | | Approximate Ratio of Returns to Cost ($ per $) |
Process	Period Over Which Expenses Incurred	Estimated Amount	Period Over Which Returns Calculated	Estimated Amount	
Burton	1909-1917	$236,000	1913-1924	$150,000,000+	600+
Dubbs	1909-1931	7,000,000+	1922-1942	135,000,000+	20
Tube and Tank	1913-1931	3,487,000	1921-1942	284,000,000+	80+
Houdry	1923-1942	11,000,000+	1936-1944	39,000,000	3.5
Fluid	1928-1952	30,000,000+	1942-1957	265,000,000+	9
TCC	1935-1950	5,000,000+	1943-1957	71,000,000+	16
Houdriflow	1935-1950		1950-1957	12,000,000	

Source: John Enos, *op. cit.*

Dubbs, and Tube and Tank processes being much greater than for the later processes. More will be said on this score below.

Since the figures in Table G-5 relate only to successful innovations, they vastly overstate the rate of return from all R and D, both successful and unsuccessful. To obtain crude estimates of the marginal rate of return from all R and D, a simple model—based on the relationship between output, on the one hand, and labor, capital, and cumulated R and D expenditures, on the other— was estimated for five major petroleum firms in 1960. The results suggested that the marginal rate of return from R and D was very high; regardless of whether technological change was assumed to be capital-embodied or disembodied, the marginal rates of return averaged from about 40 to 60 percent. However, the marginal rate of return seemed much lower in the biggest petroleum firms than in the somewhat smaller ones, indicating that, whereas the smaller firms may have been under-investing in R and D, this may not have been the case for the biggest firms.[12]

In recent years, R and D in the petroleum industry has been increasing at a decreasing rate: between 1963 and 1966, the petroleum industry's R and D expenditures increased by about one-third, between 1966 and 1969, they increased by about one-quarter, and between 1969 and 1971, they did not increase at all.[13] One reason for this slowdown in the growth of R and D was the widespread feeling that the returns from research and development were smaller than in the earlier postwar period. In the words of D.C. Baeder, Vice President of Esso Research and Engineering, "competition had intensified through the proliferation of many competent laboratories. It was not uncommon for several laboratories to announce a major new process or product within months and sometimes even days of each other. Esso Research's major response was to prune marginal research projects. Of course, this realization that R and D was costing more and more, and producing less and less on the bottom line was hardly unique to Esso Research. It hit most major companies sometime during the last decade. And the hard look was extended to basic research—that supposed fountainhead of a veritable river of innovation and profit. Clearly a significant change in research philosophy was under way."[14]

In 1966, Esso Research and Engineering carried out a retrospective study of over 100 of their inventions in the previous 20 years; the findings indicated the critical importance of real novelty in obtaining commercial success and the small returns, if any, from developing run-of-the-mill innovations based on well-known science and obvious needs. The study also suggested the significance of early market research and the ability to commercialize based on related manufacturing/sales know-how. Recognizing that the existing organizational structure was not as conducive to high-risk, high-payoff innovating as it might be, Exxon organized a new Corporate Research Laboratory oriented toward more ambitious science-based innovation. According to company officials, this laboratory has produced some interesting and worthwhile inventions in its relatively short lifetime. In any event, the events and reasoning that led to its formation provide significant insights into recent problems and trends in petroleum R and D.[15]

RESEARCH AND DEVELOPMENT IN THE BITUMINOUS COAL INDUSTRY

In contrast to the petroleum refining industry, the bituminous coal industry is not a large spender on research and development. The National Science Foundation publishes no figures on R and D spending by bituminous coal firms, but there is every reason to believe that the spending is small.[16] However, some of the biggest bituminous coal companies have done some R and D; for example, the Consolidation Coal Company, since 1966 a part of Continental Oil, has the reputation of being one of the industry's technological leaders. According to *Coal Age*, "Consol has been foremost of the coal companies in developing within its own organization new techniques of mining coal and ways of making use of the coal after it is mined."[17]

The engineering and research functions of the Consolidation Coal Company are carried out mainly by the Lee Engineering Division in McMurray, Pa. (which develops and tests mining equipment, instrumentation, and safety devices), the Research Division at Library, Pa. (which performs long-range research and experimentation concerning coal gasification, liquefaction, SO_2 scrubbing, pelletized coke, and other such matters), the Mining Research group at Continental Oil in Ponca City, Oklahoma (which studies new approaches to coal mining, such as water-jet mining), and the Scientific Systems Group (computers) in Pittsburgh, Pa. (which provides computer capability for the engineering groups and the rest of the firm). It is interesting to note that the Continental Oil Company, when it announced the acquisition of Consol, emphasized that "Consol's research and engineering program is a significant plus. We are sure you will be pleased to learn that Consol is in the forefront of research to convert coal into more versatile forms of energy—especially gasoline and high-BTU gas. The opportunities which this research may open are dramatized by coal's healthy reserve position compared with the relatively short supply of petroleum liquids and natural gas. Consol is the contractor for the U.S. government's $10 million "Project Gasoline." Under this program, Consol is building a plant at Cresap, West Virginia. Consol also has a key position in projects for producing high-BTU gas from coal."[18]

The federal government finances a substantial amount of coal research, mainly through the Office of Coal Research and the Bureau of Mines. The Office of Coal Research contracts for research and development into better methods of mining, preparing, and utilizing coal. In accord with the President's energy message of June 4, 1971, the Office has the responsibility for accelerating the coal gasification program to develop a process or processes to produce clean, high quality gas from coal on a commercial scale by 1980. About $30 million per year will be spent on this work, 2/3 to be funded by the Office, and 1/3 to be funded by industry. The Office's total estimated budget for 1974 is about $52 million. In addition, the Bureau of Mines allocated about $53 million in 1974 for mineral resource development and engineering, evaluation, and demonstration.[19]

The Office of Coal Research has helped to finance several approaches to coal gasification, including the Institute of Gas Technology's HYGAS pilot plant, the Consolidation Coal Company's CO_2 Acceptor pilot plant, Bituminous Coal Research's BI-GAS pilot plant, FMC Corporation's COED pilot plant, and others.[20] Each of these pilot plants costs millions of dollars. The Office's budget has increased very substantially in recent years: in 1972, it was about $30 million, and in 1970, about $15 million. The Bureau of

Mines also spent about $8 million in 1973 on coal gasification, most of it going to its SYNTHANE pilot plant. Recently, there have been several recommendations that such programs be accelerated further. For example, Senator Henry Jackson has called for a 10-year R and D program on coal gasification costing $660 million (60 percent to be paid for by government, 40 percent by industry), and a 12-year R and D program on coal liquefaction costing $750 million (75 percent to be paid for by government). Also, the Office of Science and Technology's Energy Advisory Panel has called for an expanded program of effort in liquefaction as part of Solvent Refined Coal effort, as well as continued emphasis on coal gasification.[21]

Besides R and D aimed at conversion of coal into high-BTU gas, there are many other approaches that are being explored to the problem of converting coal to a clean fuel. First, there is considerable work directed at the development of economical and reliable stack-gas scrubbers for the utilities, such work being done by Combustion Engineering, Bechtel, and others. To date, such work has not resulted in a commercially reliable process. Second, there is considerable work, by Westinghouse, General Electric, and others, on the conversion of coal into low-BTU gas, but such work is still at the pilot plant stage. Third, some companies are trying to refine coal into a sulfur-free solid that can be burned or heated into an oil; for example, Gulf is constructing a $19 million pilot plant financed by the Office of Coal Research. However, such work too is far from complete.[22]

SIZE OF FIRM AND INNOVATION
IN THE PETROLEUM INDUSTRY

Up to this point, we have been concerned almost exclusively with research, development, and invention: it is time that we look at the next stage of the process of technical change, innovation. An invention, when applied commercially for the first time, is called an innovation. Traditionally, economists have stressed the distinction between invention and innovation, on the ground that an invention has little or no economic significance until it is applied. This distinction becomes somewhat blurred in cases where the inventor and the innovator are the same firm. Under these circumstances, the final stages of development blend into at least a partial commitment to a market test.

To what extent have the largest petroleum firms carried out a disproportionately large share of the significant innovations? To answer this question, trade associations and trade journals were asked to list the important processes and products first introduced in the industry during 1919-58. They were also asked to rank them by importance. Then we consulted technical journals and corresponded with various firms inside and outside the industry to determine which firm first introduced each innovation commercially and when this took place. The results are shown in Table G-6. Next, based on data

Table G-6. Innovations and Innovators, Petroleum Industry, 1919-1938 and 1939-1958

Innovation	*Innovator*
1919-1938	
Burton-Clark cracking	Standard (N.J.)
Dubbs cracking	Shell
Fixed-bed catalytic cracking	Sun
Propane deasphalting of lubes	Union
Solvent dewaxing of lubes	Indian
Solvent extraction of lubes	Associated
Catalytic polymerization	Shell
Thermal polymerization	Phillips
Alkylation (H_2SO_4)	Standard (N.J.)
Desalting of crude	Ashland
Hydrogenation	Standard (N.J.)
Pipe stills and multidraw towers	Atlantic
Delayed coking	Standard (Ind.)
Clay treatment of gasoline	Barnsdall
Ammonia	Shell
Ethylene	Standard (Ind.)
Propylene	Standard (N.J.)
Butylene	Standard (N.J.)
Methanol	Cities Service
Isopropanol	Standard (N.J.)
Butanol	Standard (N.J.)
Aldehydes	Cities Service
Naphthenic acids	Standard (Calif.)
Cresylic acids	Standard (Calif.)
Ketones	Shell
Detergents	Atlantic
Odorants	Standard (Calif.)
Ethyl Chloride	Standard (N.J.)
Tetraethyl lead as antiknock agent[a]	Refiners
Octane numbers scale[a]	Ethyl
1939-1958	
Moving-bed catalytic cracking	Socony
Fluid-bed catalytic cracking	Standard (N.J.)
Catalytic reforming	Standard (Ind.)
Platforming	Old Dutch
Hydrogen-treating	Standard (N.J.)
Unifining	Union; Sohio

Table G-6 (cont.)

Innovation	Innovator
Solvent extraction of aromatics	Standard (N.J.)
Udex process	Eastern State
Propane decarbonizing	Cities Service
Alkylation (H Fl)	Phillips
Butane isomerization	Shell
Pentane and hexane isomerization	Standard (Ind.)
Molecular sieve separation	Texaco
Fluid coking	Standard (N.J.)
Sulfur	Standard (Ind.)
Cyclohexane	Phillips
Heptene	Standard (N.J.)
Tetramer	Atlantic
Trimer	Atlantic
Aromatics	Standard (N.J.)
Paraxylene	Standard (Calif.)
Ethanol	Standard (N.J.)
Butadiene	Standard (N.J.); Shell
Styrene	Shell
Cumene	Standard (Calif.)
Oxo alcohols	Standard (N.J.)
Dibasic acids	Standard (Calif.)
Carbon black (oil furnace)	Phillips
Glycerine	Shell
Synthetic rubber	Standard (N.J.)
Ethylene dichloride	Standard (N.J.)
Diallyl phthalate polymers	Shell
Epoxy resins	Shell
Polystyrene	Cosden
Resinous high-styrene copolymers	Shell
Polyethylene	Phillips

[a]Innovations excluded from Table G-7 because innovator had no crude capacity or because it was engaged primarily in another business.

Source: E. Mansfield, *op. cit.*

regarding the size (measured by daily crude capacity) of each petroleum firm in 1927 and 1947, we determined how many of these innovations were first introduced by the largest four firms. Since the more recent situation probably differed from that in the prewar era, innovations that occurred in 1939-58 were separated from those that occurred in 1919-38.

Whether or not the largest four firms introduced a disproportionately large share of the innovations depends on what one means by a disproportionately large share. If the largest firms devoted the same proportion of their resources as smaller firms both to inventive activity and to the testing and development of other people's ideas, if they could obtain applicable results as easily, and if they were as efficient and as quick to apply the results, one would expect their share of the innovations to equal their share of the market. If one computes the percentage of innovations, weighted and unweighted, carried out by the four largest firms, one finds that it exceeds their share of the market in both periods, as shown in Table G-7.

However, this is an incomplete analysis because it merely compares the performance of the largest four firms with all others. Although the largest four firms did disproportionately more innovating than all other firms taken as a group, they may not have done disproportionately more than somewhat smaller firms. To see whether this was the case, we estimated the average relationship

Table G-7. Innovations and Capacity (or output) of Largest Four Firms, Petroleum and Bituminous Coal Industries, 1919-1958

Item	Petroleum[a]		Coal[b]	
	Weighted[c]	Unweighted	Weighted[c]	Unweighted
	(percent of industry total)			
1919-38				
Process innovations	34	36	27	18
Product innovation	60	71	–	–
All innovations[d]	47	54	27	18
Capacity (or output)	33	33	11	11
1939-59				
Process innovations	58	57	30	27
Product innovations	40	34	–	–
All innovations[d]	49	43	30	27
Capacity (or output)	39	39	13	13

[a]Crude capacity is used to measure size of firm; it refers to 1927 in the earlier period and to 1947 in the later period.

[b]Annual production is used to measure size of firm; it refers to 1933 in the earlier period and to 1953 in the later period.

[c]In the columns headed "weighted," each innovation is weighted in proportion to its average rank by "importance" in the lists obtained. It was suggested that total savings be used to judge the relative importance of processes and that sales volume be used to judge the relative importance of products.

[d]The unweighted average of figures for process and product innovations.

Source: See note 4.

between the number of innovations carried out by a firm and its size, this relationship being approximated by a cubic regression. The results indicate that the size of firm where the maximum number of innovations (relative to size) occurs was about 200,000 barrels of crude capacity in 1919-38 and about 300,000 barrels of crude capacity in 1939-58; that is, it occurs at about the size of the sixth largest firm. Thus, the biggest four firms did less innovating, relative to their size, than somewhat smaller firms.

These data also allow us to compare the relative importance as innovators of firms of various sizes in the two periods: the results indicate that the smallest firms were less important sources of innovations in the later period than in the earlier one. In view of the increases in the costs of development and innovation, which we noted above, and the growing complexity of technology, this is not surprising.[23]

SIZE OF FIRM AND INNOVATION IN
THE BITUMINOUS COAL INDUSTRY

As in the case of the petroleum industry, we have a list of the significant innovations in bituminous coal in 1919-58, but this list is confined to coal preparation innovations. This list, shown in Table G-8, was derived from discussions with trade journals and government agencies. Then we consulted trade journals and corresponded with various firms inside and outside the industry to determine which firm first introduced each innovation commercially and when this took place. Next, we obtained data concerning the size (measured in terms of annual production) of each coal firm in 1933 and 1953, and determined how many innovations were first introduced by the largest four firms. Since the recent situation probably differed from that in the prewar era, we separated innovations that occurred during 1919-38 from those that occurred during 1939-58. The results are shown in Table G-8.

Judging by these crude results, the largest four coal producers carried out a disproportionately large share of the innovations in both periods, in the sense that their share of the innovations exceeded their share of the market. However, as pointed out in the previous section, this form of analysis is incomplete because it compares the performance of the largest four firms with the performance of all others. The largest four firms may not have done disproportionately more innovating than somewhat smaller firms, even though they did do disproportionately more than all other firms combined. To see whether this was the case, we estimated the relationship between the number of innovations carried out by a firm and its size, this relationship being approximated by a cubic regression.

The results indicate that the size of firm where the maximum number of innovations (relative to size) occurs was about 3,600,000 tons of coal in 1919-38 and about 7,800,000 tons of coal in 1939-58; that is, it occurs at

Table G-8. Innovations and Innovators, Bituminous Coal Preparation, 1919-1938 and 1939-1958

Innovation	*Innovator*
1919-1938	
Simon-Carves washer	Jones and Laughlin; Central Indiana
Stump air-flow cleaner	Barnes
Chance cleaner	Rock Hill
"Roto Louvre" dryer	Hanna
Vissac (McNally) dryer	Northwestern Improvement
Ruggles-Cole kiln dryer	Cottonwood
Rheolaveur	American Smelting
Menzies cone separator	Franklin County
Deister table	U.S. Steel
Carpenter dryer	Colorado Fuel and Iron
Froth flotation	Pittsburgh
1939-1958	
Raymond flash dryer	Enos
CMI drying unit	Hanna
Link-Belt separator	Pittsburgh
Bird centrifugal filter	Consolidation
Baughman "Verti-Vane" dryer	Central Indiana
Vissac Pulso updraft dryer	Northwestern Improvement
Link-Belt multilouvre dryer	Diamond; Elkhorn; Bethlehem; Eastern Gas and Fuel
Eimco filter	United Electric
Dorrco fluosolids machine	Lynnville
Parry entrainment dryer	Freeman
Heyl and Patterson fluid bed dryer	Jewell Ridge
Feldspar type jig	Northwestern Improvement
Bird-Humboldt centrifugal dryer	Clinchfield
Wemco Fagergren flotation unit	Hanna; Sevatora; Diamond
Continuous horizontal filter	Island Creek
Cyclones as thickeners[a]	Dutch State Mines

[a]Omitted from Table G-7 because innovator was not a domestic firm.
Source: E. Mansfield, *op. cit.*

about the size of the sixth largest firm. Thus, the largest four firms did less innovating, relative to their size, than somewhat smaller firms. These data also allow us to compare the relative importance as innovators of firms of various sizes in the two periods: the results indicate that the smallest firms were less important sources of innovations in the later period than in the earlier one. To some extent, this may have been due to increases in the investment required to innovate.[24]

COMPARISON WITH OTHER INDUSTRIES

To put into perspective the results presented in the previous two sections, it is helpful to look at how the relationships between innovation and firm size in the petroleum and bituminous coal industries compare with those found in other industries. Comparable data are available for three other industries: iron and steel, pharmaceuticals, and railroads. In the iron and steel industry, the largest four firms did not do a disproportionately large share of the innovating. On the contrary, the maximum number of innovations (relative to size) was found to occur among relatively small firms both in 1919-38 and 1939-58. Thus, the biggest firms in the petroleum and bituminous coal industries look better in this regard, relative to other firms in their industries, than do the biggest iron and steel firms.[25]

In the pharmaceutical industry, the largest four firms were not responsible for a disproportionately large share of the innovations in 1935-49 or 1950-62, when either unweighted data or economic weights are used. During 1935-49, the maximum number of innovations (relative to size) occurred at the size of the 10th largest firm in the industry. During 1950-62, the maximum number of innovations (relative to size) occurred at the size of the 12th largest firm in the industry. Thus, the biggest oil and coal firms look better in this regard, relative to other firms in their industries, than the biggest phramaceutical industries. However, it might be noted in passing that some observers claim that the role of the biggest drug firms has increased since 1962, due in part to increased costs of drug innovation that they attribute to the 1962 amendments to the Food, Drug, and Cosmetic Act.[26]

Finally, in the railroad industry, the largest four firms seem to have accounted for a disproportionately large share of the innovations occurring since 1920, based on the very small sample of innovations put forth by Healy. Judging by the little data that is available, the biggest oil and coal firms look no better in this regard, relative to other firms in their industries, than do the biggest railroad firms.[27]

DIFFUSION OF NEW TECHNIQUES IN
THE PETROLEUM INDUSTRY

Having discussed the R and D and innovative stages of the process of technical change, we turn next to the final stage: the diffusion of the innovation. Once an innovation is introduced, its use spreads from firm to firm and from place to place within a single firm. How rapidly an innovation spreads is obviously of great importance: for example, in the case of a process innovation, it determines how rapidly productivity increases in response to the new process. The diffusion process, like the earlier stages of the process of creating and assimilating new processes and products, is a learning process; but rather than being confined to a research laboratory or to a few firms, the learning takes place among a considerable number of users and producers. During the early stages of the diffusion process, the improvements in the new process or product may be almost as important as the new idea itself. For example, there were very important improvements in the catalytic cracking of petroleum in the period following Sun Oil Company's first introduction of the process, the original Houdry fixed-bed process being outmoded in less than a decade.[28]

Before looking at the rate of diffusion of new techniques in the petroleum industry, it is worthwhile to look briefly at how quickly petroleum inventions are applied. After all, the lag between invention and innovation is just as important as the rate of diffusion in determining how rapidly society benefits from new technology. According to Enos's estimates, the lag between invention and innovation for the major cracking processes averaged about 11 years. Schnee's results suggest that the average lag in pharmaceuticals may have been shorter than in petroleum, but Enos's results suggest that the average lag in petroleum may have been somewhat shorter than the average lag for 35 innovations in a variety of other industries. Thus, the petroleum industry seems to have been quicker than most, but not all, industries in converting inventions into innovations.[29]

Turning to the rate of diffusion, Table G-9 shows the increase over time in the percentage of American cracking capacity that was catalytic rather than thermal. As you can see, it took about 16 years (from the date of first commercial application of catalytic cracking) before one-half of U.S. cracking capacity was catalytic, and about 20 years before four-fifths of it was catalytic. The figures in Table G-9 lump together all catalytic cracking processes, and it is possible, of course, to look at the rate of diffusion of each one—Houdry, Houdriflow, T.C.C., and Fluid. For example, if we confine our attention to Fluid catalytic cracking, we find that it accounted for 2 percent of all capacity in 1943, 19 percent in 1948, 42 percent in 1953, and 62 percent in 1957. These figures provide some indication of the rate of diffusion of major new techniques in the petroleum industry.

DIFFUSION OF NEW TECHNIQUES IN
THE BITUMINOUS COAL INDUSTRY

In the bituminous coal industry, there is information concerning the rate of diffusion of three innovations—the shuttle car, trackless mobile loader, and continuous mining machine. These innovations were all of considerable importance. Figure G-1 shows the percentage of major bituminous coal firms that had introduced each of these innovations at various points in time. Two things

Table G-9. Growth of Catalytic Cracking Capacity as a Percentage of All U.S. Cracking Capacity

Year	Catalytic Cracking as a Percentage of All U.S. Cracking Capacity
1937	0.1
1938	0.6
1939	1.1
1940	5.6
1941	6.6
1942	7.0
1943	9.9
1944	21.4
1945	28.5
1946	30.0
1947	32.2
1948	34.0
1949	37.4
1950	41.7
1951	43.8
1952	46.7
1953	61.2
1954	67.5
1955	74.7
1956	79.3
1957	81.9

Source: *Oil and Gas Journal*, 1927-57.

should be noted concerning this data. First, only firms producing over 4 million tons of coal in 1956 (according to McGraw-Hill's Keystone Coal Buyers Manual) were included. Second, the data shows the percentage of firms that introduced an innovation, regardless of the scale on which they did so. Judging from Figure G-1, it took about 5 years (from the date of first commercial introduction) before half of the major coal firms had begun using shuttle cars, about 7 years before half had begun using trackless mobile loaders, and about 4 years before half had begun using continuous mining machines.[30]

According to the available data, the observed differences among these innovations in the rate of diffusion can be explained in part by differences in the profitability of using them: estimates provided by the firms seem to indicate that, among these three innovations, the use of continuous miners was most profitable and the use of trackless mobile loaders was least profitable. This

Figure G-1. Growth in the Percentage of Major Firms That Introduced Three Innovations, Bituminous Coal Industry

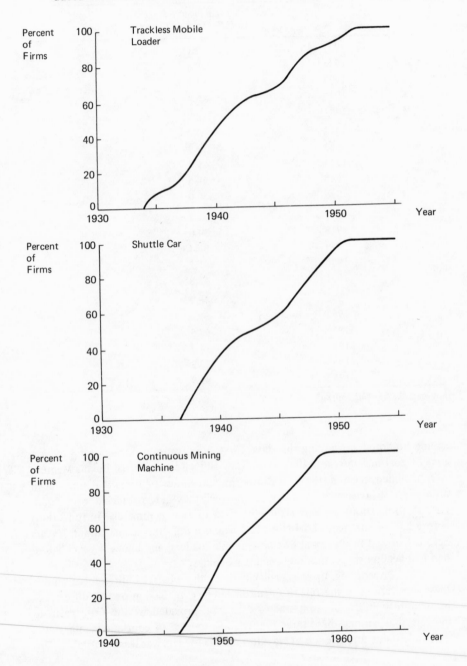

is in accord with a model of the diffusion process I put forth many years ago, and which has been tested since by Romeo, Hsia and myself.[31] Other factors included in this model, and in its more recent extensions, are the size of the investment required to introduce the innovation, the length of time the innovation has been in use in other industries, the amount of research and development carried out by the industry, and the market structure of the industry. (The effects of market structure will be described in the next section.) All of these factors can have a significant effect on the rate of diffusion of an innovation.

Based on the data for these three innovations, there seems to be some tendency in the bituminous coal industry, as in practically all other industries, for larger firms to begin using new techniques more quickly, on the average, than smaller firms. This is of some interest, but it should be emphasized that it does not mean that the larger firms are necessarily more progressive. Even if the larger firms were no more progressive, one would expect them to be quicker, on the average, than the smaller firms. To illustrate this, consider an industry with two firms, one large (70 percent of the market) and one small (30 percent of the market). If the large firm does its share of the innovating (no more, no less), it will be first in 70 percent of the cases—and it will be quicker on the average than the small firm.

It is also worth adding that, although there is some tendency for the larger firms to be quicker to begin using a new technique, there is a great deal of variation from innovation to innovation in the speed of response of a particular firm. In other words, a firm that is relatively quick to begin using one innovation may not be relatively quick to introduce another one. For example, the coefficient of correlation between how rapidly each of a sample of major coal firms introduced the continuous mining machine with how rapidly it introduced the shuttle car was −0.02, the corresponding coefficient of correlation for the continuous mining machine and the trackless mobile loader was −0.17, and the corresponding coefficient of correlation for the shuttle car and the trackless mobile loader was 0.54.[32]

INDUSTRIAL CONCENTRATION AND
THE RATE OF DIFFUSION

Since this chapter is concerned with, among other things, the effects of certain kinds of changes in market structure on the rate at which new technology is developed and utilized, it is important that we look at the relationship between the extent of concentration in an industry and the rate of diffusion of new techniques in that industry. Economists have shown considerable interest in the nature of this relationship, but until recently there has been little empirical or econometric work bearing on it. Instead, most of the discussion has been purely theoretical and speculative, many economists arguing that more competitive industries tend, all other things equal, to adopt innovations more quickly than less competitive industries, while some economists have argued that the opposite was the case.

Recently, several attempts have been made to carry out empirical work in this area. To begin with, a study was conducted of the rates of diffusion of 12 innovations in four industries: holding constant the profitability of using the innovation and the size of the investment required, there were significant interindustry differences in the rate of diffusion, and these differences seemed to be broadly consistent with the hypothesis that the rate of diffusion is higher in more competitive industries, although the results were not statistically significant.[33] Next, a study was made of the rate of diffusion of numerically controlled machine tools in the tool and die industry, the results indicating that the rate of diffusion was higher than would be expected in any of the four industries included in the previous study for an innovation of comparable profitability and size of investment. This finding, like that of the previous study, suggested that, all other things equal, innovations tend to spread more rapidly in less concentrated industries.[34]

Most recently, Romeo studied the effect of a variety of factors, including the number of firms and the inequality of firm sizes, on the rate of diffusion of numerically controlled machine tools. Based on very detailed data for ten industries, he found that—holding constant the average and interfirm variance of the profitability of the innovation, the length of time that the innovation had been used in other industries, and so on—there was a statistically significant tendency for the rate of diffusion to increase with the number of firms in the industry and to decrease with the extent of inequality of firm sizes.[35] Thus, although the empirical findings are tentative, since they are based on a relatively small number of industries and innovations, all of the empirical studies carried out to date point in the same direction, and indicate that innovations tend to spread more rapidly in less concentrated industries.

MERGER AND DIVESTITURE POLICIES: INTRAFUEL

Given the fact that relatively few studies have been made of the relationship between size of firm and invention, innovation, and diffusion in the petroleum and bituminous coal industries, as well as the equally important fact that it is difficult to move from such studies to policy conclusions, it is hard to say much with great confidence concerning the implications of existing studies for the determination of public policy regarding intrafuel mergers or divestiture. Nonetheless, it is important that we do out best to derive whatever implications we can from these studies, even though the results are bound to be tentative and subject to error, since the alternative probably is to base policies on even less reliable information than the studies at hand.

In the petroleum industry, the available information seems to suggest that, although the big oil firms frequently are not responsible for the radical, basic inventions, they have done a great deal of the development work. This work is much more expensive than the more basic research or the initial invention, and it often takes many years; since this is the case, it is not surprising that big firms have done a disproportionately large share of such work, as well as of the industry's innovating. It seems likely that the rate of technological change and innovation was more rapid in petroleum because of the existence of such firms, even though their size may have had unfortunate effects with regard to other aspects of market performance.[36]

However, this does not mean that firms must be as big as Exxon in order to carry out productive R and D or to introduce significant innovations. All of the available evidence, rough and incomplete though it may be, indicates that somewhat smaller firms perform at least as well in this regard, relative to their size, as the biggest firms. As we have seen, they spend more on R and D as a percentage of sales than the biggest firms, they devote about the same proportion of their R and D to more basic and long-term projects as the biggest firms, and they carry out a greater number of significant innovations, relative to their size, than the biggest firms. Moreover, the available evidence does not indicate that the R and D programs of the biggest petroleum firms produce more—per dollar of R and D—than somewhat smaller firms.

Thus, the available data seem to suggest that a policy designed to break up all of the reasonably large petroleum firms might well have the effect of slowing the rates of development, invention, and innovation in the petroleum industry, although its effects on the rate of diffusion of new techniques are harder to predict.[37] On the other hand, there appears to be no evidence that, if firms were allowed to merge to reach the size of the biggest firms in the industry, this would increase the rates of development, invention, and innovation in the petroleum industry. Nor does it seem likely that such an increase in concentration would increase the rate of diffusion of new techniques.[38]

Turning to the bituminous coal industry, the available data suggests that the large firms have done a disproportionately large share of the innovating, at least in the areas for which there is data. However, the industry has not invested a great deal in research and development, and much of the industry's new technology has been developed by equipment suppliers or through government financing. No doubt, one reason why the industry has devoted so little resources to inventive and innovative activity has been the fact that so many of its firms are quite small. However, this does not mean that firms must be as big as Consolidation in order to carry out productive R and D or to introduce significant innovations. The available data indicates that somewhat smaller firms perform at least as well in this regard, relative to their size, as the biggest firms.

Thus, the available data suggests that a policy designed to break up all of the reasonably large coal companies might well reduce the rate of technical change in the bituminous coal industry, and that a policy promoting the merger of some of the small coal firms into somewhat bigger units might promote the rate of technical change. But there is no evidence that, if firms were to merge into units as big as Consolidation, more rapid technical change would occur than if they were to merge into considerably smaller units. As in the case of the petroleum industry, the largest firms do not seem to have any marked edge in this regard over their somewhat smaller competitors, based on the limited information at hand.

MERGER AND DIVESTITURE POLICIES: INTERFUEL

In recent years, there has been a great deal of interest and concern with regard to the acquisition of coal firms by petroleum firms. Gulf Oil acquired Pittsburgh and Midway, which accounts for 2 percent of the nation's coal production, in 1963; Continental Oil acquired Consolidation Coal, which accounts for 12 percent, in 1966; Occidental Petroleum acquired Island Creek, which accounts for 7 percent, in 1968; and Sohio acquired Old Ben, which accounts for 2 percent, in 1968. Thus, four of the country's ten biggest coal companies were acquired by oil firms. Also, Exxon and Kerr-McGee have acquired huge amounts of coal reserves. Table G-10 shows that almost half of the 20 largest petroleum firms had coal holdings by the beginning of 1970.

The Department of Justice cleared the merger between Continental Oil and Consolidation Coal in 1965. As described by Walter Comegys, Deputy Assistant Attorney General, "after a careful study, it was concluded that the two companies were not in significant actual competition with each other in specific geographic and industry markets. It was also believed at that time that the potential for possible competition between them and for significant increased concentration in the energy market as such did not rise to the level of a reasonable probability."[39] However, in 1970, he voiced concern about the acquisition trend, which "raises long-range questions of concentration in the energy industries."[40]

There seems to be general agreement that one important reason why the oil companies have acquired the coal companies is to prepare for the time when coal can be converted economically into synthetic fuels. Some observers are worried about the possible effects of these mergers. For example, Netschert, Gerber, and Stelzer pose the following questions:

> How can the public be sure, for example, that the emergence of the synthetic fuels industries will occur at the pace which economic circumstances would, under free market forces, dictate? It could well be that the self-interest of certain companies with dominant positions, if not of the industry as a whole, would call for delaying the inauguration of a synthetic fuels industry in order to protect

Table G-10. Coal Holdings of Twenty Largest Petroleum Firms, Ranked by Assets, Early 1970

Company	1969 Assets ($ billions)	Does Company Have Coal Holdings?
Exxon	17.5	Yes
Texaco	9.3	Yes
Gulf	8.1	Yes
Mobil	7.2	No
Standard Oil (Cal.)	6.2	No
Standard Oil (Ind.)	5.2	No
Shell	4.4	Yes
Arco	4.2	Yes
Phillips	3.1	No
Continental	2.9	Yes
Sun	2.5	Yes
Union	2.5	No
Occidental	2.2	Yes
Cities Service	2.1	No
Getty	1.9	No
Sohio	1.6	Yes
Pennzoil United	1.4	No
Signal	1.3	No
Marathon	1.2	No
Amerada-Hess	1.0	No

Source: Bruce Netschert, Abraham Gerber, and Irwin Stelzer, "Competition in the Energy Markets," paper prepared for Senate Subcommittee on Antitrust and Monopoly, 1970.

existing investments in crude oil and natural gas. Of even greater concern is the fact that the energy company (and it should be borne in mind that there are already at least five major oil companies with across-the-board positions in *all* of the domestic fuel resources—oil, gas, coal, oil shale, and uranium) straddles a situation which until now has been one of intense interfuel competition. With a position in most or all of the fuels, a price rise in any of them is to the advantage of the energy company, since it makes it possible to raise the prices of all the fuels wherever they are in competition. It is all too possible that an electric utility may someday find itself facing the situation of being able to obtain each of its fuels, including uranium, from only a single supplier . . . Further, as suppliers of all the fuels used for electric power generation, the energy companies can signifciantly influence the cost of fuel to their major competitor.[41]

In this study, we are concerned only with the effects of the acquisition of coal firms by petroleum firms on the rate of technological change, not with the broader issues involved. Unfortunately, existing studies shed practically no light on this question. To try to get some idea of these effects, a number of conversations took place with R and D executives in the coal, petroleum, and electrical equipment industries, the purpose of these conversations being to determine what they felt the impact of the acquisition of coal firms by petroleum firms had been, and would be. Without exception, they felt that it would have a positive effect on the rate of development of coal gasification, liquefaction, and other such innovations, because the oil firms would devote expertise and capital that would not otherwise be available to try to solve the relevant problems. They were skeptical of the idea that the oil firms would be inclined to push these developments at a slower pace than would otherwise occur.[42]

In addition, they felt that the oil companies were not in a position to control the technological developments in these areas, since many firms outside the oil industry were heavily involved, and since much of the work being carried out by the oil firms is government-financed and available on a royalty-free basis. Without carrying out more extensive and intensive research, it is impossible to evaluate these opinions with any thoroughness or accuracy. Certainly, evidence of this sort cannot be taken at face value, since some of these people are biased, and since the sample was necessarily quite small. However, so little is known in this area that even scraps of evidence of this kind are of interest.

GAPS, LIMITATIONS, AND NEEDED RESEARCH

In conclusion, it is very important to note the many gaps in our knowledge in this area. We have summarized the available evidence concerning the relationship between size of firm and technological change in the petroleum and bituminous coal industries, and have seen what the implications are for public policy regarding mergers and divestiture. On the basis of the results, it is clear that the available data base is very skimpy and that the available models (which must be used to predict what effects changes in market structure would have) are rough at best. It would be a great mistake if anyone were to get the impression that the available evidence and models permit us to come to very precise conclusions.

NOTES

1. For definitions of research and development, see National Science Foundation, *Research and Development in Industry, 1970.*
2. National Science Foundation, "Company Funds Push Industrial R & D Spending to $18 Billion in 1971," December 13, 1972.

3. National Science Foundation, *Research and Development in Industry, 1970*, p. 94.

4. Edwin Mansfield, *Industrial Research and Technological Innovation*, W.W. Norton for the Cowles Foundation for Research in Economics at Yale University, 1968.

5. F.M. Scherer, *Industrial Market Structure and Economic Performance*, Rand McNally, 1971.

6. Edwin Mansfield, John Rapoport, Jerome Schnee, Sam Wagner, and Michael Hamburger, *Research and Innovation in the Modern Corporation*, W.W. Norton, 1971.

7. John Enos, *Petroleum Progress and Profits*, M.I.T. Press, 1962, p. 234.

8. Mansfield, Rapoport, Schnee, Wagner, and Hamburger, *op. cit.*

9. Jacob Schmookler, *Invention and Economic Growth*, Harvard, 1966.

10. E. Mansfield, *op. cit.*

11. Mansfield, Rapoport, Schnee, Wagner, and Hamburger, *op. cit.*

12. Mansfield, *op. cit.*

13. See the National Science Foundation's *Research and Development in Industry* for 1960, 1963, 1966, and 1969, as well as the reference in note 2.

14. D.L. Baeder, "Research and the Emperor's New Clothes," paper given at the Sixth Industrial Affiliates Symposium, Stanford University, May 15, 1973.

15. *Ibid.*

16. Judging from conversations with some R and D people in the bituminous coal industry, annual R & D expenditures financed by bituminous coal companies probably total well under $10 million. In addition, of course, machine manufacturers do some R and D; and, as we shall see, the government finances considerable R & D.

17. *Coal Age*, October 1972, p. 139.

18. *The Consol Story*, A Message from L.F. McCollum, Chairman of the Board, and A.W. Tarkington, President, to Stockholders of Continental Oil Company, September 19, 1966, pp. 15-16.

19. See the Budget of the United States.

20. See George Hill, "Coal Gasification Program Shifts into High Gear," paper presented at the 17th Annual Convention of the Wyoming Mining Association on June 17, 1972 (reprinted in *Coal Age*, August, 1972), and *Coal Age*, January 1973.

21. See *Coal Age*, January 1973 and April 1973.

22. With regard to the current status of scrubbers, see *Coal Age*, February 1973, p. 90. With regard to Westinghouse's program, Westinghouse received $8.2 million from the Office of Coal Research in early 1973 for initial work on a low-BTU fuel gas process. If the R and D is successful, a pilot plant will be constructed and operated by Public Service of Indiana. Besides Westinghouse, OCR, and Public Service of Indiana, Peabody Coal Co., Amax Coal Co., and Bechtel are also involved in the project, which is expected to cost $80 million, industry's share being $43 million.

23. E. Mansfield, *op. cit.*
24. E. Mansfield, *op. cit.*
25. *Ibid.*
26. Mansfield, Rapoport, Schnee, Wagner, and Hamburger, *op. cit.*
27. E. Mansfield, "Innovation and Technical Change in the Railroad Industry," *Transportation Economics*, National Bureau of Economic Research, 1965.
28. E. Mansfield, *The Economics of Technological Change*, W.W. Norton, 1968.
29. John Enos, "Invention and Innovation in the Petroleum Refining Industry," *The Rate and Direction of Inventive Activity*, National Bureau of Economic Research, 1962.
30. E. Mansfield, *Industrial Research and Technological Innovation, op. cit.*
31. Anthony Romeo, "Interindustry Differences in the Rate of Diffusion of an Innovation," unpublished Ph.D. dissertation, University of Pennsylvania; R. Hsia, "Technological Change in the Industrial Growth in Hong Kong," paper presented at the 1971 meetings of the International Economic Association at San Anton, Austria, and E. Mansfield, "Determinants of the Speed of Application of New Technology," paper presented at the 1971 meetings of the International Economic Association at San Anton, Austria.
32. E. Mansfield, *Industrial Research and Technological Innovation, op. cit.*
33. E. Mansfield, "Technical Change and the Rate of Imitation," *Econometrica*, 1961.
34. Mansfield, Rapoport, Schnee, Wagner, and Hamburger, *op. cit.*
35. A. Romeo, *op. cit.*
36. Of course, if the large oil firms had not been in existence, more process development firms like Universal Oil Products might have been formed, but it seems unlikely that they would have been able to devote as much resources to development as the large oil firms. Moreover, there frequently are advantages in combining development and production in the same organization.
37. As we noted in a previous section, the available evidence indicates that the resulting reduction in concentration would promote a more rapid rate of diffusion, but, on the other hand, the reduction in the average size of firm might reduce the rate of diffusion, since there is some evidence that the rate of diffusion decreases as the investment in the innovation—as a percent of the average assets of the firms in the industry—increases.
38. It seems unlikely that, in this range of variation of firm size, the negative effect of the increase in concentration would be offset by the positive effect of the reduction of the size of the investment in the innovation (relative to the average assets of the firms in the industry). However, much more research is needed before we can answer this question with confidence.
39. *Competitive Aspects of the Energy Industry*, hearings before the Senate Subcommittee on Antitrust and Monopoly, May 5, 6, and 7, 1970, p. 136.

40. *Ibid.*, p. 137.
41. *Ibid.*, pp. 191-2. Also see p. 136 where Comegys expresses similar words.
42. According to those queried, the oil companies have not had a great impact on the R and D programs of the coal companies (like Consol) that they have acquired, but they have stepped up their own work in this area. Of course, some of this work, like Gulf's, is financed partly by the government.

The Relationship Between Economic Structures and Political Power: The Energy Industry

Lester M. Salamon*
John J. Siegfried

This study seeks to explore systematically the relationship between economic structure and political power in the United States in an effort to lay a foundation for assessing the likely political implications of changes in the structure of the American energy industry. Such a task must necessarily be approached with some diffidence, not only because the variables to be examined are so elusive, but also because the topic has already been so thoroughly debated by so many people for so long. Indeed, from the time of Marx (who viewed the state as the "executive committee of the bourgeoisie") through that of Woodrow Wilson (who once argued that "The masters of the government of the United States are the combined capitalists and manufacturers of the United States"),[1] down to that of current liberal social scientists like Theodore Lowi (who recently published a book purporting to document the extent of expropriation of public authority by private economic interests in the United States),[2] scholars, politicians and publicists alike have portrayed economic power and political effectiveness as opposite sides of the same coin. Yet, an equally impressive list of thinkers have simultaneously documented what they claim to be numerous breaks in the chain linking economic power to political effectiveness, thanks to the "pluralistic" character of American society.

For all the attention it has attracted, however, the relationship

*Assistant Professor of Political Science and Assistant Professor of Policy Sciences, Duke University; and Assistant Professor of Economics, Vanderbilt University, respectively. This study was greatly augmented by the research assistance of James Boswell, Oliver Grawe, Kent Kraus, James Norris and Thomas Zak. Walter Mead, Milton Russell and Daniel Ogden provided extensive comments on an earlier draft which improved the clarity, logic and factual accuracy of the paper. In the usual tradition, the authors remain solely responsible for any remaining errors.

between economic and political power has still only rarely been examined through solid empirical techniques capable of confirming or denying general propositions. Anecdotal case studies—some of them quite good[3]—and vaguely supported, broad, general theories abound; but systematic empirical analyses, testing which aspects of economic structure have what political consequences under what circumstances, have been surprisingly sparse. This is particularly true, moreover, when we search for analyses that relate economic structure to policy outputs instead of to the level of political activity.[4]

This essay seeks to help fill this gap. To do so, we develop a set of hypotheses about the likely impact of economic structure on political influence and test these hypotheses against the empirical evidence in several policy areas, focusing particularly on governmental policies toward the energy industry. While we have no pretensions that our analysis constitutes the last work on this crucial subject, we do believe it throws some interesting, new, systematic, empirical evidence on a question that has too often been treated in a haphazard and imprecise way. In the process, we believe we provide some firmer ground than has been available heretofore for making judgments about the political conse-quences that flow from the peculiar structure of the American energy industry and from changes in that structure over time; and hence for making policy decisions about past and future merger activities in this industry.

INDUSTRY STRUCTURE AND POLITICAL INFLUENCE

The Theory
In general terms, two broad sets of factors shape the impact that an economic sector has on public policy: first, the nature of the political system; and second, the structure of the economic sector itself. The first of these helps define the political system's susceptibility to economic influences; the second helps determine the ability of a particular industry to take advantage of that susceptibility. While other factors—like skill, luck, timing, and leadership—also play important roles, these two set the broad parameters of the relationship between economic and political power, and therefore offer the most fruitful subjects for systematic analysis.

The Political Process
Perhaps the defining characteristic of the American political system is its permeability to outside pressures. Through voting, Congressional testi-mony, campaign finance and a host of additional avenues, the political system offers relatively easy access by citizens to the central policy-making process on a regular basis. While this relative permeability raises the possibility of broad public control of governmental policy *a la* pure democratic theory, it also raises the paradoxical possibility of translating disproportionate economic power into

disproportionate political influence in a way that can frustrate broad public control. Three aspects of the political system in particular facilitate this translation.

The first of these concerns the maldistribution of *incentives* for political action, what might be termed the "free rider problem." As numerous democratic theorists have noted, political involvement is, like any other investment, contingent upon the expectation of a reasonable rate of return on invested resources. But since each individual consumer-taxpayer typically bears only a small portion of the costs and enjoys only a small share of the benefits from public programs, he rarely has the incentive to spend the energy, time, and resources needed to influence the policy. For other political actors, like large corporations, however, the potential benefits and costs of government action are sizeable enough to make political involvement a rational investment. The consequence is a gross disparity in the incentives for political involvement that works to induce the citizen-taxpayer toward passivity while stimulating the large corporation toward political activism. In the introduction to *The Political Economy of Federal Policy*, Robert Haveman poses the free-rider problem nicely in terms of oil industry confrontations with government:

> In today's system of political economy, citizens in their capacity as taxpayers and consumers confront major obstacles in exercising their vested interest in governmental efficiency, equity and openness. Because they are unorganized, uninformed on the technical details of public issues, and without a specially designated spokesman to advance their case in the process through which tax laws are written, spending programs developed, and regulatory decisions made, their interests tend to be submerged and often ignored. For example, certain aspects of federal tax policy involving the depletion of minerals (particularly oil), are almost certainly both inefficient and highly inequitable (in that they yield huge windfalls to oil companies and their owners). The effect of this policy is to artificially raise the taxes paid by other taxpayers and to generate a misallocation of resources in the oil industry. Although this legislation has existed since 1913, those who bear the cost of this subsidy—the taxpayers— have been unable to muster the political strength to eliminate this program or even reduce it substantially. As a group they are large in number, unorganized, and generally uninformed on the intricacies of this policy. Because each of them bears only a small share of the total cost of the policy, there is little incentive for any one of them to spend the energy, time, and resources to organize opposition to the policy. As a result, in deliberations on this issue, they have no one who stands to argue their position and to lobby for it. On the other hand they are confronted by a tightly organized group of oil companies . . . each receiving a substantial share of the subsidy provided. A hired lobbyist serves as the industry spokesman at every

congressional hearing on the issue, contacts individual congressmen and senators, and makes substantial financial contributions on the industry's behalf to cooperative elected officials.[5]

Not only are the *incentives* for political involvement grossly disparate, but so is the distribution of politically relevant resources. Large-scale corporate enterprises have important political advantages by virtue of their control over sizeable quantities of several crucial political resources: money, expertise, and access to government officials. These resources are frequently direct outgrowths of the enterprises' regular economic activities, which, in the age of the "new industrial state," frequently put the enterprise in intimate contact with scores of government officials on a day-to-day basis. By contrast, the typical consumer-taxpayer has only one resource regularly at his command: the vote. While the electoral process theoretically gives the mass of unorganized citizens a mechanism to remedy whatever inequities arise from the struggle of organized groups without incurring the costs associated with more direct political action, things rarely work out so smoothly in practice. Indeed, for the electoral process to determine policy with any specificity and force, four conditions would have to be met: (1) competing candidates would have to offer clear policy alternatives; (2) voters would have to be aware of the policy stakes in elections; (3) majority preferences on these questions would have to be ascertainable in election results; and (4) elected officials would have to be bound by the positions they assumed during the campaign. Yet, as one recent American government textbook concludes on the basis of a careful review of the evidence, "none of these conditions are fulfilled in American politics," and there is "little evidence that voters can directly affect public policy through the exercise of their franchise."[6] Certainly in the case of energy policy, voters have few occasions to express a clear set of policy preferences, and would probably have difficulty formulating preferences if the occasion arose. As in his role of consumer in the corporate-controlled marketplace, so in his role as voter, the individual citizen must generally take what the prevailing structure of political interests serves up for him, although even the most powerful oligopolist will take care not to push his luck too far.

What makes this maldistribution of *incentives* and *resources* for political action so important is the highly fragmented character of the policy process in American government, which plays directly into the hands of those with organized resources focused on particular policy issues. Part of this fragmentation is dictated by constitutional provisions establishing a government of "separated institutions sharing powers."[7] But part of it grows out of the more informal arrangements these institutions have developed to process their workload. The classic example of this fragmentation is the pervasive decentralization of power in the U.S. Congress. Long ago legislators discovered that they could hold their own in the interinstitutional donnybrook that is the American political system only by channelling legislative business through specialized committees and effectively delegating legislative power to jealous subject-area potentates in the person of Committee Chairmen. Buttressed by a vibrant norm of deference to the Committee leadership, this institutional arrangement has fractionated decision-making authority into numerous relatively autonomous chunks.

What is not so clearly understood, however, is that the same fragmentation of authority that frustrates effective overall policy direction in Congress operates as well in the executive branch. Despite the deceptive symbolism of a single Chief Executive giving direction to the executive establishment, the federal bureaucracy is really a bewildering smorgasborg of institutional types with varying degrees of autonomy. Rather than unified administrative structures, Cabinet-level Departments frequently resemble loose collections of warring fiefdoms, only nominally subservient to a common sovereign. Add the numerous government corporations, the quasiindependent regulatory agencies, the special commissions, the advisory committees, the single-headed and multiheaded agencies, the institutions and institutes, and the interagency committees and the impression of unity and consistency disappears like fog under the morning sun.[8]

What produces this administrative fragmentation is the bureaucrat's need for succor in a political system that looks somewhat askance at his very existence.[9] Since Congress is the ultimate source of agency funds and authority, it is only natural that agencies should turn first to Congress for support. But since power is fragmented in Congress, it quickly becomes so in the agencies because Committee chairmen typically arrange agency structure to facilitate their own control.[10]

Agency reliance on Congressional committees and their chairmen is supplemented, however, by agency dependence on client groups, particularly those with influence at the Congressional committee stage. As one long-time government official recently noted: "Private bureaucracies in Washington now almost completely parallel the public bureaucracies in those program areas where the Federal Government contracts for services, regulates private enterprise, or provides some form of financial assistance."[11]

The consequences of this pattern of governmental decision-making are profound. Instead of a single, integrated policy process, what emerges instead is a series of individual, and largely independent, "policy subsystems" linking portions of the bureaucracy, its related congressional committees, and organized clientele groups in a symbiotic state of equilibrium.[12] The key to policy-making power, therefore, is access not to the political system generally but to these well insulated and highly structured subsystems. Voter-taxpayers can occasionally gain access to the former. However, access to the latter is typically confined to those with the expertise, resources, and influence to provide stable sources of support to key subsystem actors over extended periods of time.

Further complicating the job of the activist citizen is the predominance of administrative over legislative decision-making in modern government. The theory of separation of powers notwithstanding, the bureaucracy performs crucial legislative and judicial as well as administrative functions in American government. The vast bulk of legislation considered by Congress originates in the bureaucracy, and it is the bureaucracy that makes the most potent input into the legislative process thanks to its control of information. In addition, statute law (law passed by Congress) is overwhelmed by the far greater volume of administrative law produced by the bureaucracy in the course of interpreting Congressional intent. As a former Commissioner of the Federal Communication Commission has put it:

> While the Courts handle thousands of cases each year and Congress produces hundreds of laws each year, the administrative agencies handle hundreds of thousands of matters annually. The administrative agencies are engaged in the mass production of law, in contrast to the Courts [and Congress], which are engaged in the handicraft production of law.[13]

This predominance of administrative rule-making has important implications for the distribution of political power in American society. In particular, it puts a premium on technical expertise and on the ability to follow the complex evolution of administrative decision-making. Far from expanding Presidential power *per se*, the expansion of administrative rule-making really contributes to the influence of the private subsystem actors who alone have the resources to retain Washington legal counsels and research staffs to analyze relevant agency decisions and respond to them effectively. The structure of governmental decision-making, whether intended or not, thus facilitates the translation of economic power into political influence.

Economic Structure and
Political Influence

If the maldistribution of incentives for political action and of politically relevant resources, as well as the fragmented character of the policy process, facilitated the translation of economic power into political influence, the economic structure of particular industries helps to determine how well these industries can take advantage of this opportunity. In analyzing the relationship between economic structure and political influence it is useful to distinguish among four aspects of industry structure,[14] each of which we can hypothesize has a slightly different impact on political efficacy. By examining each of these aspects of industry structure we develop a set of hypotheses for later testing, hypotheses which, at this point, represent "hunches" based on existing literature and a priori reasoning, but which we hope to confirm or refute in the next section of this paper.

The first aspect of economic structure that affects an industry's political effectiveness is *firm size*, the absolute size of firms in the industry (measured in terms of, say, assets). In general, large firms, simply by virtue of their size, will likely have greater economic resources at their disposal than smaller firms in the same or different industries. Kaysen[15] and Edwards[16] have both suggested that economic power is an important source of political power, either directly, through campaign finance, or indirectly, through the purchase of expertise to generate information instrumental in the policy process and liaison agents to transmit this information to the relevant policy-makers. A National Industrial Conference Board study[17] found that large firms tended to represent their legislative views more frequently than did smaller firms. By the same token, Bauer, Poole and Dexter[18] found that firm size is an important determinant of the political activity of executives since only the executives of large firms could afford the luxury of hiring staffs and taking the time to inform themselves about policy issues.

What makes the absolute size of available resources, and hence firm size, so important politically is the fact that political involvement has certain fixed costs attached to it (we will call this the "threshold problem"). Like a single pain reliever commercial, corporate political activity will likely produce little payoff unless sustained over a period of time and spread over a variety of policy-making arenas. Since small firms can rarely sustain these high fixed costs, they are generally constrained to channel their political involvement more extensively through trade associations, with all the intraorganizational differences, lack of control, and consequent weakening of influence that carries with it.[19]

Access to economic resources is not the only political asset arising from larger firm size. Also important is the prestige typically accorded individuals who reach the command posts of the larger corporate enterprises, a prestige that is translatable into political influence through the numerous advisory committees and impersonal contacts that bind the worlds of business and government together. Equally important is the larger pool of expertise potentially available to the larger firms, and, on some occasions, the larger number of employees and stockholders who can be mobilized (or threatened to be mobilized) for political action. Finally, there are the reduced costs of coordinating political action. Other things being equal, therefore, we hypothesize that an industry containing large firms will have greater political influence than an industry of the same size but composed of more numerous small firms. Yet, like all true hypotheses, this one too could go the opposite way, since larger firms tend to be more visible and are more likely than smaller firms to have to contend with union officials who interfere with management's claim on worker political sentiments. Which of these interpretations holds, however, can only be settled by empirical analysis.

Closely related to *firm size* as a determinant of political influence is *industry size*, the aggregate level of economic activity represented by an industry. We would expect that large industries, whatever the average size of the firms comprising them, would have larger pools of money, talent, client support, contacts, and overall resources than smaller industries and would consequently have greater political power. We thus hypothesize a direct relationship between industry size and political influence.

In addition to firm size and industry size, a third aspect of industry structure that can influence the level of political power is the *degree of geographic dispersion* in an industry. With respect to this aspect of economic structure, opinions differ markedly about the consequences for political influence. On the one hand, an industry with all of its resources concentrated in only a few locales lacks the broad-based political support typically necessary to pass legislation and have it appropriately implemented. On the other hand, unless it has sufficient salience in some locale to attract the attention and support of at least one set of political actors, an industry can find itself ignored altogether. As one participant in a Brookings round-table discussion put it: "I can't think of anyone who has more of an impact on a congressman than a representative of a corporation having a big plant in his area."[20] On balance, we think it likely that given a certain industry size, industry political power will be greater if the industry's resources are concentrated than if they are widely dispersed. We therefore hypothesize an inverse relationship between geographical dispersion and political influence.

The fourth aspect of industry structure influencing political power is the degree of *market concentration*, the share of the total business in an industry dominated by a handful of the largest firms. Highly concentrated industries have a number of important political advantages. For example, market concentration typically yields higher profits[21] that can be channeled into political activity. A more competitive industry of roughly the same size would likely earn fewer profits and hence lack some politically relevant resources that might be available to it if it were more highly concentrated. More concentrated industries also have the advantage of avoiding the debilitating and time-consuming process of negotiating industry positions on political matters among numerous competing firms, a process that has frequently been cited as a major obstacle to trade association political effectiveness.[22] In eliminating competitors in the economic marketplace, highly concentrated industries also eliminate competition in framing industry positions in the political marketplace. In addition, the more concentrated the industry, the less severe the "free rider problem" it encounters, since firms that dominate an industry receive a larger share of whatever benefits investment in political activity brings. Hence they have a greater incentive to make this investment.

Partially counter-balancing these political advantages of economic concentration, however, are some liabilities. For example, concentrated power, particularly when combined with large firm size, is likely to enhance corporate visibility. As Bauer, Poole and Dexter put it: "A business can be too big to be politically effective along some lines. . . .nowadays many really big corporations are not eager to dance among the chickens. . . ."[23] Closely related to the visibility problem is the "wear out your welcome" problem. When a few firms must carry the entire burden of representing an industry in the public policy-making arena, they run the risk of overdrawing on their account of good-will with key policy-makers, particularly since most policy-makers can scant afford the appearance of subservience to a small group of firms. On balance, however, we believe the presumptive evidence supporting a *direct* relationship between market concentration and political power is the more persuasive for hypothesis-framing purposes.

These hypothesized relationships between industry structure and political power have obvious implications for the energy industry. First, in terms of *industry size*, the energy industry is clearly one of the largest. The petroleum sector alone, accounting for less than half of the nation's energy requirements, controlled assets in excess of $70 billion as of 1968 in its refining segment alone, making oil refining three times larger than iron and steel and two times larger than automobile manufacturing in terms of assets.[24] Not only is the industry large in terms of assets, however, but also it is large in terms of the number of people with a direct, personal stake in its prosperity. The oil industry has under lease one-fourth of the land area of the United States, much of it privately-owned land. In addition, the industry included over 196,000 service station dealers as of 1967.[25] While not easily mobilized for political action, these groups nevertheless constitute a potential source of residual support. For example, during Congressional debate over legislation to limit the use of price discrimination by suppliers among their own dealers in 1956 and 1957 the American Petroleum Institute secretly stimulated an ostensibly spontaneous groundswell of grass-roots opposition on the part of local dealers and jobbers, relieving the "majors" of the need to oppose the bill directly.[26]

Not only is the energy industry large, but also it is concentrated geographically in enough places to guarantee it rather substantial attention in numerous political jurisdictions. No fewer than 32 states have oil and gas production, sufficient to produce a majority in the United States Senate if every Senator votes to support his state's "native" industry. In 10 of these states, over 35 percent of the land area is under lease. Many of these latter states enjoy remarkable political influence in Washington as a result of the strategic positions held by their Congressional representatives. Of the 19 members of the House of Representatives with the greatest seniority in the 93d Congress, 10 came from major crude oil producing states, including four of the five *most* senior members. If we include coal as well as oil, the proportion of senior members from major producing states rises from 10 out of 19 to 13 out of 19.[27] Since seniority translates into power through the Committee system, the result is an impressive agglomeration of political muscle in precisely the most opportune spots. Texas and Louisiana alone, the two states with the largest amount of oil production, send to Congress the men who chair the House Appropriations Committee, the House Agriculture Committee, the House Armed Services Committee, the House Appropriations Committee's Subcommittee on Foreign Operations, and the Senate Finance Committee. Oklahoma, the fourth-ranking oil producing state, is represented in Congress by the Speaker of the House.

In addition to large industry size and geographic concentration, the energy industry is also characterized by relatively large *firm size*—another factor we have hypothesized relates positively to political influence. In fact, of the 25 largest U.S. corporations at the top of *Fortune*'s list of 500 major firms, 10 are major integrated oil firms.[28] What is more, while the energy industry enjoys the benefits of large firm size, it manages to avoid some of the drawbacks this involves in terms of visibility. This occurs because, despite the large size of the firms, the energy industry is still relatively unconcentrated. In 1970, the share of total production accounted for by the largest four firms was 25.3 percent for natural gas, 30.5 percent for crude oil, 33.1 percent for oil refining, 42.8 percent for regional gasoline marketing, and 30.2 percent for coal.[29] Although the oil "majors" are "in a position to dominate the industry and perhaps to control it,"[30] they can insulate themselves partially from the charges of bigness by pointing to the numerous smaller firms operating in the industry and arguing that government policies that might hurt the "majors" could hurt the small fry of the industry even more. Indeed, the "majors" are in a position to reap many of the political benefits of limited market competition while paying few of the costs. Concentrated industries, we have argued, suffer the political disadvantage of greater visibility, but gain the compensating political benefit of unity and reduced coordination costs.[31] In many respects, the petroleum "majors" enjoy the best of both worlds. The relative decentralization of their industry affords them symbolic advantages unavailable to the Big Three auto makers. Yet, through a complex array of cooperative arrangements, the majors have been able to hedge against the dangers of fragmentation, intraindustry conflict, and even "excessive" competition. These cooperative arrangements among the major oil companies take a variety of forms. For example, Mead has documented the joint ventures among the largest 32 oil companies for the purpose of bidding on oil shale resources.[32] The Federal Trade Commission has exposed the many interrelationships between major oil companies in the common ownership of domestic pipelines.[33] In addition, economists have recently turned their attention to the links between oil companies arising through their major bank associations. Some banks have members of several oil companies on their boards of directors, a practice that makes coordination of activities easier.[34]

Perhaps even more important, this pattern of interdependence among ostensible competitors is spreading from the petroleum industry alone to the energy industry as a whole. In the 1960s, there were four major oil companies that acquired leading coal producing firms,[35] while additional oil firms have acquired coal and uranium reserves without taking the highly visible merger route.[36] While it is difficult to determine precisely the economic impact of this growing web of interdependence, it is reasonable to predict that it will have significant political effects in reducing conflict within the industry and thus facilitating the formation of a united front with which to approach government.[37]

On the basis of the hypotheses we have described, there is reason to expect the energy industry to enjoy substantial political power—more so, perhaps, than most other industries. Yet before such assertions can be accepted as valid, it is necessary to subject them to systematic empirical scrutiny to determine whether the assumptions and hypotheses on which they rest find support in the evidence. It is to this all-important task of hypothesis-testing that we now turn.

FROM ECONOMIC AND POLITICAL POWER TO PUBLIC POLICY: SOME EMPIRICAL EVIDENCE

To say that the larger firms in the energy industry, or any other industry, command ample resources for political action is not yet to say that they have more effective political power. Political resources can be squandered wastefully as easily as any other resources. To evaluate the political influence of the energy industry, and the likely impact on this influence of further structural changes, it is necessary to go beyond the cataloguing of available resources and scrutinize how this industry fairs in the actual policy process.

Establishing a link between economic or political power on the one hand, and policy outcomes on the other, is no easy undertaking. The available signposts are few in number and limited in usefulness. Two works in particular deserve mention, if only to clarify the approach we adopt here. The first is Robert Dahl's *Who Governs?*,[38] a study of the role of economic notables in policy-making in New Haven, Connecticut. Dahl evaluates the influence of economic notables by studying a collection of decisions in four key policy arenas in New Haven over a ten year span. Since these notables participated infrequently in decisions in almost all of the four policy areas, Dahl concludes that economic power is not translated effectively into political power in New Haven.

The second work, Bauer, Poole and Dexter's *American Business and Public Policy*,[39] examines the role of industry groups in the formulation of American foreign trade policy—particularly in Congress—during the 1950s and early 1960s. Bauer, Poole, and Dexter scrutinized the degree of business participation in the actual policy-making arenas during consideration of trade legislation. In addition, they carefully analyzed the flow of information within the business community and its trade associations concerning foreign trade matters. On both counts, they concluded that business influence was marginal, despite what many would take to be high economic stakes in the issues. Businessmen generally had little knowledge about trade matters, and their lobby organizations proved incapable of seriously influencing consideration of the policy.

The approach we propose to pursue here differs markedly from these earlier works. We are suspicious of sole reliance on the decision-making approach to measure the political impact of economic power. As Morton Baratz and Peter Bachrach have noted,[40] some of the issues most important to economic notables may never come up for decision, and certainly not during any particular time span. In fact, the power to keep issues "settled" and therefore out of the policy-making arenas is one of the most important forms of power of all, but it is one that the decision approach can never tap.

For the purposes of this analysis, therefore, we measure power somewhat differently, as—in Bertrand Russell's terms—"the production of intended consequences."[41] The heart of our analysis consists of an effort to determine whether the pattern of policy outputs affecting the energy industry can be explained systematically in terms of the distribution of economic power reflected by industry structure.[42] Although we are interested in the full range of governmental policy outputs affecting the energy industry, both those designed to preserve industry viability by regulating the terms and conditions of competition and those redistributing costs and benefits between the industry and consumer-taxpayers generally, space and time considerations require us to restrict our attention to only a few of the myriad policies affecting this mammoth industry. In choosing among potential candidates for detailed examination, we were guided by two basic considerations: first, a decided preference for policy outputs susceptible to systematic empirical treatment so as to avoid the potential pitfalls of case study analysis; and second, a prejudice in favor of policy areas not already thoroughly explored. The federal corporation income tax thus provides a perfect candidate for scrutiny since it is both universal in its application and varied in its impact, thus offering an opportunity to assess systematically the political success of particular groups and the relationship between that success and industry structure. In addition, we also examine the pattern of state gasoline excise taxes and, on a somewhat more limited scale, three cases of regulatory policy: lead-free gasoline regulation, prorationing of crude production, and oil import quotas. In each of these areas, moreover, we are interested in two basic questions: first, how successful the energy industry has been in redistributing wealth from consumer-taxpayers generally to itself; and second, how the benefits so secured are distributed *within* the energy industry between large and small firms. Of the two questions, the first is of greatest concern to us here, since it raises most clearly the social benefit/social cost considerations that form the heart of antitrust policy.

Since our analysis is necessarily restricted to only a few of the many interactions between the energy industry and government, readers must exercise appropriate caution in drawing firm conclusions from our discussion. While we believe the policy areas selected for scrutiny are "representative" of the energy industry's political influence, we have no solid evidence with which to substantiate this claim. What is more, we are cognizant of the limitations on our analysis flowing from the relatively primitive state of the art in assessing the relationships between policy outputs and industry structure in systematic empirical terms. Pending more detailed and comprehensive analysis, therefore, we must urge readers to treat our findings as suggestive insights and resist the inclination to read firm conclusions into them.

The Federal Corporation Income Tax

The federal corporation income tax affords an unusually good opportunity to pursue a systematic, empirical analysis of the impact of economic structure on public policy through the medium of political influence, for it applies to everyone and its consequences are *relatively* easy to measure. Since there is a natural desire to shift tax burdens from one's own back and hence on to the backs of others, relative tax burdens provide a good measure of the political success of particular groups.[43]

The energy industry is the recipient of an immense tax subsidy in the form of the oil depletion allowance. But is that the only tax provision that benefits this industry? Do other industries benefit more from the same or other tax loopholes, thus supporting Wilbur Mills' comment that what makes the depletion allowance so objectionable is not that it is so unusual, but that it "sticks out alone—it's just a target that way"?[44] Most importantly, do tax burdens vary systematically with economic structure (firm size, industry size, market concentration, and geographic dispersion), as the "economic-structure-political-influence hypotheses" outlined earlier would suggest? The implications of this latter question are especially pertinent for evaluating merger activities.

To try to answer these questions, we analyze the distribution of corporate income tax burdens. The analysis falls naturally into two parts: first, an examination of the distribution of tax burdens in mining and manufacturing generally to determine the overall relationship between economic structure and tax policy;[45] and second, a more detailed appraisal of the tax burden situation in the energy industry.

Methodology. Experts have calculated the revenue loss to the Treasury caused by the "special" tax treatment of particular items of corporate income and expenditure. This "special" treatment lowers the corporate income tax paid below that which firms would be required to pay in its absence. Our measure of success at attempts to influence government policy is the effective average corporation income tax rate. This rate is defined as the ratio of tax liabilities to "true" accounting net income. "True" net income includes some income which currently is excluded from reported taxable income by the income tax laws and regulations. "True" accounting net income therefore means the cost of capital plus economic profits for each industry.

Our model is expressed most succinctly in symbols. The nominal corporation income tax rate on income in excess of $25,000 was 0.52 in 1963, the year for which we collected data. The model is:

$$0.52 - t = f(F, I, C, D, R) \tag{H.1}$$

where

t = the effective average corporation income tax rate

F = firm size (millions of dollars)

I = industry size (number of employees)

C = market seller concentration (percentage points)

D = geographical dispersion (index number based on employment)

R = the rate of "true" accounting profit plus interest on total assets (percentage points)

Our hypothesis is that the deviation of the effective average corporation income tax rate from 0.52 is affected by political factors which derive their character from economic structural variables—firm size, industry size, market concentration, geographical dispersion, and the rate of profit.

The empirical tests of the political influence hypotheses consist of regressions of firm and industry characteristics on the deviation of the effective tax rate from 0.52. The sample consists of 110 IRS "minor industries" for 1963.[46] A positive association between this deviation and firm size should help to confirm the suspicions of those supporters of an aggressive antitrust policy based on the potential political and social repercussions of increased *aggregate* concentration. A positive association between the deviation and market concentration would support the contention of trust busters who base their fear of mergers on the possible undesirable political and social consequences of increased *market* concentration.

To determine the correct functional form of the dependent variable, we employed the maximum likelihood method originally suggested by Box and Cox.[47] Examination of the likelihood function indicates that a multiplicative model is more appropriate than a linear functional form. In a multiplicative regression model the estimated coefficients are interpreted as the percentage change in the dependent variable (the reduction in the effective average corporation income tax rate) associated with a one percent increase in the value of the respective explanatory factor.

Throughout the analysis the limitations of the effective average corporation income tax rate as a measure of the effects of political influence must be kept in mind. There are substitution possibilities available to firms investing in political influence activities. In addition to alternative investments in physical and human capital, firms generally have opportunities to exert political pressure on government policy decisions other than the income tax. Yet the income tax is certainly one of the more important public policies effecting the energy industry, and therefore one likely to illustrate industry political effectiveness, so long as we exercise caution in interpreting the results as a generalized relationship between the independent variables and the overall effect of political influence.

The Variables of the Model. It is hypothesized that economic structural characteristics of the environment of firms determine the extent and character of their behavior and that this behavior, in turn, is reflected in the response of policy decisions toward achieving the participants' objectives.

Effective average corporation income tax rate. The dependent variable is the deviation from 0.52 of the effective average corporation income tax rate paid on "true" accounting profits. Reported net income is adjusted to "true" accounting profits by adding estimates of underreported profits to reported taxable net income. The source of underreported profits is a variety of "special" tax provisions which are identified by the Treasury Department.[48] Some of these "special" provisions are investment credit, special capital gains tax rates, the surtax exemption for the first $25,000 income, percentage depletion, rapid depreciation, and immediate expensing of long-lived advertising and research and development expenditures.

The tax rate variable is formulated in terms of deviations from a standard average tax rate that would evolve from a simple, basic income tax structure. This simple, basic income tax structure consists of a single proportional tax rate of 0.52 on taxable income, provides for government sharing in losses as well as profits, permits the deduction of certain costs of operation (cost of materials, repairs, bad debts, rent paid on business property, etc.), and includes the foreign tax credit provision.[49] All types of income are taxed at the same rate.

Firm size. The principal task of the analysis is to test the *firm* size-political influence assertions that seem to be the basis for much of the public's concern about conglomerates, monopoly and rising aggregate concentration. Firm size is measured by the "typical" asset size of a firm in each IRS "minor industry"; that is, the asset size of the firm which accounts for the median asset dollar in the industry. The expected coefficient of firm size is positive, since the hypothesis is that larger firms are associated with larger deviations of effective tax rates from the norm.

Industry size. The size of industries is measured by their total employment, a measure which helps us tap the political influence represented by employee dependence on industry viability. We predict a positive sign for the coefficient of industry size. Since an increase in this variable, *ceteris paribus*, comes most plausibly from the growth or addition of smaller firms, it is possible that an increase in the industry size measure is accompanied by increased difficulty in organizing the industry effectively and consequently aggravation of the "free-rider" problem. We attempt to identify this factor separately by inserting market seller concentration in the model, but to the extent that seller concentration fails to accurately measure the "free-rider" problem, the impact of industry size on the effect of political influence may well run counter to the voter patronage argument. Consequently the coefficient of industry size is subjected to a two-tail statistical significance test.

Market seller concentration. The concentration ratio for IRS "minor industries" is constructed as a weighted average (based on employment) of the total sales revenue concentration ratios of the four digit S.I.C. industries comprising each "minor industry." Four digit S.I.C. industries are more appropriate as measures of economically meaningful markets.

Since highly concentrated industries should not find the "free-rider" problem as formidable an obstacle to organizing political influence efforts as less concentrated industries, we expect the coefficient of the four firm concentration ratio to be positive.

Geographical dispersion. The effect of an industry's physical distribution on its ability to achieve favorable tax rulings and laws is a function of the scarce time and energy of elected representatives. To get a tax bill introduced, it is essential to acquire the attention and support of a member of Congress (or even more desirably, a member of the House Ways and Means Committee). Each member of Congress has many constitutent groups in his political district. If he devotes attention to their desires and needs in proportion to his perception of their impact on his impending reelection, he will direct his attention first to those bills which have the greatest impact on his political future, and continue to entertain such bills until he has exhausted his supply of time, energy, and the indulgence of his colleagues.

We hypothesize that the geographically concentrated industry will be first to have its tax burden objectives recognized by its Congressmen, *ceteris paribus*. If the incremental benefit to the elected representative of championing tax relief for a constituent group is proportional to the number of voters affected by the bill, then the incremental benefit to the elected representatives from a district including a large portion of a very concentrated industry will exceed the incremental benefit to each of the (more numerous) elected representatives from districts including parts of a less geographically concentrated industry.

A competing hypothesis with respect to geographical concentration is that widely dispersed industries have contacts with more representatives than concentrated industries, and consequently have a greater probability of finding a representative sympathetic to their interests. Also, having more individuals working for their cause in Congress may benefit widely distributed industries. Politically, when a firm with interests in many states has a problem on Capitol Hill, it can command a special kind of attention. Senator Hart has said that "when a major corporation from a state wants to discuss something with its political representatives, you can be sure he (sic) will be heard. When that same company operates in thirty states, it will be heard by thirty times as many representatives."[50]

Geographical dispersion is measured by an index based on the distribution across states of employment in each Census S.I.C. industry.[51] The dispersion index is larger the more concentrated is the employment of the industry in a few states. Therefore, a positive coefficient for the dispersion index indicates support for the high-concentration-representative-recognition hypothesis while a negative cofficient offers support for the many-contacts-many-voices hypothesis.

Profit rate. Industries earning relatively high returns to capital might behave differently from industries earning low returns to capital in their approach to influencing public policy decisions. Behavior is probably better characterized on a firm level, but industries with higher profit rates are going to be populated with firms earning, on average, higher profit rates. It is plausible that firms in more profitable industries direct their political attention principally toward protecting their attractive private market position. Industries enjoying a blockaded condition of entry, high levels of seller concentration, rapid growth in demand or reduction in costs are likely to find themselves enjoying (short-run at least) economic profits. Firms in this environment might find it advantageous to forego potential tax rate reductions if they fear that such reductions would draw attention to their industry and/or market position. Such attention could lead to unfavorable public opinion and possibly to governmental control or regulation of the industry. This argument suggests that the expected sign of the profit rate variable is negative.

Alternatively, one might argue that firms with high rates of profit have more to gain from income tax rate reductions than do firms with lower rates of profit (presuming, of course, that the firms under consideration are of the same asset size). This argument is consistent with standard profit-maximization models of economic behavior. However, recent innovations in the theory of the firm have stressed firm goals that may deviate from the pure profit maximization model. Several of these alternatives have implications for the hypothesis that firms with higher profit rates have more to gain from tax rate reductions.

If firms maximize goals including elements other than profits alone, the political activities necessary to maximize after-tax profits by influencing the tax structure may conflict with the attainment of other objectives. For example, for firms maximizing both profits and sales revenue, extensive governmental political activities might generate adverse public opinion which could hinder attainment of their sales revenue goal.[52] The political lobbying necessary to reduce tax rates could prove detrimental to the prestige and respect of executives of firms which have a managerial utility maximization goal.[53] Firms that are "satisficers" are likely to receive much less utility from reductions in the effective tax rate if they earn high rates of profit than if they earn low rates of profit.[54] These "new theories of the firm" may provide a basis for the expectation that industries enjoying high rates of return will be less likely to pursue tax rate reductions vigorously.

In view of the conflicting theories with respect to the effect of profit rates on political influence, we test the coefficient of the profit rate variable using a two-tail significance test. A finding of a positive coefficient supports the profitable-firms-have-more-to-gain-from-tax-rate-reductions hypothesis and is inconsistent with both the profitable-firms-avoid-spotlighting-themselves and profitable-firms-have-exceeded-their-minimum-profit-constraint-and-thus turn-to-satisfying-other-objectives hypotheses. The profit rate is the ratio of "true" accounting profits plus interest to total industry assets.

Summary of the Model. The basic model that is tested empirically is summarized in Equation (H.2).

$$(0.52 - t) = \alpha F^{\beta_1} I^{\beta_2} C^{\beta_3} D^{\beta_4} R^{\beta_5} e. \tag{H.2}$$

All variables are as defined previously, α and β are parameters, and e is a random disturbance term. By taking the natural logarithm of both side of Equation (H.2) we get Equation (H.3), which can be estimated directly by multiple linear regression.

$$ln(0.52 - t) = a + b_1 \, lnF + b_2 \, lnI + \tag{H.3}$$
$$b_3 \, lnC + b_4 \, lnD + b_5 \, lnR + \epsilon$$

a and b_i denote estimates of the population parameters α and β_i respectively and ϵ is a multiplicative disturbance term.

Empirical Results. The empirical results estimated from this model by least-squares regression are summarized in Equation H.4

$$ln(0.52 - t) = 3.541 + 0.084 \, lnF - 0.095 \, lnI - 0.512 \, lnC$$
$$(2.42) \qquad (-1.72) \qquad (-2.94)$$
$$-0.017 \, lnD - 0.301 \, lnR \tag{H.4}$$
$$(-0.15) \qquad (-2.46)$$

$$R^2 = 0.158 \qquad n = 110 \qquad F = 3.89$$

(*t*-ratios are shown in parentheses)

The firm size hypothesis. A one percent increase in the "typical" firm size in an industry is associated with a 0.08 percent increase in $(0.52 - t)$. The hypothesis that the elasticity of $(0.52 - t)$ with respect to firm size exceeds zero is supported at the 99 percent confidence level.

However, even though we have significant statistical support for the hypothesis, the practical importance of the finding remains to be evaluated. Without a standard for comparison it is difficult to evaluate the significance of 0.08 percent. The practical significance of the relationship depends on the loss in welfare caused by the existence of the relationship between firm size and effective tax rates and the resource costs that would be required to alter the situation. In general, the results provide support for the suspicions of those who fear that larger firms may be more successful than smaller firms in efforts to manipulate public policy to their advantage.

Industry size hypothesis. Industry size has a negative impact on $(0.52 - t)$, but an impact that is not statistically significant at usual confidence levels. This leads us to conclude that aggregate industry employment size has little, if any, net impact on tax-subsidies through its effect on the political process. Therefore we might cautiously judge, on the basis of this evidence, that larger industries find organization problems to be barriers to effective political pressure and that they suffer in the distribution of tax-subsidies as a consequence. The empirical results offer no support for domination of either the voter patronage hypothesis or the organizational problems hypothesis. The findings are consistent with the interpretation that both of these phenomena operate, but that their effects cancel each other.

Market concentration hypothesis. We hypothesized that more concentrated industries could coordinate political influence activities more easily and thus stand a better chance of waging a successful political pressure campaign. The empirical results consistently contradict this hypothesis. The regression coefficient of market concentration is negative and statistically significant at the 95 percent level in all models tested. Therefore we must accept the verdict that our a priori hypothesis concerning the organizational advantages of highly concentrated industries, independent of their size, is inconsistent with the data.

Two possible explanations might be offered for the consistently negative coefficients of the market concentration variable. The first is that firms in concentrated industries fear that the attention of zealous deconcentration advocates may be drawn to their industry if they secure favorable tax treatment. Fearing this, they might wisely avoid investments in political influence intended to affect the effective average tax rate in favor of investments which tend to increase their *before-tax* rate of return.

The second possible explanation is that concentrated industries are more likely to be highly unionized than unconcentrated industries. Unions may aspire to political goals that are inconsistent with the federal corporation income tax position of the industry. If the union is effective in achieving its goals, the highly unionized industry might be less effective in attaining lower effective average corporation income tax rates. Both of these hypotheses are *ex post* speculations and as such cannot be formally tested with our data.

Geographical Dispersion Hypotheses. We hypothesized that industries whose employment was concentrated geographically might exert more effective political pressure on elected officials. The expectation is that more geographically concentrated industries are, *ceteris paribus*, more apt to be associated with relatively lower effective average corporation income tax rates.

Our a priori hypothesis is not supported by the data. The sign of the dispersion index is negative, although not statistically significant at the 95 percent confidence level. One explanation that we suggested earlier for the negative coefficient is that a disperse distribution of physical facilities may be more advantageous than a concentrated distribution because a disperse distribution produces more access points to legislators and decision makers. This argument predicts that more geographically concentrated industries will achieve less success in avoiding corporation income taxes than widely distributed industries. The negative coefficient that evolves from the analysis seems to favor this argument over the concentration hypothesis. However, the statistical level of confidence provides little encouragement for support of the access points thesis either.

Profit rate hypotheses. Two hypotheses involving the rate of profit earned by an industry were proposed. One suggested that industries earning relatively high rates of return might abstain from activities designed to lower their effective average tax rate for fear of drawing attention to their favorable market environment that is producing the high returns. The alternative hypothesis advanced the possibility that firms earning high before-tax rates of return have more to gain from reductions in the effective corporation income tax *rate* and consequently participate in activities designed to influence the income tax structure more intensively than do firms earning lower before-tax rates of return. The regression coefficient of the profit rate variable is consistently negative and statistically significant at the 95 percent confidence level. The empirical results suggest a rejection of the higher-profits-more-to-gain hypothesis in favor of the favorable-market-environment-more-to-lose hypothesis.

Summary and Evaluation of Results.   We designed a model to study the relationship between structural characteristics of the American economy and the interindustry variation in the effect of political influence on public policy. This task required specification of a model which included the logical economic environmental factors that may affect the behavior of industries in the political arena.

In general, the statistical performance of the models employed to explain the linear variation in the logarithm of $(0.52 - t)$ is encouraging. The coefficient of determination indicates that the regression explains about 16 percent of the linear variation in the logarithm of $(0.52 - t)$. This is a reasonable statistic for cross-section regressions with more than a hundred observations. The F statistic for Equation (H.4) is statistically significant at the 99 percent confidence level, providing confidence in the overall validity of the model.

We believe our discovery of a positive association between firm size and the effective average corporation income tax rate is of substantial importance. The major competing hypotheses to the firm size influence on political success—the organizational costs argument (concentration), the voter patronage argument (industry size), and the geographical dispersion argument—do not conflict with a tentative confirmation of the firm-size-political-influence hypothesis. None of these alternatives provides an empirically supported substitute explanation for the various actual levels of federal corporation income tax rates.

Additional confidence in the discovery of a positive relationship between firm size and the effective average rate of taxation is provided by several biases remaining in the analysis. These biases tend inherently to obscure such a positive relationship. For example, the inclusion of the $25,000 surtax exemption as a "special tax provision" reduces the magnitude of the positive relationship. In view of such factors, it is surprising that any positive relationship between firm size and effective average tax rate is detected at all. While the results remain preliminary in nature, they are highly suggestive. At the very least, the tentative findings indicate that further, more comprehensive analyses of the interaction between economic variables (particularly firm size) and the distributive characteristics of public programs may prove fruitful.

Application to Merger Policy and the Energy Industry. Evaluating the political impact of mergers using the economic-structure-effective-tax-rate analysis is complicated by the fact that a particular hypothetical merger simultaneously affects more than one of the explanatory variables in the model. Unfortunately, it is difficult to provide predictions concerning the effects of energy industry mergers on effective tax burdens that go much beyond speculation. However, we are able to provide some descriptive data that lend heuristic support to a policy of caution toward such potential combinations. Information on the effect of a specific proposed merger on the economic structural variables in our model could be used to predict the change in effective average tax rates associated with the merger by our estimated model.

Because of the unique character of the special tax provisions relating to extractive industries and the enormity of the sums of tax revenue at stake, we feel that there is a distinct possibility that mergers involving petroleum companies may have an uncharacteristically strong impact on tax policy. Such a judgment finds empirical support in our data, moreover, since each of the four energy industries in the broad sample exhibits an actual tax rate lower than the predicted tax rate. This is illustrated in Table H-1, which shows both the tax rate predicted on the basis of Equation (H.4) for each of the four energy-related industries in the sample and the corresponding actual observed effective average tax rate. Only three other industries out of the 110 escaped taxation to a greater degree, as indicated by the extent to which their actual tax rates fell below those predicted by Equation (H.4). This occurs in spite of the fact that we have treated the foreign tax credit provision as a legitimate component of our norm—the basic simple corporation income tax structure.

Additional evidence that mergers in the energy industry may have effects different from those in all mining and manufacturing emerges from an examination of the differential impact of the depletion allowance on different size firms in the petroleum refining industry. According to the Internal Revenue Code, the depletion allowance for crude petroleum was 27.5 percent in 1963. However, no firm may use allowances in excess of fifty percent of net income before the depletion allowance. This limitation was far more restrictive to smaller companies than to the larger ones, who were able to expand the fifty percent of net income constraint through their diversified activities. Integrated petroleum refining firms with less than 250 million dollars of assets in 1963 were able to claim only 24.8 percent of their before depletion net income as depletion allowances. Firms 250 million dollars or greater in assets were able to deduct 41.0 percent of their net income before depletion as depletion allowances.[55]

As an added illustration, we have calculated effective average corporation income tax rates for the combined petroleum refining industries (both with and without extraction activities) by asset size class for 1967. Those firms under five million dollars of assets were excluded because many of them showed deficits in 1967 and we were not able to remove these effects from the data. Furthermore, firms of less than five million dollars assets are not representative firms in the petroleum refining industry. The tax rates for the remaining size classes are presented in Table H-2. It is clear from Table H-2 that

Table H-1. Predicted and Observed Effective Average U.S. Corporation Income Tax Rates for Four Selected Energy Related Industries for 1963

IRS Code	Industry Description	Predicted t [Equation (H-4)]	Observed t
1100	Coal mining	32.3	18.2
1310	Crude petroleum, natural gas & liquids	41.9	24.1
2911	Petroleum refining without extraction	34.5	25.0
2912	Petroleum refining with extraction (the "majors")	37.4	19.8

Table H-3. Regression of State... Coefficients and t-ratios (in parentheses)

Model	Observations	Constant	VA_r/PI	L_r	VA_c/PI	L_c			
I	All 50 states	7.658 (32.31)	-0.292 (-1.90)	-1.391 (-2.67)*	0.023 (1.10)	4.744 (0.92)	—	-0.001 (0.53)	0.17
II	All 50 states	8.33 (15.72)	-0.381 (-2.43)*	-1.395 (-2.74)*	0.042 (1.87)	3.560 (0.71)	-0.061 (-1.48)	—	0.209
III	22 non-dedication states	8.097 (23.85)	-0.017 (-0.04)	-1.929 (-2.64)*	-0.012 (-0.21)	19.858 (0.55)	—	-0.001 (-0.61)	0.127
IV	22 non-dedication states	9.087 (13.90)	-0.424 (-0.90)	-1.839 (-2.74)*	0.063 (1.09)	-0.945 (-0.03)	-0.096 (-1.84)	—	0.263
V	28 dedication states	7.587 (28.97)	-0.389 (-2.25)*	-0.494 (-0.58)	0.032 (1.35)	3.366 (0.63)	—	-0.001 (0.52)	0.110
VI	28 dedication states	6.838 (6.47)	-0.359 (-1.98)	-0.502 (-0.59)	0.023 (0.82)	3.119 (0.59)	0.058 (0.67)	—	0.117

* = statistical significance at 0.05 level.

VA_r = value added in petroleum refining (millions of dollars).

VA_c = value added in crude petroleum and natural gas extraction (millions of dollars).

PI = personal income (billions of dollars).

L_r = number of refineries with greater than 250 employees/total number of refineries.

L_c = number of crude oil and natural gas establishments with greater than 100 employees/total number of crude oil and natural gas establishments.

SR = total state government revenues (billions of dollars).

Table H-2. Effective Average U.S. Corporation Income Tax Rates for Petroleum Refining Industry (IRS 2911 + 2912) for 1967

Size Class ($1,000,000)	t_1	t_2
5-10	37.7	37.2
10-25	51.8	43.8
25-50	37.5	37.5
50-100	39.0	38.8
100-250	17.2	16.6
over 250	24.1	9.7

$t_1 = \dfrac{\text{tax liabilities}}{\text{"true" accounting profits}}$; treating foreign taxes entirely as a substitute for U.S. taxes; foreign tax credit not a "special" provision.

$t_2 = \dfrac{\text{tax liabilities-foreign tax credit}}{\text{"true" accounting profits}}$; treating the foreign tax credit as a "special" tax provision.

it is the very largest of the oil refiners, in particular the integrated "majors," that are best able to take advantage of the special provisions that affect business in general and the petroleum industry in particular. Combined with the evidence already presented, this finding, with all its limitations, gives us an empirical basis for concluding that mergers in the energy industry may have political effects that are quite different from those our evidence suggested for mining and manufacturing generally.

Motor Fuel Excise Taxes

As a second empirical test of the relationship between petroleum industry economic structure and public policy, we examine state excise taxes on motor vehicle fuels. Since state policy-making has traditionally been considered even more susceptible to private interest pressure than federal policy-making,[56] we expect significant relationships between petroleum industry economic strength in a state and state excise taxes on motor vehicle fuels. In particular, on the basis of our earlier analysis, we hypothesize that state excise taxes will vary directly with the relative size of the petroleum industry in a state and with the extent to which a state's petroleum industry is dominated by large firms. The first variable reflects an adversary model of politics, namely that since there are competing desires to capture a relatively fixed quantity of loot within any political jurisdiction, success will go to those with the greatest *share* of the resources.[57] To measure this variable, we employ the ratio of the value added of the crude oil industry and of the petroleum refining industry within each state to the personal income of the state. We separate production and refining phases of the industry since there is reason to believe refiners are relatively more interested in excise taxes than crude oil producers, both because a relatively greater proportion of refinery products go into motor fuel production, and because a much smaller proportion of refinery products are exported. Hence, crude oil producers are less likely to be interested in the motor fuel tax rate in the state in which the oil is produced.

To test our second hypothesis—that it is the relatively larger firms that exert the greater impact on political decision-making—we employ the ratio of the number of establishments greater than a fixed employment size to the total number of establishments in the industry in a state. This variable represents the dominance of large establishments within each state. The index was computed for crude petroleum using 100 employees as the cutoff size and for petroleum refining using 250 employees as the cutoff size. These sizes were chosen on the basis of data availability and to provide sufficient variation in the explanatory variables constructed from them.

The major oil companies have lobbied *against* increases in motor fuel taxes in those states which do not have a constitutional amendment requiring the dedication of motor fuel tax revenues to highway construction.[58] In those states which *do* dedicate all motor fuel tax revenues to highway construction, the larger oil companies have *not* opposed increases in motor fuel excise tax bills. This behavior is consistent with the multiple effects of motor fuel excise taxes on the gasoline market. Higher motor fuel taxes are likely to reduce total gasoline revenues if demand is elastic. If marginal cost of production is relatively low, this might result in a reduction in refinery profits. On the other hand, if higher motor fuel taxes are used to build highways, which in turn leads to an increase in vehicle use and a subsequent increase in the demand for gasoline, then refiners might enjoy higher profits as a result of increased motor fuel excise taxes. Therefore it is not surprising to learn that the industry has usually opposed increases in motor fuel taxes in nondedication states and remained relatively silent in dedication states.[59] We hypothesize that among nondedicating states, those with larger oil interests will have relatively lower motor fuel excise tax rates.

An additional factor that must be controlled in this investigation of the determinants of the level of state motor fuel excise tax rate is the "demand" for state revenues. States choose different consumption bundles; one state may choose a relatively greater share of publicly provided goods and services while another state may decide to allocate relatively more of its income to private goods and services. One important factor that may influence the relative share of the public sector in state economies is the income level. Indeed, several political scientists have established such a relationship between state income level and the level of state governmental expenditures.[60] Like other state revenues, gasoline excise taxes too may respond more to general income-related demand

state budgets to be influenced by such "demand" for state revenues considerations. We find that a relatively greater proportion of the variation in the motor fuel tax rates is explained by those models employing the relative size of the public sector as a measure of "demand" factors vis-à-vis those employing personal income. Since the relative share of the public sector is a more comprehensive measure of "demand" for state revenues, we would expect it to do a better job of "explaining" the variation in tax rate levels.

The influence of the petroleum giants in state politics might be revealed in a related, but slightly different way. A cross-classification of states by whether or not they dedicate their motor fuel excise taxes to highway construction and the relative size of the refining and crude production industries in each state is shown in Table H-4.

The data reported in Table H-4 suggested that those states where petroleum refining is a large share of state personal income are more likely to dedicate their motor fuel excise taxes to the construction of highways (78 versus 51 percent for states where petroleum refining is less than one percent of state personal income) than those states where petroleum refining is a smaller share of state personal income. This conclusion is supported by a Chi-square contingency test at a confidence level of 80 percent. A similar observation exists in the crude production industry, however the difference between large and small share states (60 versus 54 percent) in crude production is so small that it should not be interpreted to carry any significance. This observed difference between the refining and crude sectors is consistent with our findings from the earlier regression analysis.

Table H-4. Distribution of States Cross-Classified by Motor Fuel Excise Tax Dedication and Size of Industry in State

	Dedicate Motor Fuel Taxes to Highway Construction		*Do Not Dedicate Motor Fuel Taxes to Highway Construction*	
	Number of States	*Percent of Row*	*Number of States*	*Percent of Row*
Petroleum Refining				
Greater than one percent of state personal income	7	(78%)	2	(22%)
Less than one percent of state personal income	21	(51%)	20	(49%)
Crude Oil Production				
Greater than one percent of state personal income	9	(60%)	6	(40%)
Less than one percent of state personal income	19	(54%)	16	(46%)

In summary, it appears that, for those states where political efforts are made on behalf of the petroleum industry on motor fuel excise tax issues, states having relatively larger establishments tend to have relatively lower motor fuel excise tax rates and are more likely to dedicate motor fuel tax revenues to highway construction. The evidence provided by this brief study of excise tax rates is thus consistent with our findings concerning energy industry relative firm size in the analysis of income taxes.

Regulatory Policy

In addition to the tax subsidies it receives, the energy industry is also the object of a host of regulatory policies. Since governmental regulation of business is commonly perceived as a sign of the victory of public over private power, and hence of the political impotence of the affected business sector, it is important to evaluate the consequences of government regulation carefully. For example, regulatory provisions might hurt (or help) some firms in an industry more than others. Discerning the exact distribution of relative benefits and costs of a regulatory provision can tell us something important about the relative power of firms in an industry. Perhaps even more importantly, several scholars have recently argued that government regulation is rarely imposed on an industry, but is rather the product of the industry's own search for security in a pre-regulation environment characterized by costly, cutthroat competition.[61] Viewed in this way, government regulatory provisions may represent cease-fire agreements among firms in an industry reached at the expense of other economic sectors and of the public at large. The same two questions we asked about tax policies apply as well to regulatory policies: (1) How do these policies distribute benefits and costs *between* the energy industry and other economic actors? and (2) How do these policies distribute benefits and costs *within* the industry?

Answering these questions with regard to regulatory provisions is especially complicated in view of the difficulty in measuring the exact consequences of the provisions. Any regulatory measure sets off a series of ripples so widespread and complex that it frequently becomes difficult to determine which effects can appropriately be attributed to the regulation and which cannot. Therefore, rather than attempt any new analysis of the consequences of regulatory policies for the energy industry here, we propose rather to briefly review some of the major findings in the existing literature with respect to three significant regulatory measures—the lead-free gasoline regulation provisions enacted in 1970, the crude oil prorationing system, and oil import quotas. Of particular interest to us is the extent to which these regulatory provisions reflect the political influence of the energy industry, and the relationship they demonstrate between firm size and political influence.

Lead Free Gasoline Regulation. The Clean Air Act of 1970 provides a convenient starting point for this inquiry, for it seems to offer an important insight into the politics of regulation with respect to the energy industry. This bill required that automobiles comply with standards lowering emissions by 90 percent from model year 1970 cars by January 1, 1975.[62] It also contained provisions authorizing the administrator of the Environmental Protection Agency to control or prohibit the sale of any fuel additive if it was found that emissions from it would endanger public health or interfere with the passenger cars' pollution control devices.

The passage of this bill might suggest to some that the oil majors lack political effectiveness; for there are strong a priori grounds for expecting that the majors would oppose a bill like this that would be certain to raise refining costs. In fact, however, while objecting to the imposition of gasoline manufacturing *input* standards, the large integrated refiners did not express great opposition to the emission standard. Actually, several of them supported the idea. In a statement prepared for the Senate Public Works Committee's hearings prior to passage of the bill, for example, Peter N. Gammelgard, Senior Vice President for Public and Environmental Affairs, American Petroleum Institute (generally recognized as the voice of large integrated refiners) stated: "We believe that direct government regulation of manufacturing methods of gasoline composition is not in the public interest and that the federal government's role should be to establish emission standards. If compliance with those standards is left to the industry, it will encourage innovations and should result in lower costs to the public."[63] A representative of Texaco concurred, stating that "It would appear to us that the government . . . should set emission standards and permit industry in this competitive enterprise system (sic) to devise the lowest cost methods to meet such standards."[64]

A possible explanation of this apparent paradox lies in the economic structure of the petroleum industry, and particularly the majors' interest in taking advantage of a political situation that promised to allow them to improve their competitive position within the industry vis-à-vis the independent marketers and the independent refiners, while posing as stalwart defenders of the environment. Both the independent marketers and the independent refiners stood to lose disproportionately from the proposed switch to lead-free gasoline. The independent marketers, for example, were concerned that to offer customers the variety of gasoline stipulated under the bill's provisions would require gas stations to add a third pump for 91 octane low lead fuel. As William S. Jones, President of the National Oil Jobbers Council (representing 10,000 independent petroleum marketers), explained to the Senate Public Works Committee: "If wrong decisions are made, . . . , the impact upon the jobber will inflict far greater damage than that suffered by the infinitely larger concentration of capital represented by refiners and automobile manufacturers . . . if we are required to put in a third pump, it will cost the jobbers of America on the order of a quarter of a billion dollars to effect this conversion."[65] Most of the large integrated oil refiners and marketers already had more than two pump service at their stations, producing a competitive advantage that one of them readily acknowledged. "If the government does not mandate the continuation of the two-grade system," Mobil Oil's representative noted, "the adoption by the auto industry of a timetable which effectively will require the simultaneous marketing of a new low grade fuel plus the existing fuel grades for the present car population would effectively force the remaining two-grade marketer either to vacate the market for one of these grades or to make the necessary investment to market a third grade."[66]

Independent refiners had similar objections to the bill. The oil refineries would have to modify existing plants in order to produce the required low-lead fuel. But a few of the large integrated companies already were marketing low lead fuels and would therefore require much lower conversion costs. William J. Hull, Vice President of Ashland Oil and Refining Company, observed that "independent refiners whose plants are generally of small or medium size would suffer severe and immediate burdens as contrasted with the major oil companies whose plants are usually of much larger size . . ."[67]

A study commissioned by the American Petroleum Institute to explore the additional costs that would be incurred to produce the nonleaded fuel confirms Hull's conclusion.[68] This study addressed the question: "If U.S. refineries had produced unleaded gasoline in a specific year, how much added process equipment would have been needed and how much higher would operating costs have been?"[69] The researchers employed a linear programming simulation model for twelve "representative" refineries of various sizes. Each refinery was presumed to maximize efficiency. That is, the model computed the best strategy for minimizing the costs of the new lead-free restrictions for each refinery. This involved different process equipment for various size refineries.

The authors estimated that total domestic investment in new process equipment and related refinery facilities would be about $4.5 billion as a result of the air pollution bill. They projected that the increase in overall refining costs would be two cents per gallon of finished product.

But the significant finding for the purpose at hand is the variation in this increased cost with refining capacity. The relationship between refinery capacity and the increased cost per gallon caused by conversion to nonleaded fuels is shown in Figure H-1.[70] As can be seen, the greatest relative burden of the conversion to nonleaded fuel is borne by smaller refineries. It appears that a discontinuity in refinery processing exists at about 18,000 barrels per day of gasoline production. It is at this approximate capacity that a basic addition to the conversion process could be made to accommodate the lead content standards. At this size level hydrocracking becomes economically feasible, which produces significant economies for the conversion to lead-free gasoline production.

Little opposition to the bill was forthcoming from independent crude oil producers. This might be explained by the increase in the use of crude oil that would be required to process the new nonleaded fuels. This increase was projected to average 5.5 percent nationally.[71]

Figure H-1. Effect of Refinery Size on Added Manufacturing Costs For Producing Unleaded Motor Gasoline

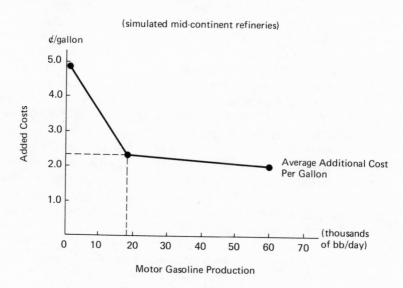

(simulated mid-continent refineries)

¢/gallon

Average Additional Cost Per Gallon

(thousands of bb/day)

Motor Gasoline Production

It is thus possible that the passage of The Clean Air Act had politically important implications for the internal structure of the energy industry. It appears that large integrated refiners and independent crude oil producers stood the most to gain. Small independent refiners and independent marketers were to bear the highest relative burden. The significance of this legislation, which was formed and considered in light of the positions of the various interested political pressure groups, may well be in its impact on the competitive position of the large refiners. Their lack of opposition to a proposal which was apparently going to cost them money may be more readily explained by the potential advantages they could expect than by their concern for the environment.

Prorationing. The prorationing system provides a different kind of example of industry regulation from that embodied in lead-free gas regulation, an example much closer to the self-regulation pattern depicted by Kolko, Lowi and Stigler. Prorationing is the oil industry's euphemism for government-sanctioned, anticompetitive price control.[72] The policy grew directly out of the chaotic conditions that existed in the early years of the oil industry. Given the highly decentralized ownership of oil wells and the technical ability to drain oil from under another's land surface, the "rule of capture," which conveys oil ownership to whoever pumps it from the ground, sent crude prices plummeting between 1920 and 1935 as a result of massive new oil discoveries. In a sense, unregulated oil production under the rule of capture is like the prisoner's dilemma of game theory. Production will occur even when receipts fail to cover costs because (1) if any producer fails to produce today, when anyone else does, he will ultimately lose due to the fugacious nature of oil and gas, and (2) his capital is of such a nature that it cannot be transferred easily to other uses, so that he must produce if he is to have any hope of recovering costs. The situation is even fiercer when one realizes that the royalty owners and the lease holders can sue the producer for alienation of property if he fails to produce, even under adverse cost conditions, when other producers continue operations. Up to a point, this situation worked to the advantage of the large refiners, who could purchase cheap crude as a consequence. But many of these same refiners were also producers. Although they sought to acquire sufficient control of crude production to stabilize the industry on their own, they were ultimately unable to do so. Under the circumstances, the oil majors reacted much the way other industry groups did during the Progressive Era—they approved government intervention to control competition in a way the merger movement could not. Historian Gabriel Kolko's description of America's economic leadership generally during this period seems to apply well to the oil industry:

> In the long run, key business leaders realized, they had no vested interest in a chaotic industry and economy in which not only their profits but their very existence might be challenged.[73]

The result is our intricate system of state production controls and well-spacing regulations buttressed by a federal law prohibiting shipment of oil produced in violation of the rations.[74]

While the prorationing system originated as a defensive measure to fend off industry collapse by regulating excessive competition, its long-term effects have been redistributive in character, at least between the petroleum industry and consumer-taxpayers generally, and perhaps within the petroleum industry itself. The energy industry as a whole gains chiefly as a consequence of the increase in the price of crude oil resulting from the prorationing system's artificial limitation on supply. One estimate placed this price inflation at $1.25 per barrel in the late 1960s.[75] A second redistributive consequence of the prorationing system results from the inefficiency the system encourages by stimulating the drilling of excessive wells. Several features of the prorationing system contribute to this inefficiency: the assignment of quotas to individual wells, instead of to entire fields; the exemption of so-called "stripper," or marginal, wells from prorationing controls; the requirement that crude buyers spread their purchases proportionally among all producers in a field; and several others. Many of these provisions were instituted to guarantee small producer political support for the prorationing controls. Taken together, they cost the public an estimated $2.15 billion annually in extra charges for oil.[76] A third redistributive impact of prorationing cited in the literature, but one about which there is more uncertainty, operates through the corporate income tax system: since the oil depletion allowance permits oil companies to deduct from income a portion of the value of crude oil produced, the higher the price of crude oil relative to the price of the final product, the larger the tax write-off.[77]

While the prorationing system's contributions to the oil industry as a whole are relatively clear, its impact on the internal structure of the industry is more difficult to unravel. For example, the system has generally strengthened the position of the "majors" vis-à-vis independent domestic refiners because the system tends to stabilize crude prices at a rather high level. Having no foreign or domestic integrated source of supply, and no ability to cross-subsidize one branch of their business with another, the independent refiners have thus been made more vulnerable to pressure from the "majors" as a consequence of prorationing.[78] At the same time, however, the system has produced real gains for some independent producers vis-à-vis the majors by requiring fair distribution of crude purchases among all wells in a field and by assigning quotas on the basis of wells, thus protecting marginal, less efficient producers. Perhaps the safest statement is that prorationing distributes benefits and costs within the industry in such a way as to maintain a positive ratio of benefits to costs for most segments of the industry without delivering the lion's share of the benefits to any one sector. In the process, however, the system exacts high costs on behalf of the industry as a whole from the public at large. That such a costly government-created, price-rigging system should prove durable in a society that touts its commitment to the doctrine of "free enterprise" is certainly testimony to the political influence of this unusual industry.

Oil Import Quotas. Were there any suspicion that the prorationing system represented merely an unusual, Depression-born quirk, and not the product of a stable constellation of political and economic power, an examination of the oil import quota should help to dispel it. In many respects, the two policies are Siamese twins, responding to similar economic needs and reflecting the same structure of economic and political power. The oil import program was a response to the challenge posed to the prorationing system by growing imports of uncontrolled, and therefore lower priced, foreign oil during the 1950s. While the majors were able to reap some benefits from imports because of their international position, the domestic independents stood to lose considerably. Realizing this, they have consistently supported tariffs and quotas on oil imports, particularly during testimony on proposed quotas before the Office of Defense Mobilization in October 1956.[79] The adoption of a quota system in 1957 over the objections of the majors seems to signify a victory for the smaller firms over the larger ones in the industry.

Yet several factors mitigate such a conclusion. In the first place, one of the major consequences of the import quota system is to protect the redistributive benefits the oil industry as a whole secures from the American public generally through the prorationing system's price controls.[80] In the second place, to secure support from the majors, proponents of import controls originally settled for a voluntary program and based "ticket" allocations on historical import levels, a formula that naturally benefited the majors the most. Within two years the voluntary oil import control program began to falter. The reasons for the breakdown of the program include noncompliance of "volunteers," newcomers to the importing business, failure to limit imports of products and unfinished oils, and opposition of the Antitrust Division.

Mandatory oil import controls were adopted in 1959. One of the major features of the mandatory program was the "sliding scale" scheme of ticket allocations to replace the historical formula of the voluntary program. The sliding scale allows smaller refiners to rely on imported oil for a larger percentage of their total inputs. Although the appearance of the sliding scale was to favor smaller refiners, the story is more complicated than that. Kenneth Dam describes the reason for the acquiescence of the majors to the "sliding scale" system:

> . . . it was in major part an effort to do equity between two classes of large companies—the traditional importers who had large foreign production holdings and the large domestic companies that had begun to "go international" in the 1950s by investment in foreign production and by beginning to import in substantial quantities. This conflict between those who were established importers in 1954 and those who began importing in substantial volumes in the 1954-1957 period (whom we may call the "latecomers") led to the adoption of the [sliding scale] . . . . Without the sliding scale the larger latecomers would . . . have received allocations very nearly equal in size to the allocations received by established importers of equal size. That this was a principal reason for adopting the sliding scale is difficult to establish from official documents, but it seems to be well understood among those involved.[81]

Moreover, although some observers have concluded from this that the sliding scale favors small refiners at the expense of large refiners since it allocates to small refiners a larger *proportional* share of imports, the larger refiners continue to enjoy the larger absolute shares of the scarcity value created by the control program because the sliding scale consists of progressively declining allocations only on the *marginal* increases in refinery capacity. Since it is the absolute level of net benefits that firms typically seek to maximize, we are reluctant to declare this battle a victory for the small refiners, even though it is fair to conclude that *changes* in the sliding scale since its initial adoption have tended to redistribute these absolute values toward the smaller refiners.[82] Perhaps nothing demonstrates the extent to which the sliding scale represents a tenable political compromise among segments of the industry more than the response that greeted a Budget Bureau proposal in 1968 to replace it with an auctioning system. Of more than

100 respondents to a Notice of Proposed Rule Making describing the auction proposal, only one was sympathetic.[83]

In short, like the prorationing system, the oil import program illustrates the oil industry's ability to redistribute wealth away from the public generally.[84] In the case of oil import quotas, the initial support for the program came from the independent crude producers, not the majors. Yet, it seems relatively clear that the form and timing of the program showed ample evidence of the majors' influence, even while conforming to more urgent independents' needs.

SUMMARY AND CONCLUSIONS

The evidence presented in this study gives systematic empirical support to the general hypothesis outlined in the first part of the paper concerning the impact of economic structure on political power, particularly for the energy industry. To the extent that political power finds reflection in actual policy outcomes, we found evidence of petroleum industry power in its success at utilizing public authority to produce outcomes of benefit to the industry. Particularly striking was (1) the discovery of a negative relationship between *firm size* and effective corporate income tax rates; (2) empirical evidence systematically linking the relative dominance of large firms in the refining industry in each state to the level of state motor fuel excise tax rates; and (3) a pattern of regulatory policies with substantial economic payoffs for the industry. In addition, we demonstrated that energy industry effective corporate income tax rates diverge farther than practically all other industries from what would be predicted if the energy industry behaved like the typical mining and manufacturing industry. These findings, incomplete though they are, thus seem to support the hypotheses about energy industry political power that were implied by our general analysis of the likely relationship between economic structure and political power: namely, that this industry is amply endowed with political resources thanks to its firm size, industry size, and pattern of geographic dispersion; and that it is simultaneously insulated from some of the political disadvantages of largeness by virtue of a market concentration pattern that gives at least a semblance of widespread competition at each of four different production stages.

To go from these empirical findings to any solid conclusions about the political implications of future mergers in the energy industry must necessarily involve a great leap of faith. Any potential merger involves such numerous simultaneous alterations in economic structure (firm size, industry size, market concentration and geographic dispersion) that it is difficult to generalize about them with any real confidence on the basis of the evidence examined here.

With these limitations in mind, however, it is possible, as a result of the analysis carried out here, to generate some more or less "informed speculations" about the political implications of possible changes in energy industry economic structure. In the first place, we find some evidence to suspect that further mergers within the petroleum industry might produce increments of political power for the industry that could result in redistributions of income and/or wealth from the general populace to owners of resources in that industry. We remain cautious in this judgment, however, because the oil industry has attained a firm size and industry size already that yields such extensive political power that it is not clear that further increases in firm size would have any effect. Predicting the implications of large scale mergers from our empirical tests is likely to take us beyond the range of our data and make such predictions relatively risky. Indeed, there is reason to believe on the basis of the evidence we have explored that the petroleum industry has already surpassed most other industries in its success at influencing government policy making.

To say that further mergers within petroleum are likely to have only marginal political impacts is not to say that the same holds true for changes in economic structure in the energy industry as a whole. To the contrary, our second "informed speculation" is that real political consequences involving both equity and efficiency implications are likely to result from further petroleum industry takeover of other energy sources. Whatever the economic consequences of such takeovers, in other words, they may have very serious *political* consequences in addition. In particular, the evolution and operation of industry-serving government regulations in the *petroleum* industry suggests a common pattern that we may very well be replicating in the current period for the *energy* industry as a whole. The pattern consists of four interrelated steps: first, competition potentially threatening to the largest firms in the industry appears; the large firms react to this competition in typical oligopolistic fashion by seeking to eliminate the competition through acquisitions, mergers, and other private anticompetitive maneuvers; because of the sheer enormity of the task and the perennial threat of antitrust action that causes delays, it becomes apparent that private action by itself will not suffice to limit the disruptive competition; at that point, the industry collectively turns to government, pointing to a serious "crisis" threatening national interests, and requests regulation to "keep the industry on a sound economic footing." The timing of this last step is naturally crucial. The proper crisis atmosphere must exist, and the major firms must be able to demonstrate that it is the "little guys" who are being hurt most. This obviously means that industry concentration must not have proceeded so far that there are too few "little guys" left to make the case credible.

We believe that this scenario may very well be the one that is now being played out, this time in the energy industry as a whole rather than in petroleum alone as has been the case in the past. The emergence in recent years of new energy sources (e.g., atomic energy, coal-gas conversion, etc.) has begun to raise threats to the entire structure of anti-competitive governmental regulations that the oil industry has managed to construct for itself over the years. In response, the majors have initiated private defensive maneuvers through mergers and acquisitions. If this analysis is correct, we may soon be approaching the politically most significant phase of all, when the industry makes its play for explicit governmental assistance in reducing competition. The more than three million dollars the industry has spent recently to advertise the "energy crisis" and thus to help generate a "crisis" atmosphere certainly lends credence to this view.[85] Though admittedly speculative, this line of argument, emerging as it does from an examination of previous industry political activities, raises troubling questions about the political implications of the recent trend toward greater concentration in the energy industry.

NOTES

1. Woodrow Wilson, *The New Freedom* (New York: Doubleday, 1913), pp. 57-8.
2. Theodore Lowi, *The End of Liberalism: Ideology, Policy, and the Crisis of Public Authority* (New York: W.W. Norton & Co., Inc., 1969), p. 102.
3. *American Business and Public Policy* (New York: Atherton Press, 1963), by Bauer, Poole, and Dexter represents possibly the most systematic and analytical contribution of this sort. While hardly "anecdotal," it is avowedly case-specific in orientation. Other contributions include Henry Kariel, *The Decline of American Pluralism* (Stanford, Calif.: Stanford University Press, 1961); Grant McConnell, *Private Power and American Democracy* (New York: Alfred A. Knopf, 1966).
4. In recent years, political scientists have made notable progress in examining systematically the link between socioeconomic and political characteristics of political systems on the one hand, and policy outputs on the other. Almost without exception, however, this research has focused on aggregate data dealing with general economic and social conditions rather than on microdata dealing wtih the structure of particular industries or economic sectors. Thus, while we can say that political systems with high levels of industrialization generally have more substantial welfare policies, regardless of the level of party competition, we are not in a position to relate policy outputs to particular aspects of industry structure. It is this gap we hope to fill. For examples of the aggregate analysis approach evident in the literature, see: Thomas R. Dye, *Economics, Politics and the Public: Policy Outcomes in the American States* (Chicago: Rand McNally, 1966); Richard Dawson and James A. Robinson, "Inter-Party Competition, Economic Variables, and Welfare Policies in the American States," *Journal of Politics*, 25 (May 1963), pp. 265-289; Charles F. Cnudde and Donald J. McCrone, "Party Competition and Welfare Policies in the American States," *American Political Science Review*, 63 (September 1969), pp. 858-66.

5. Robert H. Haveman and Robert D. Hamrin, eds., *The Political Economy of Federal Policy* (New York: Harper & Row, Publishers, Inc., 1973), pp. 6-7.

6. Thomas R. Dye and Harmon Zeigler, *The Irony of Democracy* (Belmont, Calif.: Wadsworth Publishing Co., Inc., 1970), p. 149.

7. Richard Neustadt, *Presidential Power*, Signet Edition (New York: The New American Library, 1964), p. 42.

8. For an excellent analysis of the diversity and political consequences of federal administrative fragmentation see Harold Seidman, *Politics, Position and Power* (New York: Oxford University Press, 1970).

9. For further development of this point see Lester M. Salamon and Gary L. Wamsley, "The Federal Bureaucracy: Responsive to Whom?" in Leroy Riselbach, *People vs. Government: The Responsiveness of American Institutions* (Bloomington: Indiana University Press, 1975), pp. 151-188.

10. Seidman, *op. cit.*, pp. 37, 40.

11. *Ibid.*, p. 18.

12. On the concept of "policy subsystems," see J. Leiper Freeman, *The Political Process* (New York: Random House, 1965); Lee J. Fritschler, *Smoking and Politics* (New York: Appleton-Century Crofts, 1969).

13. Quoted in Lee J. Fritschler, *op. cit.*, p. 94.

14. We are indebted here to Edwin Epstein, *The Corporation in American Politics* (Englewood Cliffs, N.J.: Prentice-Hall, Inc., 1969), particularly pp. 67-111.

15. Carl Kaysen, "The Corporation: How Much Power? What Scope?" in *The Corporation in Modern Society*, ed. by E.S. Mason (Cambridge, Mass.: Harvard University Press, 1959).

16. Corwin D. Edwards, "Conglomerate Bigness as a Source of Power," *Business Concentration and Price Policy* (Princeton, N.J.: Princeton University Press, 1955), pp. 331-60.

17. National Industrial Conference Board, *The Role of Business in Public Affairs*, Studies in Public Affairs, No. 2 (New York: National Industrial Conference Board, Inc., 1968), p. 8.

18. Raymond Bauer, Ithiel de Sola Poole, and Lewis A. Dexter, *American Business and Public Policy* (New York: Atherton Press, 1963), p. 227.

19. Commenting on the relative advantages of having a full-time firm representative in Washington and working through a trade association, Corwin Edwards notes that: "While some smaller business interests make a comparable showing through associations set up for the purpose, the experience of a Washington official is that small companies generally find out what is happening too late and prepare their case too scantily to be fully effective where their interests conflict with those of large companies." [Edwards, *op. cit.*, p. 347.]

20. Paul W. Cherington and Ralph L. Gillen, *The Business Representative in Washington* (Washington, D.C.: The Brookings Institution, 1962), p. 55.

21. For an excellent survey of the evidence relating market concentration to profit rates see Leonard W. Weiss, "Quantitative Studies of Industrial Organization," in M.D. Intrilligator, ed., *Frontiers of Quantitative Economics* (Amsterdam: North-Holland Publishing Company, 1971), pp. 362-79. Weiss concludes that "practically all observers are now convinced that there is something to the traditional hypothesis." "Almost all of the 32 concentration-profits studies except Stigler's have yielded significant relationships for years of prosperity or recessions, though they have depended on a wide variety of data and methods." (p. 371).

22. See, for example, Bauer, Poole and Dexter, *op. cit.*, pp. 332-40 and David Truman, *The Governmental Process* (New York: Alfred A. Knopf, 1958), pp. 156-87.

23. Bauer, Poole and Dexter, *op. cit.*, p. 266.

24. Thomas G. Moore, "The Petroleum Industry," in Walter Adams, ed., *The Structure of American Industry* (New York: The Macmillan Company, 1971), p. 117.

25. *Census of Business*, 1967, Vol. I.

26. Robert Engler, *The Politics of Oil* (New York: The Macmillan Co., 1961), p. 376.

27. Computed from data in *Congressional Quarterly*, Vol. XXXI (January 6, 1973), pp. 24-5.

28. *Fortune Magazine* (May 1973).

29. Thomas D. Duchesneau, *Competition in the Energy Industry*. Paper Prepared for the Ford Foundation Energy Policy Project, 1973.

30. Moore, *op. cit.*, p. 128.

31. In addition, concentrated industries may receive economic benefits that compensate for the political disadvantage of greater visibility.

32. Walter J. Mead, "The Structure of the Buyer Market for Oil Shale Resources," *Natural Resources Journal* (October 1968).

33. Federal Trade Commission, "Structure of the Petroleum Industry and its Relation to Oil Shale and Other Energy Sources" (Washington, D.C.: 1967), p. 200.

34. U.S. Congress, House Committee on Banking and Currency, *Commercial Banks and Their Trust Activities: Emerging Influence on the American Economy*, Staff Report for the Domestic Finance Subcommittee, 90th Cong., 2nd Sess., (Washington, D.C.: U.S. Government Printing Office, July, 1968).

35. Federal Trade Commission, *Large Mergers in Manufacturing and Mining: 1948-1971* (Washington, D.C.: May 1972).

36. Fred R. Harris, "Oil: Capitalism Betrayed in Its Own Camp," *The Progressive*, Vol. 37 (April 1973), p. 31.

37. We do not mean to imply by this the total absence of remaining conflicts of interest between different sections of the industry. The independent producers in the Southwest, the large "independents" (or international "minors") with their newly acquired foreign production capacity, the remaining independent refiners, and especially the

independent jobbers and dealers with their amazing patchwork of organizations continue to have real conflicts over real interests—to maintain margins, raise their allowables, increase their share of import tickets, or heighten the entry barrier against foreign crude. But most of this conflict arises because of the success of the overall energy industry in creating scarcity value—a transfer from the general taxpayer-consumer public to the energy industries. The conflict within the industry arises over dividing the spoils that have been extracted from the general public. When it comes to increasing the size of the total pie to be divided among the energy interests, unity and cooperation have a habit of suddenly appearing in the industry.

38. Robert Dahl, *Who Governs? Democracy and Power in an American City* (New Haven: Yale University Press, 1961).

39. Bauer, Poole and Dexter, *op. cit.*

40. Peter Bachrach and Morton Baratz, *Power and Poverty: Theory and Practice* (New York: Oxford University Press, 1970), pp. 3-16.

41. Bertrand Russell, *Power: A New Social Analysis* (London: Unwin, 1962), p. 25.

42. For a similar mode of analysis but focusing on quite different independent variables, see note 4, *supra*.

43. For a well written discussion of tax-subsidies, the redistributive effects of the income tax structure and the causes of its apparent perversion see Philip M. Stern, *The Rape of the Taxpayer* (New York: Random House, 1973), especially Chapters 3, "How Would You like a Special Tax Law All Your Own?," 10, " 'Tax Welfare for the Corporate Giants,' " and 11, "Ah, To Be an Oilman." Stern explains the source of a great deal of the redistribution of income resulting from the income tax laws and provides some documentation of the magnitude of this redistribution. Abundant examples from the oil industry are used as illustrations throughout the book.

44. Quoted in Ronnie Dugger, "Oil and Politics," *The Atlantic Monthly* (September 1969), p. 66.

45. An elaboration of this analysis can be found in John J. Siegfried, *The Relationship Between Economic Structure and the Effect of Political Influence: Empirical Evidence from the Corporation Income Tax Program*, unpublished Ph.D. thesis (Madison, Wisc., University of Wisconsin, 1972).

46. 1963 was chosen because of data availability and because it was a relatively "normal" year of moderate economic expansion. Our analysis seeks only to detect the existence of relationships; we do not pretend that the precise numerical magnitude of our estimates has not changed over a decade; but we do believe that the type of political relationships that we attempt to assess in this study are likely to be stable over fairly long periods of time.

47. G.E.P. Box and D.R. Cox, "An Analysis of Transformations," *Journal of the Royal Statistical Society*, Series B, Vol. XXVI (1964), pp.

211-43. The Box-Cox method transforms all variables such that $X'_i = \dfrac{X^\lambda - 1}{\lambda}$. The technique is to locate the parameter λ such that the likelihood function of the estimated equation is maximized. This occurs for $\lambda = -0.1$. A likelihood ratio test verified that the transformation using $\lambda = -0.1$ is not significantly different from that using $\lambda = 0$. The transformation $\lambda = 0$ corresponds to a simple natural logarithm transformation of all variables. Since a logarithm transformation is much easier to interpret than the maximum likelihood transformation, and since the likelihood ratio test provides intuitive support for the proposition that they are not really different, we use the logarithmic transformation throughout this paper. This is the principal stimulus for the adoption of multiplicative models. Theory also confirms the reasonableness of a multiplicative model. For further discussion of the likelihood ratio test see Paul G. Hoel, *Introduction to Mathematical Statistics* (London: John Wiley & Sons, Inc., 1962), pp. 220-25.

48. In 1968 the Treasury stipulated the following guidelines for deciding which provisions of the tax laws are "special." They stated that their analysis " . . . lists the major respects in which the current income tax bases deviate from widely accepted definitions of income and standards of business accounting and from the generally accepted structure of an income tax . . . " (*1968 Secretary of the Treasury Annual Report on the State of the Finances.*) In our study the Treasury's choice of "special" tax provisions is adopted in order to compute the underreported profits accruing to industry from unusual treatment under the corporation income tax structure. The objective is to estimate effective corporation income tax burdens from the point of view of the industry. In a contribution to a later study, "The Economics of Federal Subsidy Programs," a Staff Study prepared for the use of the Joint Economic Committee (January 11, 1972), Assistant Secretary of the Treasury, Murray L. Weidenbaum, cautioned Senator William Proxmire, Chairman of the Joint Economic Committee that "there is considerable conceptual controversy over what is and what is not a tax subsidy. In no way should the enclosed information (estimates of tax-subsidies) be interpreted as Treasury's identification of tax subsidies. Furthermore, the estimates are prepared on an individual basis for each item on the assumption that the item would be eliminated from the law without other changes in the law with respect to the other items. If two or more changes in the law are made, the aggregate revenue effect will frequently not equal the sum of the revenue effects of the individual changes." (Appendix A, p. 205).

49. The foreign tax credit provision is not treated as a special tax provision because, from the industry's viewpoint, it does not alter the tax burden to the full extent of its magnitude. This is so because if many of the foreign taxes did not exist, the industry's domestic tax rate would rise and its overall tax burden change little.

50. Senator Phillip Hart (D-Michigan), Speech to the Lawyers Club in Ann Arbor, Michigan, April 8, 1968, cited in Mark J. Green, *The Closed Enterprise System* (New York: Grossman Publishers, Inc., 1972), p. 19.

51. This index consists of the sum (across states) of the squared share of total industry employment located in each state in 1963. An index of this type is sensitive to both the number of states in which an industry has employees and the distribution of employees among these states. Fewer states or a more unequal distribution cause the index to rise. Squaring the shares weights the large employment (and hence more important) states more heavily in the index.

52. William Baumol, *Business Behavior, Value and Growth*, rev. ed. (New York: Harcourt, Brace & World, Inc., 1967).

53. Oliver Williamson, *The Economics of Discretionary Behavior: Managerial Objectives in a Theory of the Firm* (Englewood Cliffs, N.J.: Prentice-Hall, Inc., 1964).

54. It has been proposed that firms sometimes do not maximize the attainment of some well-defined objective function, but rather "satisfice." That is, they seek choices which satisfy at least minimum levels of aspiration with respect to a variety of objectives. For example, management may set a target rate of return on invested capital as its profit objective, and if that target is achieved, it turns to the satisfaction of other nonprofit objectives. See Herbert Simon, "Theories of Decision Making in Economics and Behavioral Science," *American Economic Review*, Vol. 49 (June 1959), pp. 253-83; Robert F. Lanzillotti, "Pricing Objectives in Large Companies," *American Economic Review*, Vol. 48 (December 1958), pp. 921-40.

55. Computed from data in the *Source Book of Statistics of Income, 1967* (Washington, D.C.: U.S. Treasury Department, 1968).

56. On the permeability of state legislatures to special interest pleading, see: Terry Sanford, *Storm over the States* (New York: McGraw-Hill Book Co., 1967).

57. Two alternative hypotheses are also theoretically plausible: first, that absolute industry size and not relative industry size is the key determinant of influence; and second, that the relative size of the entire special interest group sector is the relevant variable. However, we judged these hypotheses to be less probable and chose not to test them explicitly at this time.

58. Information provided by Mr. William Haga, representative of major oil company interests in the State of Tennessee. Since 1971 some major oil companies have taken a position of not opposing gasoline taxes which are used for nonhighway purposes if they are dedicated instead to mass transit purposes. Not all companies currently take this position. However, since our empirical test is for 1967, this change in policy will not affect our hypothesis. Furthermore, we are seeking to discover evidence of general relationships between eco-

nomic structure and political decision-making, and consequently the relevance of our analysis is not diminished by a change in the policy of some oil companies.

59. In 1965 twenty-eight states had constitutional amendments requiring the dedication of motor fuel tax revenues to highway construction. [National Highway Users Conference, Inc., *Good Roads Amendments: Texts of Constitutional Provisions Safeguarding Highway Revenues* (Washington, D.C.: 1965) plus mimeographed revisions since 1965.]

60. See note 4, *supra.*

61. Gabriel Kolko makes this argument most convincingly with respect to the regulatory provisions enacted during the so-called Progressive Period. See: Gabriel Kolko, *The Triumph of Conservatism: A Reinterpretation of American History* (Chicago: Quandrangle Books, 1967). For a more recent elucidation of the same point and an elaboration of the concept of "self-regulation," see Theodore Lowi, *The End of Liberalism* (New York: W.W. Norton, 1969). Merver Bernstein makes a similar point in his well-known book on regulatory commissions. Bernstein, however, argues that the tendency of regulatory commissions to serve the needs of the industry they purportedly regulate is not a result of their origins but of their latter development. Merver Bernstein, *Regulating Business by Independent Commission* (Princeton: Princeton University Press, 1966). A recent analysis by economist George Stigler, "The Theory of Economic Regulation," *The Bell Journal of Economics and Management Science*, Vol. 2 (Spring 1971), pp. 3-21, supports this argument.

62. This deadline has recently been extended until 1976.

63. U.S. Congress, Senate Committee on Public Works and Committee on Commerce, *Joint Hearings Before the Subcommittee on Air and Water Pollution*, 91st Cong., 2nd sess. (March 24, 25, 1970), p. 1109.

64. *Ibid.*, p. 1114.

65. *Ibid.*, p. 1136.

66. *Ibid.*, p. 1122.

67. *Ibid.*, p. 1115.

68. S.D. Cawson, J.F. Moore, and J.B. Rather, Jr., "A Look at the Economics of Manufacturing Unleaded Gasoline," *American Petroleum Institute Proceedings*, Vol. 47 (1967), pp. 484-517.

69. *Ibid.*, p. 486.

70. *Ibid.*, Figure 4.

71. *Ibid.*, p. 491.

72. Needless to say, industry spokesmen and regulatory officials would strenuously object to such a designation. The oft-quoted testimony of Ernest O. Thompson of the Texas Railroad Commission that "We have nothing to do with price . . . I know nothing about price" symbolizes this attitude nicely. But, as Walter J. Mead has noted, to believe that prorationing is not the equivalent of price control is to

believe that "the law of supply and demand has been repealed for oil." [Walter J. Mead, "The System of Government Subsidies to the Oil Industry," Albert E. Utton, ed., *National Petroleum Policy: A Critical Review* (Albuquerque: University of New Mexico Press, 1970), p. 123.]

73. *Op. cit.*, p. 6.

74. For a discussion of the details of the prorationing system, see: Wallace Lovejoy and I. James Pikl (eds.), *Essays on Petroleum Conservation Regulation* (Dallas: Southern Methodist University Press, 1960) and Wallace Lovejoy and Paul T. Homan, *Economic Aspects of Oil Conservation Regulation* (Baltimore: The Johns Hopkins Press, 1967).

75. Mead, *op. cit.*, pp. 123-24. Recent developments in both the domestic and international oil markets have led to a lessening in this differential.

76. Moore, *op. cit.*, p. 145.

77. Moore, *op. cit.*, p. 140.

78. John G. McLean and Robert W. Haigh, *The Growth of Integrated Oil Companies* (Boston: Graduate School of Business, Harvard University, 1954, p. 606); Moore, *op. cit.*, pp. 135-36; Engler, *op. cit.*, pp. 24-5, 137 ff.; J. Stanley Clark, *The Oil Century* (Norman, Okla.: Oklahoma University Press, 1955), p. 263.

79. Office of Defense Mobilization, *Hearings in the Matter of Petroleum*, October 22, October 24, 1956, Washington, D.C., pp. 59, 65, 66, 173, 175, 181, 182, 188, 213, 214, 216, 388, 419, 441, 479, 482, 485, 489, 496, 497, 541D, 541G, 541H, 542, 547, 551, 575-76, 578, 594, 596, 600, 603, 609. See also: Kenneth W. Dam, "Implementation of Import Quotas: The Case of Oil," *Journal of Law and Economics*, Vol. 14 (April 1971), for an excellent historical analysis of the oil import control program.

80. Readers wishing further elaboration on the oil import control program are encouraged to see the U.S. Cabinet Task Force on Oil Import Control, *The Oil Import Question*, A Report on the Relationship of Oil Imports to the National Security (Washington, D.C.: U.S. Government Printing Office, 1970). This study compares the quota system with tariffs.

81. Dam, *op. cit.*, pp. 21-2.

82. *Ibid.*, p. 17.

83. *Ibid.*, p. 56.

84. Other illustrations of the impact of political tensions within the energy industry on the oil import program include: the loosening of quotas on behalf of distributors of fuel oils through the granting of numerous temporary exceptions by the Oil Import Appeals Board; the special arrangement between the Department of the Interior and Phillips Petroleum Co. regarding oil importation into Puerto Rico; the special arrangement with Hess Oil in the Virgin Islands; and the controversial proposal by Occidental Petroleum to construct a large refinery in a "foreign trade zone" at Machiasport, Maine. This last

proposal engendered a strong political conflict between representatives from New England, who saw the Machiasport project as reducing prices to consumers in New England, and representatives from oil-producing states who feared that further special deals would cause the whole quota program to collapse. The most troublesome aspect of these special arrangements was the failure of the Interior Department to ever announce the criteria for choosing among the various applicants for the special treatment. The company chosen for special treatment gained financially from the award. Moreover, the selection of the favored companies was accompanied by all the elements of political pressure and lobbying that normally surround legislation, even though the decision was made within the Executive Branch [Dam, *op. cit.*, p. 48]. The extremely large, predictable presidential contributions by most of the major petroleum companies cannot be ignored by observers of Executive Branch allocations of valuable import tickets.

85. "The Selling of the Energy Crisis," *New York Times*, April 1, 1973.

Appendix I

Reviewer Comments

PETROLEUM TAX COMMENTS IN THE DUCHESNEAU STUDY

By the American Petroleum Institute's
Committee on Industry Statements

The Duchesneau competition study contains a number of comments on petroleum taxation which constitute either outright errors or serious misconceptions. Among these are:

> (1) "Where the full value of the [depletion] allowance is taken, a firm's federal income tax rate is reduced by half." [p. 122]

The percentage depletion allowance equals 22 percent of gross income from the producing property or 50 percent of net income, whichever is less. Thus, in situations where the depletion allowance on a particular property is not subject to the 50 percent of net limit, the reduction in tax liability ranges between 22 percent and 50 percent. Only in those situations in which operating expenses equal or exceed 56 percent of gross income from the property, such that the 50 percent of net income limitation applies, does the depletion allowance result in a 50 percent reduction in tax. These cases account for only a small part of total oil production.

> (2) " . . . the percentage depletion allowance has the greatest value to the integrated firm." [p. 100]

In fact, many of the smaller nonintegrated producers are taxed as individuals or partnerships with a 70 percent marginal rate; whereas, the integrated firms have a 48 percent marginal rate. Thus, a dollar of percentage depletion saves 70 cents of tax for the nonintegrated producer and 48 cents for the corporation. The competitive significance of percentage depletion was recently well described by the Chairman of the Senate Finance Committee:

> MR. LONG . . . In any event, what is keeping the independents in business is that depletion allowance and the deduction for intangible drilling costs . . . the depletion allowance . . . is one advantage of the independents over the majors. Look at the 70 percent personal income tax and compare it with the 48 percent rate on corporations; that depletion allowance attracts money to take a chance on drilling wildcat wells by the independents. That is the only competitive advantage they have." [*Congressional Record*–Senate, June 25, 1974, p. 11475]

Some might observe that the independent producer has other competitive advantages, e.g., flexibility of response to new opportunities. But percentage depletion is clearly not anticompetitive.

> (3) "Under current interpretations of the law, royalty payments paid to foreign governments are considered to represent tax payments." [p. 129]

American-owned oil companies pay income taxes to the foreign government in its capacity as sovereign and royalties (and land acquisition bonuses) in its capacity as landowner. The income taxes are credited against potential United States tax liability. The royalties are not; they are deductible business expenses. Royalties are never credited. (The relevance of this statement to competition in the United States oil industry is not apparent.)

> (4) "Prof. Mead notes that, in addition to the subsidy nature of the foreign tax credit, the credit provision can lead to a situation where it may be more profitable to invest in foreign countries rather than in the U.S." [p. 129]

The foreign tax credit is not a subsidy. It is not a payment from the United States Treasury to American companies which operate abroad. It is not even a forgiveness of income tax. It is recognition that the host country has primary taxing jurisdiction and that United States taxation of income which has been taxed by the host government would constitute double taxation of American companies. Such double taxation would leave American companies noncompetitive abroad, since all major countries follow some such policy in taxing foreign source income of their nationals. Some countries do not tax foreign source income at all. Other countries (including the United States) use a foreign tax credit. The United States credit mechanism leaves the American company with a tax liability equal to the greater of the American or foreign rate. In no case can the United States credit lead to a reduction in tax liability on domestic source income; hence, the credit cannot be a subsidy. It also does not cause "investment in new oil capacity to be more attractive, i.e., profitable, in foreign markets relative to the U.S." [p. 286] The attractiveness of foreign oil investment relative to United States oil investment is a function of prices and productivity of fields. There is no evidence that the oil industry has rejected economically attractive domestic exploration ventures in favor of foreign exploration ventures.

> (5) "Crude oil production . . . relative to refining and marketing, may be more profitable because of the depletion allowance. As a result, companies tend to take their profits at the crude stage." [p. 133]

In an industry characterized by low concentration and free entry—as in the oil producing business—companies have no discretion as to where they take their profits. Prices and profits are set by market forces. Moreover, even if the integrated companies had the market power to raise crude prices artificially in order to shift profits into the lower-taxed stage of the industry—a power they do *not* possess—they would have no economic incentive to do so. The reason for this is that only one or two are sufficiently integrated to make money by doing so. Raising crude prices raises royalties and state severance taxes. Unless a company has more domestic production than its refinery runs, the extra royalty and severance tax costs would offset the income tax saving. The average integration percentage of the larger companies is about 75 or 80 percent. The study, itself, confirms this on the same page: " . . . Most of the major integrated oil firms have not been crude oil self sufficient."

Index

About the Author

Thomas D. Duchesneau is Associate Professor of Economics at the University of Maine, Orono. He is a graduate of St. Anselm's College and received his Ph.D. from Boston College. Dr. Duchesneau is the author of numerous articles based on his interests in the areas of Industrial Organization, Energy Economics, Micro Theory, and Government Regulation of Business.